Signaling and Communication in Plants

Series Editors

František Baluška
Department of Plant Cell Biology, IZMB, University of Bonn, Kirschallee 1, D-53115
Bonn, Germany

Jorge Vivanco
Center for Rhizosphere Biology, Colorado State University, 217 Shepardson Building,
Fort Collins, CO 80523-1173, USA

Luis Alfonso del Río • Alain Puppo
Editors

Reactive Oxygen Species in Plant Signaling

Springer

Editors
Prof. Luis A. del Río
Estación Experimental del
Zaidín
CSIC
Depto. Bioquímica
Biología Celular y
Molecular de Plantas
Profesor Albareda, 1
E-18008 Granada
Spain

Prof. Alain Puppo
Interactions Biotiques et Santé Végétale
UMR INRA 1301/Université de Nice-Sophia
Antipolis/CNRS 6243
400, route des Chappes, BP 167
06903 Sophia Antipolis Cedex
France

ISBN: 978-3-642-00389-9 e-ISBN: 978-3-642-00390-5
DOI: 10.1007/978-3-642-00390-5
Springer Dordrecht Heidelberg London New York

Library of Congress Control Number: 2009921146

Cover design: WMXDesign GmbH, Heidelberg, Germany

Printed on acid-free paper

Springer is part of Springer Science+Business Media (www.springer.com)

Preface

Oxygen (O_2) appeared in significant amounts in the Earth's atmosphere over 2.2 billion years ago, largely due to the evolution of photosynthesis by cyanobacteria (Halliwell 2006). The O_2 molecule is a free radical, as it has two impaired electrons that have the same spin quantum number. This spin restriction makes O_2 prefer to accept its electrons one at a time, leading to the generation of the so-called reactive oxygen species (ROS). The chemical nature of these species dictates that they can create damage in cells. This has contributed to the creation of the "oxidative stress" concept; in this view, ROS are unavoidable toxic products of O_2 metabolism and aerobic organisms have evolved antioxidant defences to protect against this toxicity (Halliwell 1981; Fridovich 1998). Indeed, even in present-day plants, which are full of antioxidants, much of the protein synthetic activity of chloroplasts is used to replace oxidatively damaged D1 and other proteins (Halliwell 2006). Yet, the use of the "oxidative stress" term implies that ROS exert their effects through indiscriminate widespread inactivation of cellular functions. In this context, ROS must not be able to react with lipids, proteins or nucleic acids in order to avoid any damage to vital cellular components.

However, genetic evidence has suggested that, in planta, purely physicochemical damage may be more limited than previously thought (Foyer and Noctor 2005). Thus, the concept of "oxidative stress", which implies a state to be avoided, was re-evaluated and the term "oxidative signaling" was created (Foyer and Noctor 2005). This means that ROS production, which was originally considered as an exclusively harmful process, is also an important component of the signaling network that plants use for their development and for responding to environmental challenges. The evolution of efficient antioxidant systems has most likely enabled plant cells to overcome ROS toxicity and to use these reactive species as signal transducers (Mittler 2006).

Results obtained during the last decade have highlighted the role of ROS as signals in plants, and it is now widely accepted that ROS are key regulators of plant metabolism, morphology, and development. The role of ROS as signals for gene expression has been evidenced (Desikan et al. 2001; Vanderauwera et al. 2005), and it is now known that ROS modulate the activity of key signaling compounds such as MAP kinases (Rentel et al. 2004). Furthermore, ROS can induce protein modifications, and thiol changes have been suggested as a widespread mechanism by which ROS might affect the activity and function of proteins (Cooper et al. 2002).

The identification of ROS-generating enzymes has demonstrated that plant cells can initiate ROS production for the purpose of signaling, and that the spatio-temporal characteristics of this production are likely to play an important role in the transduction of ROS signals (Mittler 2006). It is now obvious that sophisticated processes regulate these characteristics, which lead to the generation of inter-compartmental gradients, at least of hydrogen peroxide (H_2O_2). Transport of H_2O_2 through vesicle trafficking (Leshem et al. 2006) is most probably an important component of this process. Moreover, the facilitated diffusion of H_2O_2 across membranes through specific aquaporins (Bienert et al. 2007) appears to play a crucial role in the establishment of such H_2O_2 gradients. On the other hand, it appears that oxidation of target molecules by ROS is a part of how plants perceive and respond to environmental and developmental triggers (Foyer and Noctor 2005). Among the processes where ROS involvement has been evidenced, one can first cite the plant pathogen defense. During this process, ROS are produced by plant cells via plasma membrane-bound NADPH-oxidases, cell wall-bound peroxidases and amine oxidases in the apoplast (Bolwell 1999; Grant and Loake 2000). ROS production leads to programmed cell death (Bolwell 1999) and orchestrates the plant hypersensitive disease resistance response (Levine et al. 1994). Moreover, it has been shown that H_2O_2 can mediate the systemic expression of defense-related genes (Orozco-Cardenas et al. 2001). ROS appear to also play an important role in plant development and functioning. They are essential signals in stomatal closure in response to abscisic acid and elicitors (Pei et al. 2000). They are involved in root hair growth (Foreman et al. 2003) and auxin signaling and gravitropism (Joo et al. 2001). ROS induce the expression of genes encoding proteins required for peroxisome biogenesis (López-Huertas et al. 2000). Moreover, it has recently been shown that DELLAs proteins regulate plant growth and defense processes by modulating the levels of ROS (Achard et al. 2008).

The volume of research into the roles of ROS in plants is currently growing. Thus, the purpose of this book is to present recent advances in this field. The constitutive chapters are mainly arranged around four topics: (1) the generation of ROS, their network signalling, including the retrograde signalling from the chloroplast to the nucleus and the cross-talk with hormone signaling, (2) the signaling role of ROS produced in some sub-cellular compartments, (3) the role of ROS in plant growth, development, functioning and stress acclimation, and (4) their role in biotic and abiotic interactions. In thanking the authors for their contributions, we are convinced that the forthcoming years will bring new exciting insights into this field. Unravelling the sensing and transduction of ROS, including the post-translational regulation, the modulation of their concentration at subcellular level, and the interaction networks with the reactive nitrogen species, as well as their possible role in epigenetic processes, will shed new light on ROS action in plants.

Granada, April 2009 — Luis A. del Río

Sophia-Antipolis, April 2009 — Alain Puppo

References

Achard P, Renou JP, Berthome R, Harberd NP, Genschik P (2008) Plant DELLAs restrain growth and promote survival of adversity by reducing the levels of reactive oxygen species. Curr Biol 18: 656–660.

Bienert GP, Møller ALB, Kristiansen KA, Schultz A, Møller IM, Schjoerring JK, Jahn TP (2007) Specific aquaporins facilitate the diffusion of hydrogen peroxide across membranes. J Biol Chem 282: 1183–1192.

Bolwell GP (1999) Role of active oxygen species and NO in plant defense responses. Curr Opin Plant Biol 2: 287–294.

Cooper CE, Patel RP, Brookes PS, Darley-Usmar VM (2002) Nanotransducers in cellular redox signalling: modification of thiols by reactive oxygen and nitrogen species. Trends Biochem Sci 27: 489–492.

Desikan R, A-H-Mackerness S, Hancock JT, Neill SJ (2001) Regulation of the Arabidopsis transcriptome by oxidative stress. Plant Physiol 127: 159–172.

Foreman J, Demidchik V, Bothwell JH, Mylona P, Miedema H, Torres MA, Linstead P, Costa S, Brownlee C, Jones JD, Davies JM, Dolan L (2003) Reactive oxygen species produced by NADPH oxidase regulate plant cell growth. Nature 27: 442–446.

Foyer CH, Noctor G (2005) Oxidant and antioxidant signalling in plants: a re-evaluation of the concept of oxidative stress in a physiological context. Plant Cell Environ 28: 1056–1071.

Fridovich I (1998) Oxygen toxicity: a radical explanation. J Exp Biol 201: 1203–1209.

Grant JJ, Loake GJ (2000) Role of reactive oxygen intermediates and cognate redox signaling in disease resistance. Plant Physiol 124: 21–29.

Halliwell B (1981) Toxic effects of oxygen on plant tissues. In: Chloroplast metabolism: the structure and function of chloroplasts in green leaf cells, Clarenden Press, Oxford.

Halliwell B (2006) Reactive species and antioxidants. Redox biology is a fundamental theme of aerobic life. Plant Physiol 141: 312–322.

Joo JH, Bae YS, Lee JS (2001) Role of auxin-induced reactive oxygen species in root gravitropism. Plant Physiol 126: 1055–1060.

Leshem Y, Melamed-Book N, Cagnac O, Ronen G, Nishri Y, Solomon M, Cohen G, Levine A (2006) Suppression of Arabidopsis vesicle-SNARE expression inhibited fusion of H_2O_2-containing vesicles with tonoplast and increased salt tolerance. Proc. Natl. Acad. Sci. USA 103: 18008–18013.

Levine A, Tenhaken R, Dixon R, Lamb C (1994) H2O2 from the oxidative burst orchestrates the plant hypersensitive disease resistance response. Cell 79: 583–593.

López-Huertas E, Charlton WL, Johnson B, Graham IA, Baker A (2000) Stress induces peroxisome biogenesis genes. EMBO J 19: 6770–6777.

Mittler R (2006) The roles of reactive oxygen species in plant cells. Plant Physiol 141: 311.

Orozco-Cardenas ML, Narvaez-Vasquez J, Ryan CA (2001) Hydrogen peroxide acts as a second messenger for the induction of defense genes in tomato plants in response to wounding, systemin, and methyl jasmonate. Plant Cell 13: 179–191.

Pei Z-M, Murata Y, Benning G, Thomine S, Klusener B, Allen GJ, Grill E, Schroeder JI (2000) Calcium channels activated by hydrogen peroxide mediate abscisic acid signaling in guard cells. Nature 406: 731–734.

Rentel MC, Lecourieux D, Ouaked F, Usher SL, Peterson L, Okamoto H, Knight H, Peck SC, Grierson CS, Hirt H, Knight MR (2004) OXI1 kinase is necessary for oxidative burst-mediated signalling in Arabidopsis. Nature 427: 858–861.

Vanderauwera S, Zimmermann P, Rombauts S, Vandenabeele S, Langebartels C, Gruissem W, Inze D, Van Breusegem F (2005) Genome-wide analysis of hydrogen peroxide-regulated gene expression in Arabidopsis reveals a high light-induced transcriptional cluster involved in anthocyanin biosynthesis. Plant Physiol 139: 806–821.

Contents

Reactive Oxygen-Generating NADPH Oxidases in Plants

Robert Fluhr

Abstract Reactive oxygens produced by the integral membrane protein NADPH oxidase (NOX) function in defense, development, and redox-dependent signaling. They share common structural features and are evolutionarily of ancient origin and thus ubiquitous in eukaryotes. In plants, NOX are part of a multigene family and are implicated in diverse events including innate immunity and development. Due to the fact that reactive oxygens are toxic, and in many cases short-lived, the activity of these oxidases is tightly regulated both temporally and spatially. The recent elucidation of domains for activation by calcium and Rac binding in NOX as well as its positioning on membrane lipid rafts sketches a fascinating picture of its functional dynamics. This review draws upon comparative structure–function relationships between plant and animal NOXs to portray novel aspects of their biology.

1 Diversity of the NADPH Oxidase Superfamily Basic Structure and Evolutionary Aspects

Reactive oxygen species (ROS) produced by NADPH oxidase (NOX) have been shown to play many critical roles in signaling and development in plants, including plant defense response, cell death, abiotic stress, stomatal closure, and root hair development (Foreman et al. 2003; Jones et al. 2007; Kwak et al. 2003; Torres et al. 2002; Yoshioka et al. 2003). Yet ROS species can be cytotoxic and mutagenic and for their proper function in signaling their production must be carefully controlled. The NOXs are part of a large superfamily that arose early in evolution before animal/plant divergence. Plant NOXs are called RBOH (respiratory burst oxidase homologs). *Arabidopsis* contains ten homologues that comprise a distinct phylogenetic clade

R. Fluhr
Department of Plant Sciences, Weizmann Institute of Science, P.O.B. 26, Rehovot, 76100 Israel
e-mail: robert.fluhr@weizmann.ac.il

L.A. del Río and A. Puppo (eds.), *Reactive Oxygen Species in Plant Signaling*, Signaling and Communication in Plants, DOI 10.1007/978-3-642-00390-5_1,

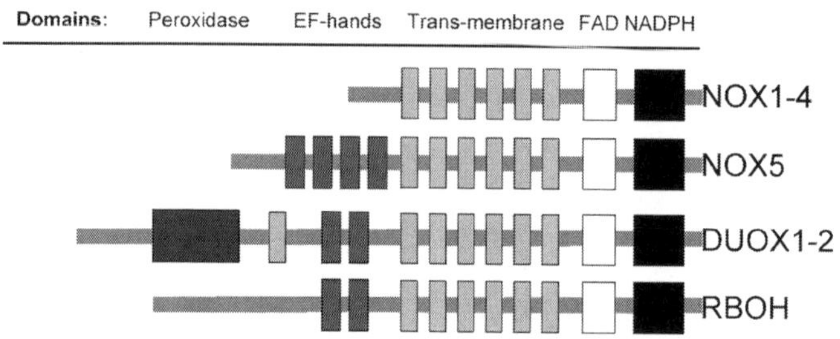

Fig. 1 Select family members of the NADPH oxidase (NOX) superfamily. The NOX/RBOH domain contains six transmembrane helices that are found in the core NOX2 structure and are found in all members of the NOX family. The NOX1-5 and DUOX1,2 are human proteins where NOX5 contains EF-hands and DUOX1,2 an additional peroxidase domain. Plant RBOH NOX homologues contain the NOX2 core and two additional EF hands

among the NOX-related genes (Sagi and Fluhr 2006). Plant RBOH are related in a structural sense to human NOX (Fig. 1) which presents a convenient paradigm for developing our understanding of RBOH biology.

In humans, there are five different NOX structural types. NOX1–3 contains the basic NOX domain and needs interaction with accessory proteins to be fully active. For example, NOX2, one of the best studied oxidases of this family, is prominent in phagocyte biology forming a complex called phox ($gp91^{phox}$, phagocyte NOX). It requires $p22^{phox}$ to form the flavocytochrome heterodimer b_{558} and additional regulatory subunits $p67^{phox}$, $p47^{phox}$, $p40^{phox}$, and the small GTPase Rac (Bedard and Krause 2007; Lambeth 2004). In contrast, the closely related NOX4 associates in epithelial cells with $p22^{phox}$ to produce large amounts of ROS constitutively and does not require accessory subunits or GTPase Rac for this activity (Martyn et al. 2006). Other structurally diverse human forms are NOX5 and DUOX1 and DUOX2 (Fig. 1). NOX5 has a N-terminal extension with EF-hand containing calcium-binding domains, while the DUOX members are characterized by a N-terminal peroxidase domain (Banfi et al. 2001; Edens et al. 2001).

Interestingly, not all animal species contain the full NOX complement found in humans, for example rat and mouse lack NOX5 having apparently lost it during their evolution. Simpler eukaryotes like yeast completely lack NOX homologues (Kawahara et al. 2007). At least three classes of NOX homologues, in a highly divergent form, are prevalent in fungi where they are thought to participate in sexual development, cellular differentiation, and appressorial formation (Aguirre et al. 2005; Giesbert et al. 2008; Takemoto et al. 2007) or play a role in fungal–host pathogenicity (Giesbert et al. 2008; Segmuller et al. 2008). Specialized NOX types continue to evolve in animals. NOX3, structurally related to NOX2, is a recent NOX derivative found only in mammals and birds. It is localized to the vestibular and cochlear sensory epithelia and spiral ganglions (Banfi et al. 2004a) apparently necessary for the formation of mineralized structures that are important for gravity perception in those organisms (Paffenholz et al. 2004).

2 RBOH Multi Gene Family

The presence of large gene families in organisms is usually an indication that each gene has specificity in its function together with a varying degree of functional overlap. Inspection of digital northern activities in *Arabidopsis* gathered from recent Affymetrix microarray slides reflects a high level of gene specialization (Table 1) (Zimmermann et al. 2004, 2005) The tissue-specific division of RBOH transcript distribution falls into three basic classes; expression throughout the plant (AtrbohD and F), in the roots (Atrboh A-G, I), and in a pollen-specific manner (AtrbohH and J). The tissue-specific expression is reflected in the phylogenetic distribution in which H and J form a small sub-clade (Sagi and Fluhr 2006). The most common abiotic inducers of Atrboh transcript accumulation include conditions

Table 1 Tissue specificity and response activities of atrboh transcripts

RBOH	AGI	Localization[a]	Induction characteristics[b]
A	At5g07390	Root, elongation zone	Hypoxia/salt stress, genotoxic, nitrogen starvation, fad2 mutant. Repressed: in *oxt6* (redox gene regulator)
B	At1g09090	Root, elongation zone	Anoxia, hypoxia, methyl jasmonate, UV-B, elevated in rbohC mutant. Repressed: ABA, cold, zeatin cycloheximide, myb 50, leaf1 mutant, repressed by miR164b, *oxt6*
C	At5g51060	Root, elongation zone	Most highly expressed transcript of the RBOH class. *B. cinerea, P. syringae*: *Agrobacterium*, ozone. Repressed: cycloheximide, H_2O_2, 6-benzyl adenine
D	At5g47910	All plant parts, less in roots	Cycloheximide, anoxia, H_2O_2, chitin, ozone, $AgNO_3$, methyl jasmonate, *max4* (inhibits shoot branching), *F. occidentalis, P infestens, P, syringae*. Repressed: ABA, high CO_2, fls2-17
E	At1g19230	Cell suspension, root, seeds	*Agrobacterium*, nitrogen starvation,, miR156b genotoxic. Repressed: senescence, miR172a, *oxt6*
F	At1g64060	All plant parts	*Agrobacterium*, brassinolide. Repressed: isoxaben, *mpk4* (constitutive SAR)
G	At4g25090	Root, elongation zone	Low nitrogen, salicylic acid, glucose sucrose miR172a. Repressed: myb61.2
H	At5g60010	Stamens, pollen	Repressed *oxt6*
I	At4g11230	Roots, elongation zone	Anoxia, cycloheximide, norflurazone, HSP90. Repressed: sfr6.4, *oxt6*
J	At3g45810	Stamens, pollen	Upregulated in rhd2 background, sfr6.1. Repressed: in camta1-2, *oxt6*

[a]Based on 3,110 microarray database compiled in Genevestigator V3. Results shown are significantly higher than background ($p \leq 0.06$). Experiments are summarized in https://www.genevestigator.ethz.ch. (Zimmermann et al. 2005)

[b]Induction of more than twofold or where indicated repressed by 1/2-fold and above 200 in the relative signal value.

of anoxia/hypoxia and nitrogen stress where AtrbohC and F are also induced by a variety of biotic stress. AtrbohC was specifically identified as playing a role in root hair development (Foreman et al. 2003), while AtrbohD is the major constitutively active form and AtrbohF is a biotic stress induced form (Torres et al. 2002). Both AtrbohD and F contribute to innate immune defense and ABA dependent guard cell opening, but only AtrbohF appears to be important in ABA-dependent root shortening (Kwak et al. 2003; Torres et al. 2002). Inspection of the transcriptional behavior of the RBOH family shows that these genes are in some cases regulated in concert. Thus, many of the RBOH family in *Arabidopsis* (6/10, Table 1) are suppressed in the *oxt6* mutant background (oxidative stress tolerant) that displays reduced sensitivity to a catalase inhibitor 3-amino-1, 2, 4-triazole (AT) and buthionine S,R-sulfoximine (BSO), an inhibitor of glutathione synthesis. The *oxt6* mutation involves inactivation of the specificity factor in the processing of transcript cleavage and polyadenylation (Zhang et al. 2008). Transcriptional profiling of oxt6 mutants detected elevated constitutive upregulated expression of a subset of genes that encode proteins containing thioredoxin- and glutaredoxin-related domains that represent alternative ROS-scavenging pathways. Thus, it is of note that oxt6 represses, in a reciprocal manner, members of the RBOH family, i.e., that are potential sources of ROS, and points to a common control mechanism for those genes.

3 Structurally Conserved Aspects of NOXs

Membrane proteins are notoriously difficult to crystallize and so far no crystal-based structure of a NOX is available. Thus, our functional understanding of its structure is based on decoding sequence motifs, analysis of the effects of mutations, and biochemical studies. The core NOX-like domain is characterized by six trans-membrane helices that support two heme groups bound to conserved histidines as shown in Fig. 2. The C-terminal contains topologically distinct hydrophilic domains for NADPH and FAD. The NADPH interaction sites have been identified in human NOX2 by competition of the photoaffinity ligand 2 azido-NADPH by NADPH (Segal et al. 1992). The amino acid sequences have further been identified by direct comparison to the NADPH-binding domains of spinach ferrodoxin NADPH reductase as established by crystal structure (Rotrosen et al. 1992). A comparison of these motifs in relation to AtrbohD sequence is shown in Fig. 3.

Other important domains can be discerned by identifying those mutations in human NOX2 that confer dysfunction and are also conserved in the RBOH sequence. The classic examples have been gleaned from mutations that cause chronic granulomatous disease (CGD), a genetic immune disorder caused by point mutations in NOX2. People with CGD are susceptible to ingress of exotic microorganism not usually encountered in people with normal immune systems, e.g., *Aspergillus* growth. The recurrent infections can cause tumor-like masses called granulomas. Some of the point mutations are conserved throughout all NOX types and cause instability of the polypeptide or its inactivity. In this manner, localization of the

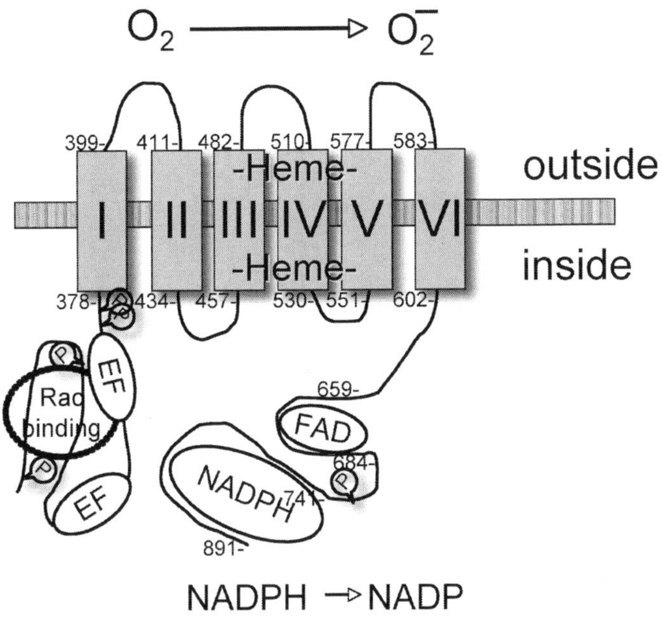

Fig. 2 Diagram of AtrbohD membrane localization. The conversion of NADPH to NADP will result in vectoral transfer of electrons with the participation of FAD and heme to an oxygen acceptor yielding superoxides. The core NOX-like domain is characterized by six transmembrane helices labeled I-VI. The two heme groups bound to conserved histidines that are present in helices III and V. The predicted "cytoplasmic side" contains binding cavities for FAD and the NADPH substrate. The presence of EF-hands and a putative Rac ineteracting site are depicted in the N-terminal region. Putative amino acid locations for transmembrane spanning approximations are based on the inspection of TMpred – Prediction of Transmembrane Regions and Orientation (http://www.ch.embnet.org/software/TMPRED_form.html) of the AtrbohD sequence and were further modified by manual alignment comparison to NOX2 and NOX5 sequence (Kawahara et al. 2007). *P* indicates putative phosphorylated amino acids and *Rac* indicates Rac/Rop small GTPases putative binding as discussed in the text

FAD binding region was predicted from studies using plasma membranes from a CGD patient with a point mutation at His-338. In that case, the NOX2 showed low FAD content in the plasma membrane and failed to produce superoxide (Yoshida et al. 1998). His-338 is located in the conserved amino acid motif ^{338}HPFT of human NOX2 which differs from the region suggested previously (Rotrosen et al. 1992). The equivalent conserved motif in AtrbohD is ^{660}HPFS. In human gp91^{phox}, mutants that are unable to reduce oxygen still retain diaphorase activity as measured by the reduction of iodonitrotetrazolium violet by NADPH (Cross et al. 1995).

Electrons from reduced FAD are transferred to two asymmetrical hemes that are ligated to histidine residues found in the transmembrane domains III and V. The exact positions of the histidines were established by alignment of gp91^{phox} to the yeast iron reductase. Yeast and plant iron reductases function to reduce exogenous ferric ions before their uptake. Both membrane proteins show a great deal of homology to

Gene	NADP pyrophoshate		NADP ribose		Adenine ring		C-4 atom of Nicotinamide
FNR	^{167}MLGTGTGIAPF	23	^{201}FLGVP	28	^{234}SREQTNEKGEKMYIQ	22	^{270}YMCG
NOX2	^{406}LVGAGIGVTPF	25	^{442}YWLCR	57	^{504}GLKQ----KTLYGR	21	^{535}FLCG
AtrbohD	^{741}LVGLGIGATPM	29	^{780}YWVT	72	^{856}GTRVK–SHFAKP	32	^{888}FYCG

Fig. 3 Spacing of NADPH interacting domains in AtrbohD compared to NOX2 and spinach ferrodoxin NADPH reductase (FNR) as defined by crystallography of FNR. The amino acid position is indicated at the start of each subdomain. The number of amino acids between each subdomain is indicated

other animal NOXs. The histidines in the yeast iron reductase were sequentially mutated and the resultant heme spectral activity monitored to reveal which mutations disrupted heme ligation capability (Finegold et al. 1996). Based on this work, the conserved AtrbohD histidines that serve as ligands for the hemes can be identified as follows: the cytoplasmic-proximal heme is ligated to histidines H459 and H561 and the outer heme ligation is to histidine H473 and H574 (Fig. 2). Incorporation of heme is also an important checkpoint in NOX biosynthesis. Indeed, blocking heme synthesis by application of succinyl acetone completely inhibited heterodimer formation of p22phox and gp65 which together form the core human NOX (DeLeo et al. 2000). Thus, it appears that insertion of the heme group is necessary for stabilization of the gp65 precomplex. The animal NOX2 oxidases are highly glycosylated indicating their passage through the Golgi, although the inhibition of glycolysation by application of tunicamycin does not affect the formation of stable heterodimer (DeLeo et al. 2000).

Additional domains of importance for the function of NOX2 that are highly conserved include: H209R, a conserved histidine binding residue (AtrbohD, 473); G389E, situated between the FAD and NADPH domains (AtrbohD, G713); and L420P (AtrbohD, L755) and W516R (AtrbohD, 869) that are in the NADPH domain. In comparison to the human NOX2, the NOX5 and RBOH proteins have shorter A and E extracellular loops (between transmembrane domains I and II and V and VI, respectively), as well as variable insertions between the FAD and NADPH domains (Fig. 2) (Kawahara et al. 2007). These conserved structural features are invaluable orientation guides that facilitate the transfer of structure-function relationships within the NOX superfamily.

4 Chemical Highlights of Superoxide Formation

Our knowledge of plant NOX chemistry is garnered from studies conducted in the animal NOX family (Bedard and Krause 2007; Vignais 2002). NOXs are transmembrane redox chain proteins that connect the electron donor, nicotinamide adenine dinucleotide phosphate (NADPH), on the cytosolic side of the membrane with the electron acceptor, oxygen, on the outer side of the membrane (Fig. 2). NADPH transfers two electrons (1.1) that culminates in the reduction of two oxygen molecules (1.2).

$$\mathrm{NADPH} -> \mathrm{NADP}^{+} + \mathrm{H}^{+} + 2\mathrm{e}^{-} \quad (1.1)$$

$$\mathrm{NADPH} + 2\mathrm{O}_2 -> \mathrm{NADP}^{+} + 2\mathrm{O}_2^{-} + \mathrm{H}^{+} \quad (1.2)$$

In the first step, electrons are transferred from NADPH to FAD by a two-electron transfer. Reduced FAD is then reoxidized in sequential single-electron transfers to the iron center of the inner heme. Oxidase activity is lost when the FAD is removed, e.g., by solubilization in detergents, but can be restored when FAD is added back (Light et al. 1981). Since the iron of the heme can only accept one electron at a time, the inner heme must donate its electron to the outer heme before the second electron can be accepted from the now partially reduced flavin, FADH. Low-potential flavoproteins may redox cycle as one-electron carriers. In that case, the oxidized form of FAD would cycle between the semi-quinone form (FADH·) and the fully reduced form ($FADH_2$). This scenario would require dimerization of two NOXs for functional NOX activity (Mayhew et al. 1996; Vignais 2002). Transfer of the electron from the inner heme to the outer heme is actually against the measured electromotive force between these two groups; hence, to create an energetically favorable state, oxygen must be bound to the outer heme to accept the electron. The transfer of the electron from the heme group to oxygen is not well understood. EPR studies indicate that all iron coordination sites are normally occupied, a situation that would not leave room for oxygen binding (Fujii et al. 1995; Isogai et al. 1995). However, upon addition of arachidonic acid, an activator of $gp91^{phox}$, a different EPR signal indicative of a pentacoordinated configuration of the heme ion was obtained, suggesting that under conditions of stimulation the heme group may direly interact with oxygen (Doussiere et al. 1996). Conformational changes in the NOX are anticipated to be important for switching the molecule from an inactive to an active form. In its active form, the phagocyte NOX is an electron transfer chain that results from a biological assembly of cytosolic regulatory factors ($p67^{phox}$, $p47^{phox}$, $p40^{phox}$, and Rac) on the membrane cytochrome b_{558} formed by two subunits: $gp91^{phox}$ and $p22^{phox}$ (Cross and Segal 2004; Nauseef 2004). It is assumed that initiation of electron flow is the consequence of a conformational change in $gp91^{phox}$ induced by its interaction with $p67^{phox}$ (Mizrahi et al. 2006). Mechanisms that bring about conformational changes in the NOX structure are a recurrent theme in its regulation, e.g., during activation of NOX5 and Atrboh by calcium as will be discussed below.

5 NOX Inhibitors

Ongoing oxidative basal metabolism, photosynthesis and many enzymatic activities produce ROS. Thus, specific inhibitors have been sought with the ability to identify the contribution of NOX activity to the redox milieu. However, at present, no specific NOX inhibitors exist, although the most commonly used NOX inhibitor, iodonium-derivative diphenylene iodonium (DPI), has great utility if its limitations are taken into account. Human neutrophil NOX is not inhibited by DPI pretreatment in the absence of an activating stimulus. It is likely that a reduced redox center in the oxidase could serve as electron donor to DPI to form a DPI radical. Inhibition occurs after direct phenylation of the redox cofactor or of other reactive groups in the area (Odonnell et al. 1993). Consistent with this, adducts were formed between photoreduced flavin and the inhibitor DPI. Furthermore, in pig neutrophil membranes, the quantity of recoverable intact flavin was greatly diminished when NADPH was present indicating that the flavin may be the site of DPI activation. However, it has been shown that iodonium compounds decrease the absorbance of the heme Soret peak in neutrophil membranes incubated with NADPH. The decrease was correlated with the loss of oxidase activity (Doussiere et al. 1999). Thus, the heme groups may be a secondary site of DPI action. Importantly, as suggested by the mechanism of action, DPI is a nonspecific inhibitor of many different electron transporters. It inhibits not only all the NOX isoforms, but also nitric oxide synthase (Stuehr et al. 1991), plant molybdenum-cofactor flavin-containing enzymes, xanthine dehydrogenase (XDH) (Yesbergenova et al. 2005), xanthine oxidase, mitochondrial complex I (Hutchinson et al. 2007), and cytochrome *P*-450 reductase (Zhukov and Ingelman-Sundberg 1999). The use of multiple inhibitors is thus necessary to get an indication of the relative contribution of NOXs to a particular milieu of reactive oxygen production (Yesbergenova et al. 2005).

6 Calcium-Binding Domains in RBOH Proteins

The canonical EF-hand consists of two alpha helices positioned roughly perpendicular to one another and linked by a short loop region of about 12 amino acids. The loop structure of the EF-hand binds calcium ions with residues positioned at 1, 3, 5, 7, 9, and 12 (X, Y, Z, -X, -Y, -Z). The positive calcium ion is complexed by charge interaction through negatively charged oxygen-containing side chains, such as aspartate and glutamate, that appear invariantly in the twelfth position and provide two oxygens for calcium binding (bidentate ligand). The sixth residue is invariantly a glycine that facilitates conformational bending requirements of the backbone. The remaining residues are typically hydrophobic and support a hydrophobic core that stabilizes the two helices.

All RBOH contain two EF-hand motifs where the EF-hand located closest to the N-terminal motif tends to be canonical. AtrbohI is exceptional, lacking canonical EF-hands in both structures (Fig. 4). The second EF-hand motif in all Atrboh is non-canonical lacking the invariant aspartate or glutamate at position 12 (-Z) as illustrated in Fig 4 (Kawahara et al. 2007). AtrbohF was shown to bind calcium (Keller et al. 1998) and calcium was shown to activate partially purified plant membrane fractions (Sagi and Fluhr 2001). Further insight into the binding properties of RBOH was obtained by examining a recombinant polypeptide of 83 amino acid size of the N-terminal region of AtrbohD. Calcium-dependent conformational changes were followed by fluorescence spectroscopy taking advantage of the presence of

	EF-1		EF-II[1]
	X Y Z-Y-X -Z		X Y Z-Y-X -Z
AtrbohA	^{234}DKDSDGRLNEAE	32	DPYHYGYIMIEN
AtrbohB	^{184}DKNLDGRITGDE	32	DRDNLGYIELHN
AtrbohC	^{239}DKDADGRLTEDE	32	DPDNIGYIMLES
AtrbohD	^{266}DKDEDGRVTEEE	32	DPDNAGFIMIEN
AtrbohE	^{281}DSNEDGKITREE	32	DPENFGYIELWQ
AtrbohF	^{277}DKNEDGRITEEE	32	DPERLGYIELWQ
AtrbohG	^{189}DKDSDGRLTEDE	32	DPDHMGYIMMES
AtrbohH	^{208}DKNGDGKLTEEE	32	DPDHKGYIEMWQ
AtrbohI	^{265}CYQLSSNLVKHI	32	APDGLYSQYIEL
AtrbohJ	^{218}DKDGDGKLTEEE	32	DPNEQGYIEMWQ
	*		* ^

Fig. 4 Atrboh EF-hand motif alignment in Atrboh genes. The sequences are deduced by comparison to canonical EF-hand domains and the NOX5 sequence. The positions of critical calcium binding amino acids are indicated by *X,Y,Z-Y-X* and *-Z*, and the amino acid position and spacing between domains are indicated. The invariant glycine at position 6 is indicated by * and the non-canonical -Z residue in the second EF-hand is indicated by ^

resident tyrosine residues as indicators of such changes (Ogasawara et al. 2008). The results showed that a level of 0.1 mM of calcium induced 50% of the observable conformational change. Considering that cellular calcium levels are orders of magnitude lower, the EF hands in this context (i.e., a short polypeptide) indicate only moderate sensitivity to calcium. When the non-canonical asparagine in the second EF-hand at the twelfth position was mutated to the acidic amino acid (N321E), a higher affinity to calcium was obtained indicating a difference in the affinity of each domain for calcium. However, when the EF-hands are examined in the context of the complete polypeptide, i.e., in a human cell line transfected with the full AtrbohD, a higher sensitivity to calcium was obtained. Indeed, in those cell lines, an interesting synergistic activation in the presence of the phosphatase inhibitor, calyculin A, and the calcium ionophore, ionomycin, was observed. Calyculin A leads to a higher phosphorylation state of the polypeptide. As the level of cellular calcium under these conditions was measured at 10^{-7} M this indicates a higher affinity to calcium due to either the phosphorylation of the protein, or that EF hands in the context of the full protein bind calcium more efficiently.

The NOX5 prototype could provide an important paradigm for understanding plant RBOH biology. NOX5 contains three canonical EF-hands whilst the fourth at the N-terminus is non-canonical. In a cell-free system, its activity is entirely dependent on calcium (Banfi et al. 2004b). Flow dialysis experiments established binding for all four sites with a tenfold difference in affinity assigned to the non-canonical structure. Interestingly, when tyrosine fluorescence was used to measure conformational change, such change occurred when both the N and C terminal fragments were together, indicating the necessity of their interaction. The researchers took advantage of the fact that a single cysteine was present in the alpha helical region flanking EF-III and noted that its titratable reduction by DTNB was accelerated in the presence of calcium, suggesting a more accessible conformation in that state. Indeed, when TNS, a hydrophobic probe, was used to measure the degree of hydrophobic exposure in the polypeptide, a large increase in its interaction was observed after calcium addition. Changes in hydrophobicity were further established by showing calcium-dependent binding of the protein melittin (Banfi et al. 2004b). Melittin, a 26 amino acid peptide, makes up 50% of the dry weight of bee venom; its hydrophobicity acts to destroy the victims blood cells by breaking up their membranes. Melittin was shown to form a complex with the NOX5 N-terminal fragment only in the presence of calcium. The calcium-dependent formation of a hydrophobic core is reminiscent of calcium binding to calmodulin that results in exposure of a deep hydrophobic cavity essential for recognition of calmodulin target protein. Interestingly, the binding of melittin was shown to inhibit NOX5 ROS production implying that the N-terminal region could interact with the 'business end' of NOX5. Indeed, recombinant N-terminal fragment could be shown to bind in a calcium-dependent manner to recombinant C-terminal containing the FAD and NADPH domains (Banfi et al. 2004b). Thus, the amino terminus of NOX5 operates in a manner analogous to calmodulin by promoting interaction with the C-terminal domain which then facilitates electron delivery and ROS production. The C-terminus of NOX5 was shown to contain a canonical calmodulin binding site of the so-called 1-14 CaMBD subclass that contains bulky hydrophobic residues in positions 1, 5, and 8 with a basic residue in position 13 (Fig. 5). This domain interacts with the N-terminal region and appears to increase the sensitivity of NOX5 enzymatic activity to calcium (Tirone and Cox 2007). Inspection of the RBOH plant sequences reveals that only AtrbohI sequence fills these criteria. Interestingly, only AtrbohI completely lacks canonical EF-hands sequence in its N-terminus, which may indicate alternative modes of regulation for this protein.

```
.........1...5..8...13.....NADPH
NOX5-      92WSKVFQKVAAEKKGK-VQVFFCG
AtrbohA-   WRSVFKRIAVNHPKTRVGVFYCG
AtrbohB-   WRSVFKHVAVNHVNQRVGVFYCG
AtrbohC-   WRNVYKRIAMDHPNTKVGVFYCG
AtrbohD-   WRQVYKKIAVQHPGKRIGVFYCG
AtrbohE-   WKEVFSSIARKHPNSTVGVFYCG
AtrbohF-   WKKVLTKLSSKHCNARIGVFYCG
AtrbohG-   WKNVYKQIAMDHPGANVGVFYCG
AtrbohI-   WKKVLSKISTKHRNARIGVFYCG
AtrbohJ-   WRKVFSELSNKHETSRIGVFYCG
```

Fig. 5 Putative calmodulin binding motifs in human NOX5 and Atrboh genes. Evidence for calmodulin binding in NOX5 is presented in the text and putative homologous regions are indicated for Atrboh genes. The sequence of AtrboH lacks homology to the other Atrboh in this region and is not included

7 Phosphorylation of RBOH Potentiates Calcium Responses

The relatively high levels of calcium required to activate NOX5 raised the question of its physiological relevance. To address this, COS-7 cells that do not normally experience elevated calcium levels, were transformed by NOX5 and examined for activity (Jagnandan et al. 2007). In this case, the addition of 12-myristate 13-acetate (PMA) was necessary to activate ROS production. The activity was greatly potentiated by the addition of ionomycin in a calcium-dependent manner. PMA was shown to induce phosphorylation of NOX5. A cluster of potential PKC substrate residues was identified at S486, S490, T494, and S498 all localized to the FAD domain region. Site-specific mutagenesis to alanine was carried out, and PMA activation and potentiation of the response to calcium was examined. While S486A showed no effect, mutagenesis of the other potential sites showed additive inhibition of the response. Indeed, mutations of T494E/S498E produced a phosphomimetic response in the absence of PMA. Interestingly, at the equivalent NOX5 T494 position, a conserved serine is present in AtrbohB, D, and F.

Arabidopsis leaves with functional FLS2 receptor rapidly produce ROS in response to the application of flagellin elicitor, flg22 (Gomez-Gomez et al. 1999). In a proteomic phosphorylation study, amino acid sites were recovered that show differential phosphorylation after xylanase or flg22 application these include; ^{35}GAFpSGPLGRPK, ^{341}ILpSQMLSQK, and ^{161}TSpSAAIHALK and indicate potential RBOH regulatory sites (Benschop et al. 2007). Inspection of the data in supplementary Table III (Benschop et al. 2007) reveals that at least in the case of xylanase treatment, an additional peptide ^{697}PPpTAGKpS showed differential phosphorylation. This peptide serine is at a conserved position similar to NOX5 T494. Whether it serves the same function in Atrboh is unknown. The sites in AtrbohD, S343 and S347, were confirmed to undergo phosphorylation after treatment with flg22 elicitor (Nuhse et al. 2007). These sites are conserved in several other Atrboh members and are indicated in Fig 2. Importantly, when they were mutated to alanine and transformed into the AtrbohD mutant, the leaf strips no longer responded to flg22 (Nuhse et al. 2007). None of these sites overlap with the phosphorylation site found for NOX5. Thus, AtrbohD has been shown to undergo elicitor-augmented phosphorylation at conserved and unique sites that contribute to its activity.

Kinases that may be directly responsible for RBOH regulation have been identified in potato. Using an anti-phosphopeptide antibody, Ser-82 and Ser-97, that are present in the N terminus of potato StrbohB, were identified as potential phosphorylation sites and Ser-82 was phosphorylated by pathogen signals in planta (Kobayashi et al. 2007). Two calcium-dependent protein kinases, StCDPK4 and StCDPK5, were shown to specifically phosphorylate these serines in a calcium-dependent manner. Significantly, in *N. benthamiana* the over-expression of the constitutively active mutant of StCDPK5 phosphorylated Ser-82 of StrbohB and induced ROS accumulation that depended on the presence of the homologous gene NbrbohB. In leaves where NbrbohB expression was abrogated, the loss of function could be complemented by expression of wild-type potato StrbohB but not by a mutant (S82A/S97A). These results suggest that calcium-dependent protein kinases are conserved and responsible for phosphorylation of the StrbohB N-terminal region and that this activity can regulate the oxidative burst. The positions of these residues fall on conserved serines present in Atrboh (S127 and S148) that have not been shown to be phosphorylated after elicitor treatment (above). The lack of identity in phosphorylated residues may be a result of comparison of non-orthologous genes or indicate inter-specific differences. However, in both *Arabidopsis* and potato, it seems evident that phosphorylation of the N-terminus serves to potentiate RBOH activity.

8 Small G Proteins Interact with RBOH

Full activity of human Nox1, -2, and -3 requires the direct action of a Rac GTPase. A Rac interaction site was discovered in human NOX2 juxtaposed to the first NADPH interacting site within amino acids 419–430 of NOX2 (Fig. 3) (Kao et al. 2008). Human NOX5 does not show a requirement for Rac and all the amino acids that are critical for Rac interaction in NOX2 are absent. It is therefore unlikely that this region is used for this purpose in plant RBOH. Rac/Rop small GTPases are prevalent in the plant kingdom and are part of multigene families (11 in *Arabidopsps*; 9 in rice) (Gu et al. 2004; Wong et al. 2007). OsRac1, which is located in the plasma membrane, is involved in Osrboh-dependent ROS production and cell death initiation during defense signaling (Kawasaki et al. 1999; Ono et al. 2001). Indeed, constitutively active and dominant negative forms of the small GTP-binding protein were shown to directly modulate ROS production. Furthermore, ROS scavengers such as metallothionein were reciprocally downregulated by Rac showing that OsRac1 plays a critical role in regulating the cellular redox milieu (Wong et al. 2004). The constitutively active (CA) form of OsRac2 and OsRac7, but not the dominant negative (DN) form, interacted with four different rice and two potato RBOH N-terminal fragments (Wong et al. 2007). Furthermore, transient agroinfiltration of *N. benthamiana* leaves produced ROS when the CA form was used. More evidence for the involvement of small GTPases in RBOH upregulation was provided by the phosphatidic acid-dependent induction of ROS in *Arabidopsis* cells (Park et al. 2004). In contrast to the results that Rac plays a positive role in RBOH induction, tobacco cells transformed with sense constructs of Ntrac5 or the constitutive active GTP form, Ntrac5V15, showed a decrease in the production of ROS after elicitation with cryptogein (Morel et al. 2004). Hence, Ntrac5 could be considered as a negative regulator of NtrbohD. Taken together, there is compelling evidence that Rac modulates RBOH and we may expect to find that different Rac types either induce or inhibit RBOH activity.

In animal NOX 1, 2, and 4, Rac binding was localized to the NADPH interacting region. In contrast, in rice, both N-terminal EF-hands were essential for Rac interaction and the full two hybrid interaction was obtained only when a considerable length of N-terminal region was used. In vitro examination showed that OsrbohB (residues 138–313) binds the GTP-binding form of OsRac1 (in the presence GMPPCP a nonhydrolyzable Rac substrate), apparently the binding of Rac to Osrboh is very sensitive to the character of its bound cofactor. In a fluorescent resonance energy transfer (FRET) system in which both interacting partners were fused to different fluochromes on the same polyprotein, a strong FRET signal was obtained with the CA but not DN OsRAC1 form. When calcium was added, this FRET signal decreased suggesting a negative interaction between calcium and OsRac binding (Wong et al. 2007). The authors therefore suggested a scenario in which calcium, at low levels, indirectly activates RBOH through CDPK phosphorylation, subsequently producing a conformational status that induces Rac binding and RBOH activation. Finally, continued cellular calcium buildup would disrupt Rac binding and return RBOH binding to a quiescent state. This scenario is seemingly in conflict with other evidence presented above, including direct activation of human NOX5 by calcium (Banfi et al. 2004b), the fact that calcium serves as a direct positive activator for partially purified tobacco membranes (Sagi and Fluhr 2001), and that in a heterologous expression system recombinant AtrbohD is activated by calcium (Ogasawara et al. 2008). It is thus likely that both calcium and Rac are positive – although mutually exclusive – activators permitting multiple inputs to impact on RBOH activity.

9 Membrane Microdomain Milieu of RBOH

Cellular fractionation of plant tissue indicates that RBOH proteins are found in plasmalemma membrane (Sagi and Fluhr 2001; Simon-Plas et al. 2002). However, the plasmalemma is not a uniform entity. Thus, NtrbohD was found to be enriched in tobacco BY2 cells on chemically distinct membrane microdomains called lipid rafts (Mongrand et al. 2004). Such rafts form a distinct lipid phase within the membrane and can be isolated due to their insolubility in nonionic detergents, e.g. Triton X-100. Plant lipid rafts are similar to animal lipid rafts and include a four- to fivefold enrichment of sterols and sphingolipids as their major components but are depleted for phospholipids (Borner et al. 2005; Grennan 2007). An exhaustive survey of tobacco BY2 cells identified 145 proteins in the detergent resistant membrane (DRM) fraction (Morel et al. 2006). This work confirmed that NtrbohD was present in lipid rafts as well as its possible regulator, the small G protein Rac5. Other proteins associated with hypersensitive response and pathogen resistance, such as HIN9, HIN18, and members of the gene family HIR, were also detected (Nadimpalli et al. 2000; Takahashi et al. 2004). In contrast, in an analysis of the *Medicago* root plasma membrane, RBOH was not detected, instead a slew of redox-generating proteins were recovered which could fill the role of ROS generation in root tissue (Lefebvre et al. 2007). Lipid rafts seem to be the focus of particular cellular events, i.e., while transport proteins are under-represented, others, like proteins that coordinate signaling, are enriched, including cellular trafficking, cell wall metabolism, and RBOH.

The localization of RBOH into lipid rafts can have far-meaning ramifications in RBOH biology. This can be discerned from following the dynamic changes in positioning of human NOX in membranes during its activation. In general, the NOX proteins are resistant to solubilization in Triton X-100. Following phagocyte activation, cholesterol-enriched microdomains act to recruit and/or to organize the cytosolic NOX factors in the assembly of the active NOX (Vilhardt and van Deurs 2004). Kinetic analysis of the activation of NOX in neutrophils implies that the onset, but not the maximal rate, of enzyme activity is determined by its presence on lipid rafts (Shao et al. 2003). However, different NOXs or cell types may alter the microdomain localization. Thus, NOX4 is highly expressed in cardiovascular tissue and ROS is implicated in cardiovascular diseases, including hypertension (Ellmark et al. 2005). The D1-like receptor agonist, fenoldopam, used for hypertension treatment dispersed NOX subunits of human renal proximal tubule (hRPT) cells that are found within lipid rafts and resulted in the reduction of ROS production, suggesting a novel mode of action for this drug. In contrast, cholesterol depletion caused the translocation of NOX subunits out of the lipid rafts and increased measurable ROS. Thus, in human renal proximal tubule cells, lipid rafts maintain NOX in a quiescent state suggesting that NOX micro-localization is important for its regulation (Han et al. 2008). In contrast, in endothelial cells, the activation by TNF-α stimulates lipid raft clustering of NOX2 components. This is important, as NOX2 positioning facilitates endothelial cell migration in the process of tissue healing (Ikeda et al. 2005; Ushio-Fukai 2006). In this context, caveolae invaginations of the plasma membrane of endothelial cells recruit NOX1 under unstimulated conditions and undergo activation by further recruitment of Rac1 after stimulation. The dynamic properties of lipid rafts that contribute to NOX activation have important ramifications for understanding RBOH biology.

10 Extracellular and Intracellular Localization of NOX and RBOH

Spatial, temporal, and quantitative components of ROS appearance dictate its biological significance during signaling. Therefore, the localization of NOX activity is crucial for its function. Cellular fractionation of plant tissue indicates that RBOH proteins are found in the plasmalemma membrane (Sagi and Fluhr 2001; Simon-Plas et al. 2002). Furthermore, recombinant GFP polypeptides of potato StrbohA and StrbohB localized to onion peel plasmalemma membranes during transient transformation (Kobayashi et al. 2006). The distribution on the membrane can be asymmetric. For example, RBOH activity in AtrbohC-dependent ROS signaling in root hair growth is distributed mainly to the tip (Carol et al. 2005; Foreman et al. 2003), and in RBOH involvement in xylem differentiation to one particular cell side (Barcelo 2005).

When situated on the plasmalemma, RBOH enzymatic activity would mediate the vectoral transfer of electrons to produce short-lived superoxides extracellularly. The locally produced signals do not propagate more then a few cell lengths from its source (Allan and Fluhr 1997). While extracellular functions for such ROS are evident (e.g. cell wall cross-linking) it is likely that the ROS signals generated by RBOH are processed intracellularly (Sagi et al. 2004; Sagi and Fluhr 2006). The relationship between externally produced superoxide product to modulation of the intracellular ROS signal are fundamental questions that remain to be answered. Given the importance of membrane topology in ROS production and subsequent redox signaling a thorough understanding of where O_2^- is converted to H_2O_2 is necessary. For example, the superoxide product is membrane impermeable in animals due to its negative charge in ambient conditions of pH (*p*Ka of superoxide is 4.8, e.g. blood pH is 7.4). However, under conditions of exceptionally low pH, the superoxide can be protonated and as such has been shown to be capable of crossing a yeast membrane (Wallace et al. 2004). In plants, the physiological range of extracellular pH is 5; in this case 16% of the superoxide would be in the membrane permeable hydroperoxyl (HO_2•) form. Thus, the external pH status could moderate the compartmentalization of superoxides produced by RBOH outside the membrane. Once having entered, the cytoplasmic SOD would convert the incoming superoxide to hydrogen peroxide. This scenario does not seem to provide efficient use of the NOX product. Alternatively, the superoxide could be converted extracellularly by an unknown SOD and the resultant product, H_2O_2, would then be freely permeable (Fig. 6). Evidence in support of the presence of extracellular SOD was found in secreted tobacco nectar that contains large amounts of hydrogen peroxide. The hydrogen peroxide is supplied in-part by the tandem action of RBOH and an extracellular germin-like superoxide dismutase protein called Nectarin I (Carter and Thornburg 2004). Other extracellular SOD activities are known, including a high isoelectric point-SOD isoform (hipI-SOD) that has been identified in the secondary cell wall of *Zinnia elegans* and pine (Karlsson et al. 2005; Karpinska et al. 2001).

Evidence from human NOX biology shows that they are not situated solely in the outer membrane; indeed, NADPH subunits have been identified in diverse intracellular locations. Within the cell, NOX5, the form most structurally related to RBOH, was detected in detergent-resistant microdomains of the endoplasmic reticulum. Generally, a strong perinuclear staining, including an extensive network of branching tubules that are reminiscent of localization to the endoplasmic reticulum as well as the plasma membrane, was observed. In the case of NOX5 over-expressing cells, it is difficult to distinguish between ER retention due to NOX5 over-expression and a true ER localization. (Jagnandan et al. 2007; Serrander et al. 2007).

The phagocyte-NOX2 expressed in endothelial cells differs from the neutrophil enzyme in that it exhibits activity even in the absence of agonist stimulation. When the subcellular location of oxidase subunits and activity were followed, 50% of the total activity and protein were found by fractionation to be in a 'low g' nuclear fraction (Li and Shah 2002). These results indicate that, in contrast to the neutrophil enzyme, a substantial proportion of the NOX of unstimulated endothelial cells exists as a preassembled intracellular complex associated with the cytoskeleton. The use of GFP-labeled recombinant proteins has provided additional visible support for internal, i.e., endoplasmic reticulum, localization for active NOXs particularly in the nuclear and ER regions (Ambasta et al. 2004; Li and Shah 2002; Martyn et al. 2006; Van Buul et al. 2005).

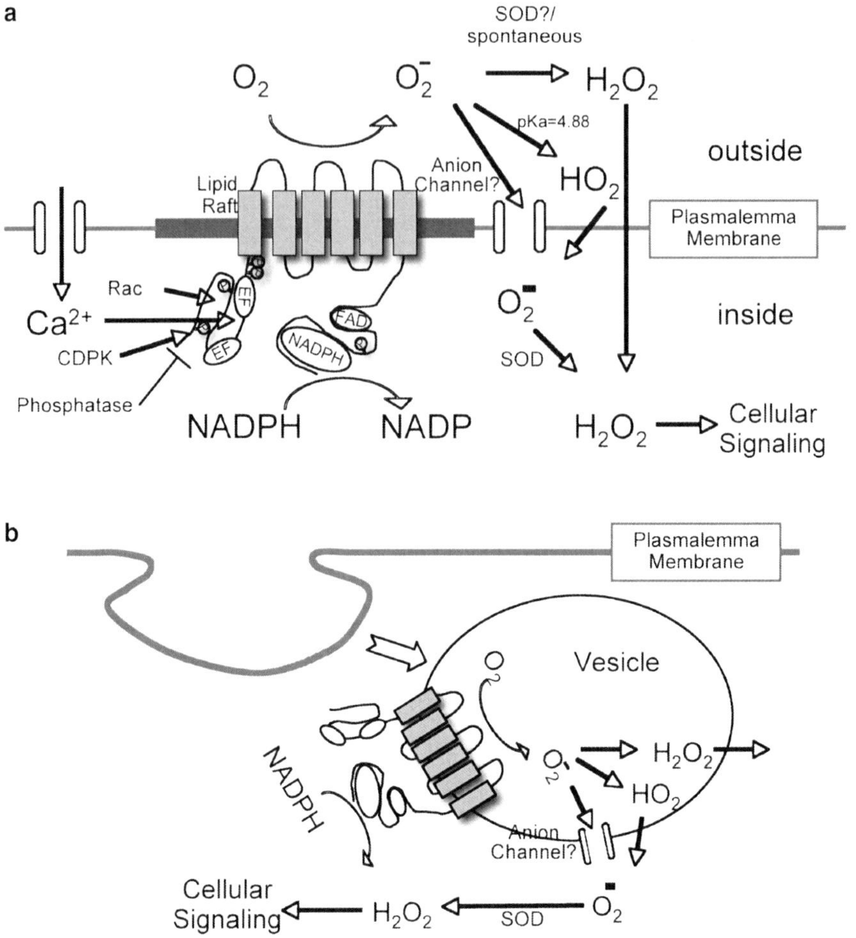

Fig. 6 Mode of RBOH activity. RBOH is shown situated on the plasmallema (**a**) or on a vesicle (**b**). In each case, the superoxide signal must diffuse into the intracellular space to induce redox-dependent signaling. RBOH interacts with calcium ions by binding through the EF hands as well as with Rac by binding within the N-terminal region. CDPK-like kinases phosphorylate the N-terminal region or together with other kinases at other regions, leading to potentiation of RBOH activity. The superoxide product will dismutate spontaneously to H_2O_2 or diffuse through the membrane via an anionic channel or diffuse through the membrane in a protonated form. Within the cell, the ROS product facilitates cellular signaling

Endocytosis is associated with downregulation of receptor-mediated signaling at the plasma membrane. However, it can also play an important role in the specificity and amplification of membrane-initiated receptor signals (Sorkin and von Zastrow 2002). NOXs are known to be internalized during activation by endocytosis. For example, H_2O_2-mediated redox change is a key mediator in IL-1β and TNF-α activation of NF-κB. The source of H_2O_2 was found to be provided by NOX2 that is situated in activated endosomes formed after stimulation by IL-1β (Li et al. 2006).

In this scenario, the local increases in H_2O_2 facilitates the redox-dependent association of TRAF6 with the receptor complex on ligand-activated endosomes, leading to activation of downstream IKK kinases and ultimately of NF-κB. Seemingly, upon membrane invagination, topologically the NOXs would produce superoxides that tend to be trapped in such endosomes. However, in mammary epithelial tumor cells that produce superoxides in their endosomes, an anion channel was found to play a role in endosome superoxide transport, thus providing an avenue for superoxide escape. In addition, activated endosomes recruited human SOD1 suggesting the rapid generation of H_2O_2 from exiting dismutating O_2^- at the endosomal surface (Mumbengegwi et al. 2008). It remains to be seen if such scenarios for ROS membrane translocation and efficient superoxide processing are prevalent elsewhere.

Only indirect evidence exists in plants for internalization of RBOH. Cryptogein, a proteinaceous elicitor, produced by the pathogenic fungus *Phytophthora cryptogea*, induces a hypersensitive response in tobacco (Binet et al. 2001; Ricci et al. 1989). An unidentified plasma membrane receptor leads to a cascade of events, including calcium influx, potassium and chloride efflux, cytosolic acidification, and plasma membrane depolarization that lead to RBOH activation (Bourque et al. 1999; Pugin et al. 1997; Simon-Plas et al. 2002). The production of apoplastic ROS in the oxidative defense-related burst promoted by cryptogein and other pathogens is biphasic: a rapid and transient production of ROS, which occurs within minutes of the perception of pathogens, and a second much later phase of ROS production within hours, which is specific to hypersensitive response-inducing pathogens (Grant and Loake 2000). The use of Amplex Red reagents, that can report real-time H_2O_2 accumulation in tobacco BY-2 cell populations, showed an internal signal for H_2O_2 that developed more rapidly than the external apoplastic signal. Thus, the ROS signal was actually initiated within the cells (Ashtamker et al. 2007). Furthermore, the fluorescent probe for H_2O_2, 2′,7′-dichlorofluorescin diacetate (DCF), showed signal development first in the nuclear region and only after a short delay in the cell periphery. Those results are consistent with ROS originating at intracellular sites. Indeed, subcellular accumulation of ROS was evident in the cytoplasm, endoplasmic reticulum (ER), and the nuclear region, implying diverse localization for tobacco RBOH (Ashtamker et al. 2007). These results are consistent with the observation that cryptogein-stimulated clathrin-mediated endocytosis in BY-2 cells was correlated with ROS production (Leborgne-Castel et al. 2008). Similarly, in *Arabidopsis* roots, RBOH-dependent hydrogen peroxide is produced during salt stress. During salt stress, ROS-loaded vesicles were noted to undergo trafficking to the central vacuole (Leshem et al. 2006).

11 Conclusion

NOXs are tethered to perform multiple signaling tasks. Some of what we have learned about the exquisite temporal and spatial control is depicted in Fig. 6. RBOH is localized to dynamic lipid rafts in the plasmalemma membrane and likely also in different cell types or special physiological conditions on the endomembrane system. There is evidence that the initial activation of RBOH is via a calcium signal. RBOH binds to the calcium ions with EF hands domains located in the N-terminal region which can also interact with Rac. As calcium and Rac bind to the same region, they may be mutually exclusive; however, it is likely that they are both positive activators of RBOH activity. In NOX5, the N-terminal region was shown to interact with the C-terminus in a calcium-dependent manner; it remains to be seen if this is true for RBOH. CDPK-like activities phosphorylate the N-terminal region of RBOH. Either alone or together with additional kinases, other sites are phosphorylated in an elicitor-dependent manner. In addition, Rac interaction domains are present and, together with phosphorylation, lead to potentiation of RBOH activity. It is likely that a cellular phosphatase activity serves to negatively moderate this potentiation. Once produced, the superoxide product can diffuse through the membrane in a protonated form or dismutate extracellulary to H_2O_2 which would then diffuse through the membrane. Alternatively, the superoxide may be transferred via an anionic channel as was found to be the case in mammary epithelial endosomes.

The unique ephemeral properties of superoxides and their rapid permutation to hydrogen peroxide enable them to play a role in influencing local redox responses. Recent elucidation of the structure function relationships of plant RBOH as summarized above are impressive first steps in understanding the biology of RBOH.

Acknowledgement The author acknowledges the support of the Dr. Josef Cohn MINERVA Center for Biomembrane Research.

References

Aguirre J, Rios-Momberg M, Hewitt D, Hansberg W (2005) Reactive oxygen species and development in microbial eukaryotes. Trends Microbiol 13:111–118

Allan AC, Fluhr R (1997) Two distinct sources of elicited reactive oxygen species in tobacco epidermal cells. Plant Cell 9:1559–1572

Ambasta RK, Kumar P, Griendling KK, Schmidt HHHW, Busse R, Brandes RP (2004) Direct interaction of the novel nox proteins with p22phox is required for the formation of a functionally active NADPH oxidase. J Biol Chem 279:45935–45941

Ashtamker C, Kiss V, Sagi M, Davydov O, Fluhr R (2007) Diverse subcellular locations of cryptogein-induced reactive oxygen species production in tobacco bright yellow-2 cells. Plant Physiol 143:1817–1826

Banfi B, Malgrange B, Knisz J, Steger K, Dubois-Dauphin M, Krause KH (2004a) NOX3, a superoxide-generating NADPH oxidase of the inner ear. J Biol Chem 279:46065–46072

Banfi B, Molnar G, Maturana A, Steger K, Hegedus B, Demaurex N, Krause KH (2001) A Ca^{2+}-activated NADPH oxidase in testis, spleen, and lymph nodes. J Biol Chem 276:37594–37601

Banfi B, Tirone F, Durussel I, Knisz J, Moskwa P, Molnar GZ, Krause KH, Cox JA (2004b) Mechanism of Ca^{2+} activation of the NADPH oxidase 5 (NOX5). J Biol Chem 279:18583–18591

Barcelo AR (2005) Xylem parenchyma cells deliver the H_2O_2 necessary for lignification in differentiating xylem vessels. Planta 220:747–756

Bedard K, Krause KH (2007) The NOX family of ROS-generating NADPH oxidases: physiology and pathophysiology. Physiol Rev 87:245–313

Benschop JJ, Mohammed S, O'Flaherty M, Heck AJR, Slijper M, Menke FLH (2007) Quantitative phosphoproteomics of early elicitor signaling in Arabidopsis. Mol Cell Proteomics 6:1198–1214
Binet MN, Humbert C, Lecourieux D, Vantard M, Pugin A (2001) Disruption of microtubular cytoskeleton induced by cryptogein, an elicitor of hypersensitive response in tobacco cells. Plant Physiol 125:564–572
Borner GHH, Sherrier DJ, Weimar T, Michaelson LV, Hawkins ND, MacAskill A, Napier JA, Beale MH, Lilley KS, Dupree P (2005) Analysis of detergent-resistant membranes in Arabidopsis. Evidence for plasma membrane lipid rafts. Plant Physiol 137:104–116
Bourque S, Binet MN, Ponchet M, Pugin A, Lebrun-Garcia A (1999) Characterization of the cryptogein binding sites on plant plasma membranes. J Biol Chem 274:34699–34705
Carol RJ, Takeda S, Linstead P, Durrant MC, Kakesova H, Derbyshire P, Drea S, Zarsky V, Dolan L (2005) A RhoGDP dissociation inhibitor spatially regulates growth in root hair cells. Nature 438:1013–1016
Carter C, Thornburg RW (2004) Is the nectar redox cycle a floral defense against microbial attack? Trends Plant Sci 9:320–324
Cross AR, Heyworth PG, Rae J, Curnutte JT (1995) A variant X-linked chronic granulomatous disease patient (X91+) with partially functional cytochrome b. J Biol Chem 270:8194–8200
Cross AR, Segal AW (2004) The NADPH oxidase of professional phagocytes – prototype of the NOX electron transport chain systems. Biochim Biophys Acta Bioenergetics 1657:1–22
DeLeo FR, Burritt JB, Yu LX, Jesaitis AJ, Dinauer MC, Nauseef WM (2000) Processing and maturation of flavocytochrome b_{558} include incorporation of heme as a prerequisite for heterodimer assembly. J Biol Chem 275:13986–13993
Doussiere J, Gaillard J, Vignais PV (1996) Electron transfer across the O_2^- generating flavocytochrome b of neutrophils. Evidence for a transition from a low-spin state to a high-spin state of the heme iron component. Biochemistry 35:13400–13410
Doussiere J, Gaillard J, Vignais PV (1999) The heme component of the neutrophil NADPH oxidase complex is a target for aryliodonium compounds. Biochemistry 38:3694–3703
Edens WA, Sharling L, Cheng GJ, Shapira R, Kinkade JM, Lee T, Edens HA, Tang XX, Sullards C, Flaherty DB, Benian GM, Lambeth JD (2001) Tyrosine cross-linking of extracellular matrix is catalyzed by Duox, a multidomain oxidase/peroxidase with homology to the phagocyte oxidase subunit gp91^{phox}. J Cell Biol 154:879–891
Ellmark SHM, Dusting GJ, Fui MNT, Guzzo-Pernell N, Drummond GR (2005) The contribution of Nox4 to NADPH oxidase activity in mouse vascular smooth muscle. Cardiovasc Res 65:495–504
Finegold AA, Shatwell KP, Segal AW, Klausner RD, Dancis A (1996) Intramembrane bis-heme motif for transmembrane electron transport conserved in a yeast iron reductase and the human NADPH oxidase. J Biol Chem 271:31021–31024
Foreman J, Demidchik V, Bothwell JHF, Mylona P, Miedema H, Torres MA, Linstead P, Costa S, Brownlee C, Jones JDG, Davies JM, Dolan L (2003) Reactive oxygen species produced by NADPH oxidase regulate plant cell growth. Nature 422:442–446
Fujii H, Johnson MK, Finnegan MG, Miki T, Yoshida LS, Kakinuma K (1995) Electron-Spin-Resonance studies on neutrophil cytochrome b_{558} – evidence that low-spin heme iron is essential for O_2^- generating activity. J Biol Chem 270:12685–12689
Giesbert S, Schurg T, Scheele S, Tudzynski P (2008) The NADPH oxidase Cpnox1 is required for full pathogenicity of the ergot fungus *Claviceps purpurea*. Mol Plant Path 9:317–327
Gomez-Gomez L, Felix G, Boller T (1999) A single locus determines sensitivity to bacterial flagellin in *Arabidopsis thaliana*. Plant J 18:277–284
Grant JJ, Loake GJ (2000) Role of reactive oxygen intermediates and cognate redox signaling in disease resistance. Plant Physiol 124:21–29
Grennan AK (2007) Lipid rafts in plants. Plant Physiol 143:1083–1085
Gu Y, Wang Z, Yang Z (2004) ROP/RAC GTPase: an old new master regulator for plant signaling. Curr Opin Plant Biol 7:527–536
Han WX, Li HW, Villar VAM, Pascua AM, Dajani MI, Wang XY, Natarajan A, Quinn MT, Felder RA, Jose PA, Yu PY (2008) Lipid rafts keep NADPH oxidase in the inactive state in human renal proximal tubule cells. Hypertension 51:481–487

Hutchinson DS, Csikasz RI, Yamamoto DL, Shabalina IG, Wikstrom P, Wilcke M, Bengtsson T (2007) Diphenylene iodonium stimulates glucose uptake in skeletal muscle cells through mitochondrial complex I inhibition and activation of AMP-activated protein kinase. Cell Signal 19:1610–1620

Ikeda S, Yamaoka-Toji M, Hilenski L, Patrushev NA, Anwar GM, Quinn MT, Ushio-Fukai M (2005) IQGAP1 regulates reactive oxygen species-dependent endothelial cell migration through interacting with Nox2. Arterioscler Thromb Vasc Biol 25:2295–2300

Isogai Y, Iizuka T, Shiro Y (1995) The mechanism of electron donation to molecular-oxygen by phagocytic cytochrome b_{558}. J Biol Chem 270:7853–7857

Jagnandan D, Church JE, Banfi B, Stuehr DJ, Marrero MB, Fulton DJR (2007) Novel mechanism of activation of NADPH oxidase 5 – Calcium sensitization via phosphorylation. J Biol Chem 282:6494–6507

Jones MA, Raymond MJ, Yang ZB, Smirnoff N (2007) NADPH oxidase-dependent reactive oxygen species formation required for root hair growth depends on ROP GTPase. J Exp Bot 58:1261–1270

Kao YY, Gianni D, Bohl B, Taylor RM, Bokoch GM (2008) Identification of a conserved Rac-binding site on NADPH oxidases supports a direct GTPase regulatory mechanism. J Biol Chem 283:12736–12746

Karlsson M, Melzer M, Prokhorenko I, Johansson T, Wingsle G (2005) Hydrogen peroxide and expression of hipI-superoxide dismutase are associated with the development of secondary cell walls in *Zinnia elegans*. J Exp Bot 56:2085–2093

Karpinska B, Karlsson M, Schinkel H, Streller S, Suss KH, Melzer M, Wingsle G (2001) A novel superoxide dismutase with a high isoelectric point in higher plants. Expression, regulation, and protein localization. Plant Physiol 126:1668–1677

Kawahara T, Quinn MT, Lambeth JD (2007) Molecular evolution of the reactive oxygen-generating NADPH oxidase (Nox/Duox) family of enzymes. BMC Evol Biol 7:109

Kawasaki T, Henmi K, Ono E, Hatakeyama S, Iwano M, Satoh H, Shimamoto K (1999) The small GTP-binding protein Rac is a regulator of cell death in plants. Proc Natl Acad Sci USA 96:10922–10926

Keller T, Damude HG, Werner D, Doerner P, Dixon RA, Lamb C (1998) A plant homolog of the neutrophil NADPH oxidase gp91^{phox} subunit gene encodes a plasma membrane protein with Ca^{2+} binding motifs. Plant Cell 10:255–266

Kobayashi M, Kawakita K, Maeshima M, Doke N, Yoshioka H (2006) Subcellular localization of Strboh proteins and NADPH-dependent O_2^--generating activity in potato tuber tissues. J Exp Bot 57:1373–1379

Kobayashi M, Ohura I, Kawakita K, Yokota N, Fujiwara M, Shimamoto K, Doke N, Yoshioka H (2007) Calcium-dependent protein kinases regulate the production of reactive oxygen species by potato NADPH oxidase. Plant Cell 19:1065–1080

Kwak JM, Mori IC, Pei ZM, Leonhardt N, Torres MA, Dangl JL, Bloom RE, Bodde S, Jones JDG, Schroeder JI (2003) NADPH oxidase AtrbohD and AtrbohF genes function in ROS-dependent ABA signaling in Arabidopsis. EMBO J 22:2623–2633

Lambeth JD (2004) Nox enzymes and the biology of reactive oxygen. Nat Rev Immuno 4:181–189

Leborgne-Castel N, Lherminier J, Der C, Fromentin J, Houot V, Simon-Plas F (2008) The plant defense elicitor cryptogein stimulates clathrin-mediated endocytosis correlated with reactive oxygen species production in Bright Yellow-2 tobacco cells. Plant Physiol 146:1255–1266

Lefebvre B, Furt F, Hartmann MA, Michaelson LV, Carde JP, Sargueil-Boiron F, Rossignol M, Napier JA, Cullimore J, Bessoule JJ, Mongrand S (2007) Characterization of lipid rafts from *Medicago truncatula* root plasma membranes: A proteomic study reveals the presence of a raft-associated redox system. Plant Physiol 144:402–418

Leshem Y, Melamed-Book N, Cagnac O, Ronen G, Nishri Y, Solomon M, Cohen G, Levine A (2006) Suppression of Arabidopsis vesicle-SNARE expression inhibited fusion of H2O2 containing vesicles with tonoplast and increased salt tolerance. Proc Natl Acad Sci USA 103:18008–18013

Li JM, Shah AM (2002) Intracellular localization and preassembly of the NADPH oxidase complex in cultured endothelial cells. J Biol Chem 277:19952–19960

Li QA, Harraz MM, Zhou WH, Zhang LN, Ding W, Zhang YL, Eggleston T, Yeaman C, Banfi B, Engelhardt JF (2006) Nox2 and Rac1 regulate H_2O_2-dependent recruitment of TRAF6 to endosomal interleukin-1 receptor complexes. Mol Cell Biol 26:140–154

Light DR, Walsh C, O'Callaghan AM, Goetzl EJ, Tauber AI (1981) Characteristics of the cofactor requirements for the superoxide-generating NADPH oxidase of human polymorphonuclear leukocytes. Biochemistry 20:1468–1476

Martyn KD, Frederick LM, von Loehneysen K, Dinauer MC, Knaus UG (2006) Functional analysis of Nox4 reveals unique characteristics compared to other NADPH oxidases. Cell Signal 18:69–82

Mayhew SG, OConnell DP, OFarrell PA, Yalloway GN, Geoghegan SM (1996) Regulation of the redox potentials of flavodoxins: Modification of the flavin binding site. Biochem Soc Trans 24:122–127

Mizrahi A, Berdichevsky Y, Ugolev Y, Molshanski-Mor S, Nakash Y, Dahan I, Alloul N, Gorzalczany Y, Sarfstein R, Hirshberg M, Pick E (2006) Assembly of the phagocyte NADPH oxidase complex: chimeric constructs derived from the cytosolic components as tools for exploring structure-function relationships. J Leukocyte Biol 79:881–895

Mongrand S, Morel J, Laroche J, Claverol S, Carde JP, Hartmann MA, Bonneu M, Simon-Plas F, Lessire R, Bessoule JJ (2004) Lipid rafts in higher plant cells – Purification and characterization of triton X-100-insoluble microdomains from tobacco plasma membrane. J Biol Chem 279:36277–36286

Morel J, Claverol S, Mongrand S, Furt F, Fromentin J, Bessoule JJ, Blein JP, Simon-Plas F (2006) Proteomics of plant detergent-resistant membranes. Mol Cell Proteomics 5:1396–1411

Morel J, Fromentin J, Blein JP, Simon-Plas F, Elmayan T (2004) Rac regulation of NtrbohD, the oxidase responsible for the oxidative burst in elicited tobacco cell. Plant J 37:282–293

Mumbengegwi DR, Li Q, Li CH, Bear CE, Engelhardt JF (2008) Evidence for a superoxide permeability pathway in endosomal membranes. Mol Cell Biol 28:3700–3712

Nadimpalli R, Yalpani N, Johal GS, Simmons CR (2000) Prohibitins, stomatins, and plant disease response genes compose a protein superfamily that controls cell proliferation, ion channel regulation, and death. J Biol Chem 275:29579–29586

Nauseef WM (2004) Assembly of the phagocyte NADPH oxidase. Histochem Cell Biol 122:277–291

Nuhse TS, Bottrill AR, Jones AME, Peck SC (2007) Quantitative phosphoproteomic analysis of plasma membrane proteins reveals regulatory mechanisms of plant innate immune responses. Plant J 51:931–940

Odonnell VB, Tew DG, Jones OTG, England PJ (1993) Studies on the inhibitory mechanism of iodonium compounds with special reference to neutrophil Nadph Oxidase. Biochem J 290:41–49

Ogasawara Y, Kaya H, Hiraoka G, Yumoto F, Kimura S, Kadota Y, Hishinuma H, Senzaki E, Yamagoe S, Nagata K, Nara M, Suzuki K, Tanokura M, Kuchitsu K (2008) Synergistic activation of the Arabidopsis NADPH oxidase AtrbohD by Ca^{2+} and phosphorylation. J Biol Chem 283:8885–8892

Ono E, Wong HL, Kawasaki T, Hasegawa M, Kodama O, Shimamoto K (2001) Essential role of the small GTPase Rac in disease resistance of rice. Proc Natl Acad Sci USA 98:759–764

Paffenholz R, Bergstrom RA, Pasutto F, Wabnitz P, Munroe RJ, Jagla W, Heinzmann U, Marquardt A, Bareiss A, Laufs I, Russ A, Stumm G, Schimenti JC, Bergstrom DE (2004) Vestibular defects in head-tilt mice result from mutations in Nox3, encoding an NADPH oxidase. Genes Dev 18:486–491

Park J, Gu Y, Lee Y, Yang ZB, Lee Y (2004) Phosphatidic acid induces leaf cell death in Arabidopsis by activating the Rho-related small G protein GTPase-mediated pathway of reactive oxygen species generation. Plant Physiol 134:129–136

Pugin A, Frachisse JM, Tavernier E, Bligny R, Gout E, Douce R, Guern J (1997) Early events induced by the elicitor cryptogein in tobacco cells: Involvement of a plasma membrane NADPH oxidase and activation of glycolysis and the pentose phosphate pathway. Plant Cell 9:2077–2091

Ricci P, Bonnet P, Huet JC, Sallantin M, Beauvaiscante F, Bruneteau M, Billard V, Michel G, Pernollet JC (1989) Structure and activity of proteins from pathogenic fungi *Phytophthora* eliciting necrosis and acquired-resistance in tobacco. Eur J Biochem 183:555–563

Rotrosen D, Yeung CL, Leto TL, Malech HL, Kwong CH (1992) Cytochrome-B_{558}: the flavin-binding component of the phagocyte NADPH Oxidase. Science 256:1459–1462

Sagi M, Davydov O, Orazova S, Yesbergenova Z, Ophir R, Stratmann JW, Fluhr R (2004) Plant respiratory burst oxidase homologs impinge on wound responsiveness and development in *Lycopersicon esculentum*. Plant Cell 16:616–628

Sagi M, Fluhr R (2001) Superoxide production by plant homologues of the $gp91^{phox}$ NADPH oxidase. Modulation of activity by calcium and by tobacco mosaic virus infection. Plant Physiol 126:1281–1290

Sagi M, Fluhr R (2006) Production of reactive oxygen species by plant NADPH oxidases. Plant Physiol 141:336–340

Segal AW, West I, Wientjes F, Nugent JHA, Chavan AJ, Haley B, Garcia RC, Rosen H, Scrace G (1992) Cytochrome b_{-245} Is a flavocytochrome containing FAD and the NADPH-binding site of the microbicidal oxidase of phagocytes. Biochem J 284:781–788

Segmuller N, Kokkelink L, Giesbert S, Odinius D, van Kan J, Tudzynski P (2008) NADPH Oxidases are involved in differentiation and pathogenicity in Botrytis cinerea. Mol Plant Microbe Interactions 21:808–819

Serrander L, Jaquet V, Bedard K, Plastre O, Hartley O, Arnaudeau S, Demaurex N, Schlegel W, Krause KH (2007) NOX5 is expressed at the plasma membrane and generates superoxide in response to protein kinase C activation. Biochimie 89:1159–1167

Shao DM, Segal AW, Dekker LV (2003) Lipid rafts determine efficiency of NADPH oxidase activation in neutrophils. FEBS Lett 550:101–106

Simon-Plas F, Elmayan T, Blein JP (2002) The plasma membrane oxidase NtrbohD is responsible for AOS production in elicited tobacco cells. Plant J 31:137–147

Sorkin A, von Zastrow M (2002) Signal transduction and endocytosis: Close encounters of many kinds. Nat Rev Mol Cell Biol 3:600–614

Stuehr DJ, Fasehun OA, Kwon NS, Gross SS, Gonzalez JA, Levi R, Nathan CF (1991) Inhibition of macrophage and endothelial-cell nitric-oxide synthase by diphenyleneiodonium and its analogs. FASEB J 5:98–103

Takahashi Y, Berberich T, Yamashita K, Uehara Y, Miyazaki A, Kusano T (2004) Identification of tobacco HIN1 and two closely related genes as spermine-responsive genes and their differential expression during the Tobacco mosaic virus-induced hypersensitive response and during leaf- and flower-senescence. Plant Mol Biol 54:613–622

Takemoto D, Tanaka A, Scott B (2007) NADPH oxidases in fungi: Diverse roles of reactive oxygen species in fungal cellular differentiation. Fungal Genet Biol 44:1065–1076

Tirone F, Cox JA (2007) NADPH oxidase 5 (NOX5) interacts with and is regulated by calmodulin. FEBS Lett 581:1202–1208

Torres MA, Dangl JL, Jones JDG (2002) Arabidopsis gp91(phox) homologues AtrbohD and AtrbohF are required for accumulation of reactive oxygen intermediates in the plant defense response. Proc Natl Acad Sci USA 99:517–522

Ushio-Fukai M (2006) Redox signaling in angiogenesis: Role of NADPH oxidase. Cardiovasc Res 71:226–235

Van Buul JD, Fernandez-Borja M, Anthony EC, Hordijk PL (2005) Expression and localization of NOX2 and NOX4 in primary human endothelial cells. Antioxid Redox Signal 7:308–317

Vignais PV (2002) The superoxide-generating NADPH oxidase: structural aspects and activation mechanism. Cell Mol Life Sci 59:1428–1459

Vilhardt F, van Deurs B (2004) The phagocyte NADPH oxidase depends on cholesterol-enriched membrane microdomains for assembly. EMBO J 23:739–748

Wallace MA, Liou LL, Martins J, Clement MHS, Bailey S, Longo VD, Valentine JS, Gralla EB (2004) Superoxide inhibits 4Fe–4S cluster enzymes involved in amino acid biosynthesis - Cross-compartment protection by CuZn-superoxide dismutase. J Biol Chem 279:32055–32062

Wong HL, Pinontoan R, Hayashi K, Tabata R, Yaeno T, Hasegawa K, Kojima C, Yoshioka H, Iba K, Kawasaki T, Shimamoto K (2007) Regulation of rice NADPH oxidase by binding of Rac GTPase to its N-terminal extension. Plant Cell 19:4022–4034

Wong HL, Sakamoto T, Kawasaki T, Umemura K, Shimamoto K (2004) Down-regulation of metallothionein, a reactive oxygen scavenger, by the small GTPase OsRac1 in rice. Plant Physiol 135:1447–1456

Yesbergenova Z, Yang GH, Oron E, Soffer D, Fluhr R, Sagi M (2005) The plant Mo-hydroxylases aldehyde oxidase and xanthine dehydrogenase have distinct reactive oxygen species signatures and are induced by drought and abscisic acid. Plant J 42:862–876

Yoshida LS, Saruta F, Yoshikawa K, Tatsuzawa O, Tsunawaki S (1998) Mutation at histidine 338 of gp91phox depletes FAD and affects expression of cytochrome b558 of the human NADPH oxidase. J Biol Chem 273:27879–27886

Yoshioka H, Numata N, Nakajima K, Katou S, Kawakita K, Rowland O, Jones JDG, Doke N (2003) *Nicotiana benthamiana* gp91^{phox} homologs NbrbohA and NbrbohB participate in H2O2 accumulation and resistance to *Phytophthora infestans*. Plant Cell 15:706–718

Zhang J, Addepalli B, Yun KY, Hunt AG, Xu R, Rao S, Li QQ, Falcone DL (2008) A polyadenylation factor subunit implicated in regulating oxidative signaling in Arabidopsis thaliana. PLoS ONE 3:e2410

Zhukov A, Ingelman-Sundberg M (1999) Relationship between cytochrome P450 catalytic cycling and stability: fast degradation of ethanol-inducible cytochrome P450 2E1 (CYP2E1) in hepatoma cells is abolished by inactivation of its electron donor NADPH-cytochrome P450 reductase. Biochem J 340:453–458

Zimmermann P, Hennig L, Gruissem W (2005) Gene-expression analysis and network discovery using Genevestigator. Trends Plant Sci 10:407–409

Zimmermann P, Hirsch-Hoffmann M, Hennig L, Gruissem W (2004) GENEVESTIGATOR. Arabidopsis microarray database and analysis toolbox. Plant Physiol 136:2621–2632

Integration of ROS and Hormone Signaling

Izumi C. Mori, Yoshiyuki Murata, and Misugi Uraji

Abstract Plant hormones are major determinants of regulation of development and stress response of plants. Hormone signal networks are integrated in order to render an appropriate response to internal and external stimuli. Recently, emerging evidence suggests the role of reactive oxygen species (ROS) in hormone signal integration processes. In this chapter, we review the involvement of ROS in plant hormone responses and discuss the function of ROS in hormone signal integration in plant cells.

1 Auxin

1.1 Oxidative Degradation of Auxin, ROS Generation, and Cell Wall Loosening

Classically, it was reported that the auxin, indole-3-acetic acid (IAA), was degraded by spontaneous oxidation with hydrogen peroxide (H_2O_2), and thus H_2O_2 negatively affects auxin signaling by decreasing the concentration of auxin in root tissue (Omran 1977). Later, many papers have been published demonstrating that auxin is oxidized by molecular oxygen (O_2), but not H_2O_2, in a reaction catalyzed by peroxidases (POX) (Nakajima and Yamazaki 1979; Savitsky et al. 1999). Degradation of auxin by POX-mediated oxidation may sound a mode of negative regulation of auxin by decreasing concentration of auxin in tissue. Paradoxically, recent studies suggest that the degradation of IAA by cell wall-associated POX is rather a mechanism that induces elongation and gravitropism of roots, elongation of coleoptile, and hypothetically, phototropism (Joo et al. 2001; Schopfer 2001; Liszkay et al. 2004).

I. C. Mori (✉)
Research Institute for Bioresouces, Okayama Univeristy, 710-0046, Kurashiki, Japan

Y. Murata and M. Uraji
Graduate School of Natural Science and Technology, Okayama University, Okayama, 700-8530, Japan

L.A. del Río and A. Puppo (eds.), *Reactive Oxygen Species in Plant Signaling*, Signaling and Communication in Plants, DOI 10.1007/978-3-642-00390-5_2,

Auxin induces cell elongation via hydroxyl radical (OH•)-mediated cell wall loosening (Schweikert et al. 2000; Schopfer 2001); that is, cell elongation is thought to be a consequence of cell wall loosening. OH• nonspecifically cleaves cell wall polysaccharides resulting in cell wall loosening, suggested by the inhibitory effect of OH• scavengers on coleoptile elongation (Schopfer 2001). OH• immediately reacts with most organic compounds in the proximity of the production site due to its extreme reactivity. In cell walls, OH• can be formed from H_2O_2 by the Haber–Weiss reaction. It has also been shown that H_2O_2 and superoxide anion (O_2•⁻) scavengers inhibit coleoptile elongation of maize (Schopfer 2001), indicating serial generation of multiple ROS.

Many reports, mainly of interest in chemistry as well as biology, have shown that IAA reacts with dioxygen (O_2) catalyzed by peroxidase (POX) resulting in production of O_2•⁻ (Smith et al. 1982; Kawano 2003). Generated O_2•⁻ is readily transformed to H_2O_2 by a reaction called dismutation or disproportionation, spontaneously or catalyzed by superoxide dismutase (SOD).

During IAA-induced ROS-dependent cell elongation, OH• can easily be produced from H_2O_2 and in turn cleave polymer in cell walls. The peroxidase, which is involved in this process, is hypothesized to be located in the cell wall. It was shown in the production of OH• and the presence of POX in the maize root elongation zone (Liszkay et al. 2004). The results also showed that the inhibitors of NADPH oxidase (NOX) affect elongation, indicating the contribution of NOX as well as cell POX. Since the NOX inhibitor used in the study, diphenylene iodinium (DPI), as the authors noticed, can also inhibit other redox enzymes, the involvement of NOX remains to be further examined.

Studies reported that auxin induced ROS production in root gravitropism (Joo et al. 2001) and in coleoptile elongation (Schopfer 2001) of maize. The root gravitropism study showed asymmetric ROS generation upon gravi-stimulus at specific areas, whereas induction of ROS by salicylic acid in the root showed no specific localization (Joo et al. 2001).

As mentioned above, auxin induces a series of ROS in the root elongation zone (Schopfer 2001). Joo et al. (2005) showed an elevation of the H_2O_2 level in maize root upon auxin and gravi-stimulation. Interestingly, inhibitors of phosphatidylinositol 3-kinase (PI3-K), LY294002 and wortmannin suppressed auxin and gravi-stimuli-induced ROS accumulation in maize root (Joo et al. 2005), indicating the involvement of phospholipid metabolism in upstream of ROS generation. It should be noticed that PI3-K is reported to function in the activation of NOX in mammalian leucocytes (Ellson et al. 2006) and ABA-induced ROS generation in guard cells (Park et al. 2003) as discussed in Sect. 2.2.1.

1.2 ROS Generation by Cell Wall-Associated Peroxidases, a Hypothesis

Kawano (2003) proposed the model that includes a parallel IAA signal pathway leading to elongation growth (Fig. 1). Besides the signaling pathway mediated by

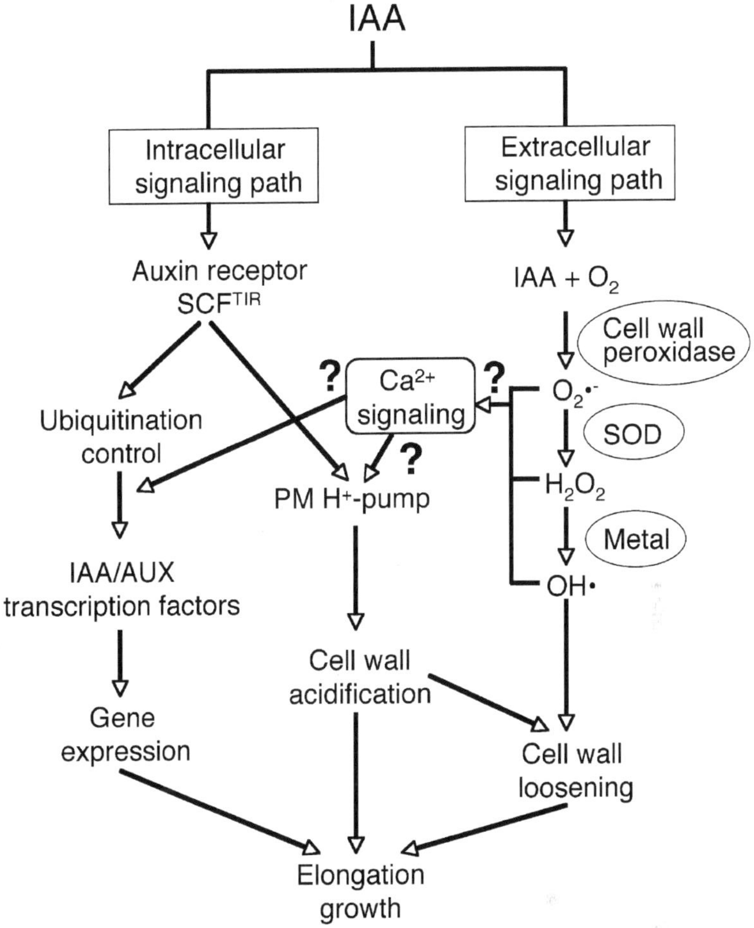

Fig. 1 A hypothetical model of parallel auxin signal pathways leading to elongation growth (modified from Kawano 2003). Indole-3-acetic acid (IAA) is received intracellularly and extracellularly. IAA binds to the auxin receptor, F-box protein, that consists of ubiquitine ligase E3, SCF^{TIR}. The downstream IAA/Aux transcription regulators are controlled by the ubiquitin ligase-mediated protein degradation leading to expression regulation of auxin-inducible genes. Possibly parallel to the gene expression pathway, plasma membrane associated H^+-pump is activated resulting in cell wall acidification. Extracellular IAA is oxidized by cell wall peroxidase reacting with dioxygen (O_2) and produces superoxide ($O_2{\bullet}^-$). H_2O_2 is formed from $O_2{\bullet}^-$ catalyzed by plasma membrane associated superoxide dismutase (SOD), followed by hydroxyl radical (OH•) formation in the presence of metals. OH• cleaves cell wall polysaccharide to loosen cell walls. ROS activates the universal ROS/Ca^{2+} signal cassette (Mori and Schroeder 2004; Demidchik et al. 2007) that might modify the intracellular signaling pathway with Ca^{2+} signaling

the F-box auxin receptor, the SCF^{TIR} that induces gene expression through AUX/IAA transcription factors (Dharmasiri et al. 2005; Kepinski and Leyser 2005) (note that when Kawano's model was proposed in 2003, the auxin receptor, SCF^{TIR} had not yet been reported), the peroxidase-mediated reaction produces $O_2{\bullet}^-$ coupled with the IAA breakdown in the cell wall. The $O_2\bullet^-$ or H_2O_2 generated possibly elicit cellular Ca^{2+} signaling through plasma membrane Ca^{2+}-permeable channels, as proposed in guard cells, that affects intracellular Ca^{2+} signaling. Parallel to this, OH• cleaves polysaccharides, hence loosens the cell wall (Kawano 2003).

Kawano (2003) also pointed out the similarity of the peroxidase (POX)-mediated salicylic acid (SA)-induced stomatal closure model (Mori et al. 2001) to the IAA-POX associated ROS production mechanism. The modified Mori's model (Kawano 2003) predicts the interaction of POX-mediated SA and chitooligosaccharide, and NOX-mediated ABA and peptide elicitor ROS signaling pathways. A recent study also suggests the contribution of NOX in MeJA signaling (Suhita et al. 2004). This model is constructed to explain the signal integration system in guard cells, but not for cell elongation. The physiological significance of SA or chitooligosaccharide-induced ROS generation in cell wall loosening, as proposed for IAA response, remains to be clarified. However, hypothetically POX-mediated $O_2\bullet^-$ production coupled with following H_2O_2 and OH• production could be a common mechanism in several ROS-dependent plant cell signaling.

2 Abscisic Acid

Abscisic acid (ABA) is a plant hormone that primarily mediates the induction of plant tolerance to a range of abiotic stresses and regulates growth and development, such as the closure of stomata upon drought stress, induction of a series of stress responsive genes, delay of vegetative growth, and promotion of seed maturation (Giraudat 1995, Finkelstein et al. 2008). While not all the details of the involvement of ROS in ABA-relating physiology are known, in this section a few examples are described, i.e., the ABA-induced stomatal closure and the induction of ROS-scavenging gene expression. The function of ABA in the inhibition of germination is discussed in Sect. 2.5.

2.1 ABA-induced ROS Generation in Guard Cells

Plants control stomatal aperture to optimize the balance of water loss and carbon dioxide uptake in response to environmental stresses. Abscisic acid is the drought-responsive plant hormone that induces stomatal closure. Turgor pressure of guard cells that surround stomatal pore is the motive force of stomatal movement. Abscisic acid changes the electrophysiological characteristics of ion transporters, such as H^+-pump, inward-, and outward-K^+ channels, anion channels, Ca^{2+} channels, and hypothetically water channels, so that guard cells lose turgor pressure. It was reported that hydrogen peroxide activates plasma membrane Ca^{2+}-permeable channels (Pei et al. 2000; Köhler et al. 2003), which is involved in creating Ca^{2+} signature (Allen et al. 2000, 2001). Kwak et al. (2003) demonstrated that the NOX catalytic subunits, AtrbohD and AtrbohF were involved in the ABA-induced ROS generation.

Plant NOX reduces O_2 using the reduction potential of NADPH or NADH, resulting in extracellular formation of superoxide anion ($O_2{\bullet}^-$). Accumulation of H_2O_2 in the cytosol was proved by the H_2O_2-specific fluorescent dye, H_2DCF (Pei et al. 2000; Zhang et al. 2001). However, it is not yet clear how the generation of extracellular $O_2{\bullet}^-$ leads to an H_2O_2 increase in the cytosol. At the same time, most probably OH• is formed from H_2O_2 by Fenton or Haber-Weiss reactions. It is known that Ca^{2+} channel activation might be a next downstream event after ROS generation (Murata et al. 2001). We will discuss the role of ROS to activate plasma membrane Ca^{2+}-permeable channels in the Sect. 2.4.1.

The mechanism of NOX activation in ABA signaling is not clarified. In mammalian neutrophil cells, NOX consists of a protein complex formed by such as two membrane-bound elements ($gp91^{phox}$ and $p22^{phox}$), three cytosolic regulatory components ($p67^{phox}$, $p47^{phox}$, and $p40^{phox}$), and the Rac2 small G protein (Babior 2004). Binding of phosphatidylinositol-3-phosphatase (PI3P) to a NOX subunit, $p40^{phox}$ is essential for the activation (Ellson et al. 2006). Unlike mammalian cells, the plant genome lacks regulatory components of NOX including $p40^{phox}$ (see Chap. 1). However, interestingly, a pharmacological study showed that phosphatidylinositol 3-kinase (PI3-K) and PI3P are involved in an ABA-induced stomatal closure in *Vicia faba* (Jung et al. 2002). PI3-K inhibitors, wortmannin and LY294002, inhibit ABA-induced ROS accumulation and stomatal closure (Park et al. 2003), suggesting a function for PI3P and PI3K in the ABA signaling upstream of ROS production.

Genetic analyses of ABA signaling using *Arabidopsis* mutants revealed that protein phosphorylation and dephosphorylation is associated upstream of ROS generation. ABA-induced ROS generation in guard cells is impaired in *open stomata 1* (*ost1*) protein kinase and *abscisic acid insensitive 1* (*abi1*) protein phosphatase 2C mutants (Murata et al. 2001; Mustilli et al. 2002). However, the details of the signal pathway that leads to the activation of NOX and accumulation of ROS are still unclear. The interaction of methyl jasmonate and ABA signaling in guard cells is discussed in Sect. 2.4.1.

2.2 *Induction of Antioxidant Enzymes by ABA in Response to Abiotic Stress*

ABA is involved in drought resistance of plants. Recent reports indicate that ROS elevation occurs in the extracellular space of ABA-treated maize leaves. The ROS-dependent ABA signal pathway induces the gene expression and activity of ROS-scavenging enzymes, such as superoxide dismutase (SOD), chloroplastic and cytosolic ascorbate peroxidases (APXs), and chloroplastic and cytosolic glutathione reductase (GR) (Hu et al. 2005, 2006, 2007). Using $CeCl_3$-staining, it was shown that ABA induces H_2O_2 accumulation exclusively in the intercellular space (cell wall) (Hu et al. 2005, 2006). The inhibition of the ABA-induced elevation of antioxidant enzyme activity by the NOX inhibitor, DPI, and ROS scavengers, Tiron and dimethylthiourea, indicates roles of NOX-dependent

ROS production in ABA-induced antioxidant defense response (Hu et al. 2005). In the ABA-deficient maize mutant *vp5*, water stress did not induce H_2O_2 accumulation, suggesting a physiological significance of the upregulation of antioxidant enzymes in ABA-dependent drought response of maize.

As we will discuss in gibberellin-induced aleurone cell death in the following section, ABA prevents the downregulation of the antioxidation system in aleurone cells (Sect. 2.5). Those results imply the common function of ABA to induce the upregulation of antioxidant enzymes in several aspects of ABA responses.

3 Methyl Jasmonate, Salicylic Acid, and Ethylene

3.1 Methyl Jasmonate and Jasmonate

Methyl jasmonate (MeJA) is a volatile plant hormone that mediates various plant defense responses (Liechti and Farmer 2002; Turner et al. 2002). Signal interaction between MeJA and ABA is discussed in Sect. 2.4.1. The interactions of jasmonate (JA), a MeJA-relating hormone, and SA and ethylene signaling are discussed in Sects. 2.3.2 and 2.3.3.

JA induces H_2O_2 accumulation and cell death at very high concentrations (>50 μM). At lower concentrations, JA induces PR-proteins expression and nitric oxide (NO) generation. Low concentrations of JA apparently do not induce ROS. JA-induced NO production counteracts the SA response. The physiological significance of JA-induced ROS production and its role in signal integration is still controversial.

3.2 Salicylic Acid

3.2.1 Systemic Acquired Resistance

Salicylic acid (SA) induces systemic acquired resistance (SAR) and hypersensitive response cell death in response to pathogen attack (Loake and Grant 2007). H_2O_2 was suggested to be a second messenger of SA in SAR induction (Chen and Klessig 1991; Chen et al. 1993a, b). In tobacco, the SA-binding protein was identified as catalase (Chen et al. 1993a, b). SA inhibited catalase activity in vitro and induced H_2O_2 generation in tobacco tissue (Chen et al. 1993b). H_2O_2 induced the expression of pathogenesis-related (PR) proteins that is a consequence of SAR (Chen et al. 1993b). These observations led to the hypothesis, that H_2O_2 accumulation is a second messenger for induction of SAR. Later reports disagreed with this hypothesis, showing that H_2O_2 was not accumulated during the induction of SAR in TMV-infected tobacco (Neuenschwander et al. 1995; Ryals et al. 1995). The expression of bacterial salicylate hydroxylase, which artificially cancels SA accumulation in the tissue, demonstrated that the H_2O_2-induced PR protein accumulation is SA-dependent, suggesting that SA acts downstream of H_2O_2 generation (Bi et al. 1995; Neuenschwander et al. 1995). The inhibition of catalase activity in the tissue after a challenge of pathogen was not observed (Bi et al. 1995; Summermatter et al. 1995).

3.2.2 SA-induced Cell Death

Significant numbers of studies have shown that SA induces ROS in plants (Chen et al. 1993b; Rao et al. 2000). SA induces oxidative damage, such as lipid peroxidation and protein oxidation (Rao et al. 1997), and the expression of the glutathione S-transferase gene at relatively low concentration (Chen and Singh 1999). At relatively high concentration, SA induces SOD gene expression and represses CAT, APX, and POX genes expression (Rao et al. 1997), suggesting the involvement of SA in reduction/oxidation regulation of the cells under biotic stressed conditions. As discussed above, SA induces a H_2O_2 increase, and meanwhile H_2O_2 induces SA accumulation. To understand the relationship of the H_2O_2-induced rise of the SA concentration and the SA-induced elevation of H_2O_2 concentration in cell signaling, further investigations are necessary.

The elevation of the ROS level is hypothesized to be an executor of the PCD in gibberellin (GA)-response of aleurone cells. High concentrations of SA induce hyper-responsive cell death and antioxidant enzyme downregulation similar to GA (Delaney et al. 1994). Hypothetically, suppression of antioxidant gene expression by SA can act as the cell death executor as well as GA. Future comparative studies of hormone-induced cell death will shed light on a common function and mechanism of downregulation of antioxidant genes.

3.2.3 Similarity of Mechanisms of ROS Accumulation by SA and Auxin

It was demonstrated that SA-induced Ca^{2+} elevation involves ROS production in tobacco suspension culture (Kawano and Muto 2000). SA induces stomatal closure to prevent pathogen invasion through the aperture (Manthe et al. 1992; Lee 1998; Mori et al. 2001). Mori et al. (2001) demonstrated that ROS scavengers, catalase and Tiron, and a peroxidase inhibitor, salicylhydroxamic acid, inhibited SA-induced stomatal closure suggesting possible involvement of the POX-mediated SA signaling model in stomatal closure. Interaction of cell wall-associated POX-mediated and NOX-mediated ROS generation in stomata is discussed in Sect. 2.4.2.

SA radical, that is formed by reaction of SA with the compound II reaction intermediate of peroxidase, spontaneously reacts with O_2 resulting in $O_2\bullet^-$ production (Durner and Klessig 1995; Rao et al. 1997; Kawano 2003). The involvement of compound II suggests that SA is oxidized by H_2O_2, but not by O_2. Contrary to SA, IAA is oxidized by O_2 to produce $O_2\bullet^-$ (Nakajima and Yamazaki 1979; Savitsky et al. 1999). Those two proposed hormone-induced $O_2\bullet^-$ production systems likely use cell wall-associated POXs, indicating partly shared signal pathways by the two hormones. It should be noticed that both SA and IAA signal cannot be explained solely by ROS production. There might be parallel signal paths. To clarify the mechanism of SA-induced ROS accumulation may need further investigation, since some other mechanisms might function in the process, such as downregulation of scavenger enzymes, direct inhibition of CAT and POX at very high concentrations, and NOX-mediated mechanisms.

3.3 *Ethylene*

Ethylene is a gaseous plant hormone that is involved in many important physiological processes including stress-induced programmed cell death. Roles of ROS in hormone-induced cell death processes were mentioned in Sects. 2.3.2 and 2.5 in respect to SA and GA, respectively. Relationship of ethylene and programmed cell death (PCD) has been well studied, but the detailed function of ROS in the ethylene-induced PCD is relatively unclear. Ethylene and ROS are reported to coregulate PCD (Overmyer et al. 2000; Rao et al. 2002). Overmyer et al. (2000) showed that ozone treatment induced the formation of lesions, expression of ethylene synthesizing enzyme and an increment of ethylene concentration. Further, application of the precursor of ethylene, 1-aminocyclopropane-1-carboxylic acid (ACC), enhanced ozone-induced cell damage. ACC did not apparently enhance the damage in the ethylene insensitive mutant, *ein2*. The ROS generation mechanism is suggested to be NOX according to the observation of the partial inhibition by DPI on ozone-induced lesion formation. As discussed in Sect. 2.4.1, the sole result of DPI is not enough to predict a ROS generation mechanism. A function for mitochondria in the ROS-dependent PCD process was also suggested (Lam et al. 2001).

Jasmonic acid (JA) is known to halt expansion of ethylene-induced PCD lesion (containment of lesion) (Overmyer et al. 2003). JA also shows antagonistic action to SA-induced ROS-dependent cell death through NO production. In aleurone cell, ABA counteracts GA-induced PCD (Fath et al. 2002). While signaling mechanisms of ROS induction and PCD in ethylene are not fully understood, analogy to cell death in SA and GA responses suggests a relationship between the downregulation of antioxidant gene expression and the counteraction of NO in ethylene-induced PCD.

4 Hormone Signal Integration in Guard Cells

4.1 *Signal Integration in Guard Cells, Methyl Jasmonate, and Abscisic Acid*

It has been reported that cytosolic free Ca^{2+} concentration ($[Ca^{2+}]_{cyt}$), nitric oxide (NO) and reactive oxygen species (ROS) function as second messengers in ABA-induced stomatal closure. ABA induces repetitive elevation of $[Ca^{2+}]_{cyt}$ in guard cells (McAinsh et al. 1996; Allen et al. 1999). Ca^{2+} is mobilized from extracellular space via plasma membrane Ca^{2+}-permeable channels and internal Ca^{2+} storage (Schroeder and Hagiwara 1990; McAinsh et al. 1996; Pei et al. 2000).

Genetic and pharmacological approaches revealed that ABA induces an increase of ROS produced by plasma membrane NOX (Pei et al. 2000; Murata et al. 2001; Kwak et al. 2003). NOX produce $O_2\bullet^-$, most likely in extracellular space (Sumimoto 2008; see Chap. 1). Observation using a membrane-permeable H_2O_2-sensitive fluorescent indicator, H_2DCF-DA, showed that guard cells accumulate intracellular H_2O_2 in response to ABA (Pei et al. 2000; Zhang et al. 2001), suggesting successive production of H_2O_2 after NOX-mediated $O_2\bullet^-$ generation. Since $O_2\bullet^-$ and its protonated form, $HO_2\bullet$ readily react with lipids and proteins, $O_2\bullet^-$ and $HO_2\bullet$ may not enter the cell through the plasma membrane. H_2O_2 might be formed extracellularly from $O_2\bullet^-$ and enter into the cell probably through aquaporins (Henzler and Steudle 2000; Bienert et al. 2007), to increase the H_2O_2 level in the cytosol of guard cells.

A whole-cell patch-clamp experiment demonstrated that application of H_2O_2 activated plasma membrane Ca^{2+}-permeable channels in guard cell protoplasts of *Vicia faba* and *Arabidopsis*, which suggests a tandem interaction of two second messengers, ROS and $[Ca^{2+}]_{cyt}$ elevations in ABA signaling (Pei et al. 2000; Köhler et al. 2003). However, it is not yet clear if other ROS, $O_2\bullet^-$, and/or $OH\bullet$ are involved in the activation of Ca^{2+}-permeable channels of guard cells. Two molecules of $O_2\bullet^-$ and/or its protonated form, $HO_2\bullet$, react to form H_2O_2 and O_2 spontaneously or enzymatically via superoxide dismutase (SOD). The Haber–Weiss–Fenton reaction produces $OH\bullet$ from H_2O_2. Thus, the actual chemical species among ROS that activate the Ca^{2+} channels is still controversial.

It might be noticed that $OH\bullet$ activates similar plasma membrane Ca^{2+}-permeable channels in the elongation zone of *Arabidopsis* roots (Demidchik et al. 2003). Interestingly, it is suggested that two distinguishable Ca^{2+}-permeable channels, which were activated by either $OH\bullet$ or H_2O_2, coexist in the epidermis of the root elongation zone (Demidchik et al. 2007). The similarity of Ca^{2+} channels in tip growth of root hair, germination of algal spore, cell elongation of root, and turgor regulation of guard cells is discussed elsewhere (Mori and Schroeder 2004). On the other hand, $OH\bullet$ immediately reacts with target molecules in the proximity of the production site and thus is rather a reaction intermediate than signal transducer, as discussed by Mori and Schroeder (2004). The ABA-insensitive *Arabidopsis* mutants, *gca2* and *abi2-1*, exhibited impairment of ROS activation of the plasma membrane Ca^{2+}-permeable channels (Pei et al. 2000; Murata et al. 2001), providing genetic evidence for a central role of the activation of the Ca^{2+} channels in ROS-dependent ABA signaling in guard cells (Fig. 2).

Methyl jasmonate (MeJA) induces stomatal closure mediated by ROS-dependent signaling pathway as well as ABA (Gehring et al. 1997; Suhita et al. 2003, 2004). MeJA is a derivative of jasmonic acid (JA), which is involved in plant growth, development, and defense, and is accumulated in plant when it is exposed to wounding, pathogen infection, water deficit (Creelman and Mullet 1997), and ozone exposure (Rao et al. 2000). MeJA and ABA signal pathways are transduced separately upstream of ROS generation and become integrated at the level of ROS generation

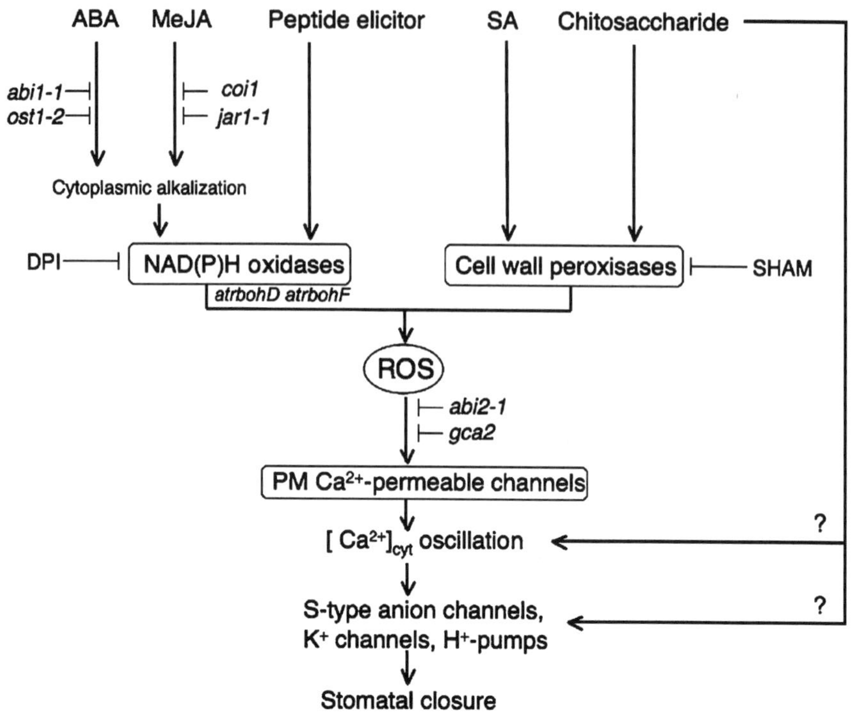

Fig. 2 Possible signal transduction networks for stomatal closure induced by ABA, MeJA, SA, and elicitors. ABA and MeJA increase the ROS level in guard cells via NAD(P)H oxidase (NOX) pathway that was preceded with cytoplasmic alkalization. Peptide elicitor also increases ROS level through NAD(P)H oxidase pathway. The upstream of cytoplasmic alkalinization is distinguishable in *abi1-1*, *ost1-2*, *coi1*, and *jar1-1* mutants. NOX, which function in ABA, MeJA, and peptide elicitor in guard cells were identified by the inhibitor, diphenylene iodinium (DPI), and loss-of-function mutant of *atrbohD* and *atrbohF*. Contrary to the NOX pathway, salicylic acid (SA) and chitosaccharide signal are transduced via cell wall peroxidase pathway that is inhibited by a peroxidase inhibitor, salicyl hydroxamate (SHAM). The following signaling pathway, hypothetically, shares ROS generation, plasma membrane Ca^{2+}-permeable channel activation, and control of a series of ion transporters that contribute to turgor and stomatal aperture regulation

in the pathway (Fig. 2). MeJA, like ABA, induce the signal path in guard cells, including elevation of ROS, activation of plasma membrane Ca^{2+}-permeable channels, activation of S-type anion channels, and stomatal closure. Recent studies showed that activation of the plasma membrane Ca^{2+}-permeable channels and S-type anion chanenls by MeJA treatment of patched protoplast was mediated by ROS generation (Suhita et al. 2004; Munemasa et al. 2007). The jasmonate insensitive *Arabidopsis* mutants, *jar1-1* (Suhita et al. 2004) and *coi1* (Munemasa et al. 2007) lack the ability to close stomata in response to MeJA, whereas these mutants possess responsiveness to ABA. Elevation of ROS, activation of plasma membrane Ca^{2+}-permeable and S-type anion channels induced by MeJA were not observed in *coi1* mutant

(Munemasa et al. 2007). Meanwhile, MeJA did not increase ROS levels in *jar1-1* guard cells (Suhita et al. 2004). On the other hand, *ost1-2* guard cells are responsive to MeJA in stomatal closure and ROS accumulation, but insensitive to ABA (Suhita et al. 2004). These observations indicate a function of ROS production in MeJA signaling like that in ABA signaling (Fig. 2).

It was shown that ABA- and MeJA-induced ROS generation in guard cells were catalyzed by NOX (Kwak et al. 2003). There are 6–10 genes encoding the NOX catalytic subunit in the *Arabidopsis* genome. Among those genes, the double gene disruption mutant of *atrbohD* and *atrbohF* demonstrated impairment of ABA-induced stomatal closure and plasma membrane Ca^{2+}-permeable channel activation (Kwak et al. 2003). Elevation of ROS level induced by ABA in guard cells was also impaired in the *atrbohD atrbohF* double mutant. It was shown that MeJA-induced ROS elevation in guard cells was also impaired in the *atrbohD atrbohF* double mutant, indicating that signaling pathways of two hormones share the ROS production and the downstream process, and signals are integrated upstream of ROS generation (Suhita et al. 2004; Saito et al. 2008) (Fig. 2).

4.2 *ROS Signal in Other Stimuli in Guard Cells*

ROS generation is reported to mediate many stomatal closing signals. In guard cells of tomato and *Commelina*, oligogalacturonic acid and chitosan, which function as an elicitor, induce ROS accumulation and stomatal closure (Lee et al. 1999). Salicylic acid also induces extracellular generation of $O_2\bullet^-$ leading to stomatal closure (Mori et al. 2001). Klüsener et al. (2002) demonstrated that ROS activation of the plasma membrane Ca^{2+}-permeable channels is involved in elicitor-induced stomatal closure as well as in ABA and MeJA.

5 Gibberellin and Abscisic Acid in Seed Gemination

Many plant hormones such as auxin, gibberellin (GA), cytokinin, and brasinolide regulate elongation and division of plant cells, hence determine architecture of plants. Involvement of ROS in auxin signaling has already been mentioned above (Sect. 2.1). In this section, ROS production in GA response is described. Historically, GA was found to function in internodal elongation. Gibberellin was initially isolated and identified as the toxin that caused "Bakanae" (foolish seedling) disease of rice (Tamura 1990). Infection of Bakanae fungus, *Gibberella fujikuroi*, renders overdose of GA resulting in excessive internode elongation, pale yellow leaves, and low crop yield. However, as far as we know, a function for ROS in the GA-induced internodal elongation has not been demonstrated. This section focuses on GA-induced seed germination and counteraction of ABA.

5.1 *ROS Production in Programmed Cell Death of Aleurone Cells*

In response to GA, aleurone cells execute programmed cell death (PCD) following the excretion of amylases and proteases for providing necessary energy source and nutrients to the germinating embryos in cereal seeds. PCD of aleurone cells during germination is not specific to cereal seeds, but common in monocotyledons and dicotyledons (Fath et al. 2001; Bethke et al. 2007). It should be noted that induction of amylase secretion from aleurone cells is not regulated by ROS generation, but by PCD.

ROS are thought to function as cell death executers during germination (Fath et al. 2002). Fath et al. (2002) demonstrated that GA downregulates catalase (CAT), ascorbate peroxidase (APX), and superoxide dismutase (SOD) gene expression, resulting in elevation of ROS level leading to PCD of aleurone cells. In other words, moderately high levels of CAT, APX, and SOD maintain the ROS level not so high as to initiate PCD. However, once GA initiates suppression of gene expression of antioxidant enzymes, ROS level in tissue increases and in turn oxidizes proteins, DNA, and lipids, and ends up with cell death. A study using chemical ROS scavengers, ROS generated in the embryo and seed coat of radish was reported to consist of, at least, O_2•$^-$, H_2O_2, and OH• (Schopfer 2001; Schopfer et al. 2002). Fath et al. (2002) hypothesized that beta-oxidation in the glyoxysome is a major ROS source in barley aleurone cells. On the other hand, Schopfer et al. (2001) based on pharmacological evidence suggested a model consisting of a cascade of two mechanisms of ROS metabolism, i.e., plasma membrane-associated NOX and cell wall-POX.

5.2 *Antagonistic Action of Abscisic and Nitric Oxide Against Gibberellic Acid*

Antagonistic to GA, ABA sustains the level of CAT, APX, and SOD in aleurone cells and, hence, prevents or delays PCD (Fath et al. 2002). The detail of signal transduction process of ABA in aleurone cells is not well understood. Thus, the signaling pathway, which leads to cancellation of GA-induced gene suppression by ABA is unrevealed. ABA and GA signals may not be integrated at the level of second messengers, but rather at the level of gene expression in ROS metabolism-relating (antioxidant) enzymes. In the previous Sect. 2.5.1, we mentioned possible candidates of ROS source in GA signaling in aleurone cells. Two models have been proposed for the ROS accumulation in germinating seeds, i.e., downregulation of scavenging enzymes and the cascade of NOX/peroxidase (Schopfer 2001; Fath et al. 2002). Both mechanisms might contribute to the ROS accumulation process. The contribution ratio can be variable species to species.

Nitric oxide (NO) also functions to counteract GA in PCD of aleurone cells (Beligni et al. 2002; Bethke et al. 2007). In guard cells, ABA induces NO production that precedes stomatal closure (Desikan et al. 2002; Bright et al. 2006; Munemasa et al. 2007). In analogy between aleurone cells and guard cells, one can hypothesize that ABA induces NO accumulation during suppression of the GA-induced PCD by ABA. However, the NO induction by ABA in aleurone cells has not been demonstrated. NO and ABA might regulate GA action independently.

A series of studies by Jones and colleagues suggests that an interaction between ABA and GA occurs at least through gene expression of CAT, APX, and SOD in the process of PCD of alleurone cells (Fath et al. 2001, 2002; Beligni et al. 2002; Bethke et al. 2007). Apparent evidence suggested that amylase excretion from aleurone cells is inhibited by ABA as well as by PCD, but not by NO (Beligni et al. 2002; Bethke et al. 2007). This indicates that NO is not an adequate factor in ABA signaling in aleurone cells, unlike guard cells.

6 Cytokinin

6.1 *Oxidative Degradation of Cytokinine*

Cytokinins function in many aspects in plant development; however, the ROS generation in response to cytokinin and the involvement of ROS in cytokinin signaling has not been well assessed. Here, we briefly review the degradation of cytokinin by $O_2{\bullet}^-$ and O_2.

Intracellular signaling processes play a central role in signal integration of plant hormones. Besides cellular signaling, hormone signal network can also be achieved indirectly through catabolism of plant hormones. Oxidative degradation of cytokinin occurs by reaction with either molecular oxygen (O_2) or $O_2{\bullet}^-$. It was reported that endogenous zeatin riboside was oxidized to form adenine by spontaneous reaction with $O_2{\bullet}^-$ in soybean seeds (Gidrol et al. 1994). Cytokinins are also enzymatically catabolized by the dual functional cytokinin oxidase/dehydrogenase reacting with O_2, not with $O_2{\bullet}^-$ (Kopecny et al. 2005). ABA and MeJA are reported to induce the generation of $O_2{\bullet}^-$ catalyzed by NOX in guard cells (Kwak et al. 2003; Suhita et al. 2004). SA is also reported to elicit production of $O_2{\bullet}^-$ catalyzed by cell wall POX. Upon a pathogen-response, SA induces a two-phase ROS generation. The latter ROS generation is associated with hyperresponse leading to necrotic cell death (Delaney et al. 1994). Gidrol et al. (1994) suggested that environmental stresses induce ROS accumulation and in turn degrade cytokinin in seed tissue. ABA, MeJA, and SA, as well as other ROS-inducing stimuli, might hypothetically down-regulate cytokinin signal via the $O_2{\bullet}^-$-mediated degradation mechanism. Interaction of ABA/MeJA/SA and cytokinin through ROS in a single cell type is yet to be examined to elucidate the function of ROS in cytokinin, and interaction between cytokinin and other hormone signals. As mentioned in the previous section, IAA also generates ROS via a POX-catalyzed mechanism. Cytokinin and auxin counteract in lateral root formation. However, it remains unclear that interaction of auxin and cytokinin via degradation of cytokinin with ROS evoked by IAA. The involvement of ROS generation in lateral root formation that depends on auxin function is still unclear.

7 Conclusion

The importance of ROS accumulation signals in hormone response is unambiguous. The involvement of ROS accumulation in hormone signal integration is highly feasible. However, solid evidence of the signal integration process is still lacking. It is shown that methyl jasmonate (MeJA) and abscisic acid (ABA) signaling pathways use the same NOX for ROS production in *Arabidopsis* guard cells. Studies using some hormone insensitive mutants revealed that MeJA and ABA signaling share signal events downstream of ROS production. Salicylic acid, ethylene, and gibberellin induce ROS accumulation that is associated with cell death, in which downregulation of antioxidant enzyme expression may be involved. Auxin and salicylic acid induce ROS accumulation through cell wall peroxidases. Auxin-induced ROS production leads to cell wall loosening. SA induces cell death and stomatal closure, and probably boosts PR-protein induction. No evidence has yet been shown concerning possible roles of ROS in brassinolide signaling. The existence of intracellular ROS receptors in hormone signaling is still unclear. The identification of ROS receptors would further clarify the role of ROS generation in hormone signal integration.

References

Allen GJ, Chu SP, Harrington CL, Schumacher K, Hoffmann T, Tang YY, Grill E, Schroeder JI (2001) A defined range of guard cell calcium oscillation parameters encodes stomatal movements. Nature 411:1053–1057

Allen GJ, Chu SP, Schumacher K, Shimazaki CT, Vafeados D, Kemper A, Hawke SD, Tallman G, Tsien RY, Harper JF, Chory J, Schroeder JI (2000) Alteration of stimulus-specific guard cell calcium oscillations and stomatal closing in *Arabidopsis det3* mutant. Science 289:2338–2342

Allen GJ, Kwak JM, Chu SP, Llopis J, Tsien RY, Harper JF, Schroeder JI (1999) Cameleon calcium indicator reports cytoplasmic calcium dynamics in Arabidopsis guard cells. Plant J 19:735–747

Babior BM (2004) NADPH oxidase. Curr Opin Immunol 16:42–47

Beligni MV, Fath A, Bethke PC, Lamattina L, Jones RL (2002) Nitric oxide acts as an antioxidant and delays programmed cell death in barley aleurone layers. Plant Physiol 129:1642–1650

Bethke PC, Libourel IG, Aoyama N, Chung YY, Still DW, Jones RL (2007) The *Arabidopsis* aleurone layer responds to nitric oxide, gibberellin, and abscisic acid and is sufficient and necessary for seed dormancy. Plant Physiol 143:1173–1188

Bi YM, Kenton P, Mur L, Darby R, Draper J (1995) Hydrogen peroxide does not function downstream of salicylic acid in the induction of pr protein expression. Plant J 8:235–245

Bienert GP, Moller AL, Kristiansen KA, Schulz A, Moller IM, Schjoerring JK, Jahn TP (2007) Specific aquaporins facilitate the diffusion of hydrogen peroxide across membranes. J Biol Chem 282:1183–1192

Bright J, Desikan R, Hancock JT, Weir IS, Neill SJ (2006) ABA-induced NO generation and stomatal closure in *Arabidopsis* are dependent on H_2O_2 synthesis. Plant J 45:113–122

Chen W, Singh KB (1999) The auxin, hydrogen peroxide and salicylic acid induced expression of the Arabidopsis GST6 promoter is mediated in part by an OCS element. Plant J 19:667–677

Chen Z, Klessig DF (1991) Identification of a soluble salicylic acid-binding protein that may function in signal transduction in the plant disease-resistance response. Proc Natl Acad Sci USA 88:8179–8183

Chen Z, Ricigliano JW, Klessig DF (1993a) Purification and characterization of a soluble salicylic acid-binding protein from tobacco. Proc Natl Acad Sci USA 90:9533–9537

Chen Z, Silva H, Klessig DF (1993b) Active oxygen species in the induction of plant systemic acquired resistance by salicylic acid. Science 262:1883–1886

Creelman RA, Mullet JE (1997) Biosynthesis and action of jasmonates in plants. Annu Rev Plant Physiol Plant Mol Biol 48:355–381

Delaney TP, Uknes S, Vernooij B, Friedrich L, Weymann K, Negrotto D, Gaffney T, Gut-Rella M, Kessmann H, Ward E, Ryals J (1994) A central role of salicylic acid in plant disease resistance. Science 266:1247–1250

Demidchik V, Shabala SN, Coutts KB, Tester MA, Davies JM (2003) Free oxygen radicals regulate plasma membrane Ca^{2+}- and K^{+}-permeable channels in plant root cells. J Cell Sci 116:81–88

Demidchik V, Shabala SN, Davies JM (2007) Spatial variation in H_2O_2 response of *Arabidopsis thaliana* root epidermal Ca^{2+} flux and plasma membrane Ca^{2+} channels. Plant J 49:377–386

Desikan R, Griffiths R, Hancock J, Neill S (2002) A new role for an old enzyme: Nitrate reductase-mediated nitric oxide generation is required for abscisic acid-induced stomatal closure in *Arabidopsis thaliana*. Proc Natl Acad Sci USA 99:16314–16318

Dharmasiri N, Dharmasiri S, Estelle M (2005) The F-box protein TIR1 is an auxin receptor. Nature 435:441–445

Durner J, Klessig DF (1995) Inhibition of ascorbate peroxidase by salicylic acid and 2,6-dichloroisonicotinic acid, two inducers of plant defense responses. Proc Natl Acad Sci USA 92:11312–11316

Ellson C, Davidson K, Anderson K, Stephens LR, Hawkins PT (2006) Ptdins3p binding to the px domain of $p40^{phox}$ is a physiological signal in NADPH oxidase activation. EMBO J 25:4468–4478

Fath A, Bethke P, Beligni V, Jones R (2002) Active oxygen and cell death in cereal aleurone cells. J Exp Bot 53:1273–1282

Fath A, Bethke PC, Jones RL (2001) Enzymes that scavenge reactive oxygen species are down-regulated prior to gibberellic acid-induced programmed cell death in barley aleurone. Plant Physiol 126:156–166

Finkelstein R, Reeves W, Ariizumi T, Steber C (2008) Molecular aspects of seed dormancy. Annu Rev Plant Biol 59:387–415

Gehring CA, Irving HR, McConchie R, Parish RW (1997) Jasmonates induce intracellular alkalinization and closure of *Paphiopedilum* guard cells. Ann Bot (Lond) 80:485–489

Gidrol X, Lin WS, Degousee N, Yip SF, Kush A (1994) Accumulation of reactive oxygen species and oxidation of cytokinin in germinating soybean seeds. Eur J Biochem 224:21–28

Giraudat J (1995) Abscisic acid signaling. Curr Opin Cell Biol 7:232–238

Henzler T, Steudle E (2000) Transport and metabolic degradation of hydrogen peroxide in *Chara corallina:* Model calculations and measurements with the pressure probe suggest transport of H_2O_2 across water channels. J Exp Bot 51:2053–2066

Hu X, Jiang M, Zhang A, Lu J (2005) Abscisic acid-induced apoplastic H_2O_2 accumulation up-regulates the activities of chloroplastic and cytosolic antioxidant enzymes in maize leaves. Planta 223:57–68

Hu X, Jiang M, Zhang J, Zhang A, Lin F, Tan M (2007) Calcium-calmodulin is required for abscisic acid-induced antioxidant defense and functions both upstream and downstream of H_2O_2 production in leaves of maize (*Zea mays*) plants. New Phytol 173:27–38

Hu X, Zhang A, Zhang J, Jiang M (2006) Abscisic acid is a key inducer of hydrogen peroxide production in leaves of maize plants exposed to water stress. Plant Cell Physiol 47:1484–1495
Joo JH, Bae YS, Lee JS (2001) Role of auxin-induced reactive oxygen species in root gravitropism. Plant Physiol 126:1055–1060
Joo JH, Yoo HJ, Hwang I, Lee JS, Nam KH, Bae YS (2005) Auxin-induced reactive oxygen species production requires the activation of phosphatidylinositol 3-kinase. FEBS Lett 579:1243–1248
Jung JY, Kim YW, Kwak JM, Hwang JU, Young J, Schroeder JI, Hwang I, Lee Y (2002) Phosphatidylinositol 3- and 4-phosphate are required for normal stomatal movements. Plant Cell 14:2399–2412
Kawano T (2003) Roles of the reactive oxygen species-generating peroxidase reactions in plant defense and growth induction. Plant Cell Rep 21:829–837
Kawano T, Muto S (2000) Mechanism of peroxidase actions for salicylic acid-induced generation of active oxygen species and an increase in cytosolic calcium in tobacco cell suspension culture. J Exp Bot 51:685–693
Kepinski S, Leyser O (2005) The Arabidopsis F-box protein TIR1 is an auxin receptor. Nature 435:446–451
Klüsener B, Young JJ, Murata Y, Allen GJ, Mori IC, Hugouvieux V, Schroeder JI (2002) Convergence of calcium signaling pathways of pathogenic elicitors and abscisic acid in Arabidopsis guard cells. Plant Physiol. 130:2152–2163
Köhler B, Hills A, Blatt MR (2003) Control of guard cell ion channels by hydrogen peroxide and abscisic acid indicates their action through alternate signaling pathways. Plant Physiol 131:385–388
Kopecny D, Pethe C, Sebela M, Houba-Herin N, Madzak C, Majira A, Laloue M (2005) High-level expression and characterization of *Zea mays* cytokinin oxidase/dehydrogenase in *Yarrowia lipolytica*. Biochimie 87:1011–1022
Kwak JM, Mori IC, Pei ZM, Leonhardt N, Torres MA, Dangl JL, Bloom RE, Bodde S, Jones JD, Schroeder JI (2003) NADPH oxidase AtrbohD and AtrbohF genes function in ROS-dependent ABA signaling in *Arabidopsis*. EMBO J 22:2623–2633
Lam E, Kato N, Lawton M (2001) Programmed cell death, mitochondria and the plant hypersensitive response. Nature 411:848–853
Lee JS (1998) The mechanism of stomatal closing by salicylic acid in *Commelina communis* l. J Plant Biol 41:97–102
Lee S, Choi H, Suh S, Doo IS, Oh KY, Choi EJ, Schroeder Taylor AT, Low PS, Lee Y (1999) Oligogalacturonic acid and chitosan reduce stomatal aperture by inducing the evolution of reactive oxygen species from guard cells of tomato and *Commelina communis*. Plant Physiol 121:147–152
Liechti R, Farmer EE (2002) The jasmonate pathway. Science 296:1649–1650
Liszkay A, van der Zalm E, Schopfer P (2004) Production of reactive oxygen intermediates $O2^{\bullet -}$, H_2O_2, and •OH by maize roots and their role in wall loosening and elongation growth. Plant Physiol 136:3114–3123; discussion 3001
Loake G, Grant M (2007) Salicylic acid in plant defence – the players and protagonists. Curr Opin Plant Biol 10:466–472
Manthe B, Schulz M, Schnabl H (1992) Effects of salycylic acid on growth and stomatal movements of *Vicia faba* l: Evidence for salicylic acid metabolization. J Chem Ecol 18:1525–1539
McAinsh MR, Clayton H, Mansfield TA, Hetherington AM (1996) Changes in stomatal behavior and guard cell cytosolic free calcium in response to oxidative stress. Plant Physiol 111:1031–1042
Memelink J, Lingthorst HJM, Schilperoort RA, Hoge HC (1990) Tobacco genes encoding acidic and basic isofroms of pathogenesis-related proteins display different expression patterns. Plant Mol Biol 14:119–126
Mori IC, Schroeder JI (2004) Reactive oxygen species activation of plant Ca^{2+} channels. A signaling mechanism in polar growth, hormone transduction, stress signaling, and hypothetically mechanotransduction. Plant Physiol 135:702–708
Munemasa S, Oda K, Watanabe-Sugimoto M, Nakamura Y, Shimoishi Y, Murata Y (2007) The *coronatine-insensitive 1* mutation reveals the hormonal signaling interaction between abscisic

acid and methyl jasmonate in *Arabidopsis* guard cells. Specific impairment of ion channel *activation and second messenger production. Plant Physiol* 143:1398–1407
Murata Y, Pei ZM, Mori IC, Schroeder J (2001) Abscisic acid activation of plasma membrane Ca^{2+} channels in guard cells requires cytosolic NAD(P)H and is differentially disrupted upstream and downstream of reactive oxygen species production in *abi1-1* and *abi2-1* protein phosphatase 2C mutants. Plant Cell 13:2513–2523
Mustilli AC, Merlot S, Vavasseur A, Fenzi F, Giraudat J (2002) *Arabidopsis ost1* protein kinase mediates the regulation of stomatal aperture by abscisic acid and acts upstream of reactive oxygen species production. **Plant** Cell 14:3089–3099
Nakajima R, Yamazaki I (1979) The mechanism of indole-3-acetic acid oxidation by horseradish peroxidases. J Biol Chem 254:872–878
Neuenschwander U, Vernooij B, Friedrich L, Uknes S, Kessmann H, Ryals J (1995) Is H_2O_2 a second messenger of salicylic acid in systemic acquired resistance? Plant J 8:227–233
Omran RG (1977) The direct involvement of hydrogen peroxide in indoleacetic acid inactivation. Biochem Biophys Res Commun 78:970–976
Overmyer K, Brosche M, Kangasjarvi J (2003) Reactive oxygen species and hormonal control of cell death. Trends Plant Sci 8:335–342
Overmyer K, Tuominen H, Kettunen R, Betz C, Langebartels C, Sandermann H Jr, Kangasjarvi J (2000) Ozone-sensitive arabidopsis rcd1 mutant reveals opposite roles for ethylene and jasmonate signaling pathways in regulating superoxide-dependent cell death. Plant Cell 12:1849–1862
Park KY, Jung JY, Park J, Hwang JU, Kim YW, Hwang I, Lee Y (2003) A role for phosphatidylinositol 3-phosphate in abscisic acid-induced reactive oxygen species generation in guard cells. Plant Physiol 132:92–98
Pei ZM, Murata Y, Benning G, Thomine S, Klusener B, Allen GJ, Grill E, Schroeder JI (2000) Calcium channels activated by hydrogen peroxide mediate abscisic acid signalling in guard cells. Nature 406:731–734
Rao MV, Lee H, Creelman RA, Mullet JE, Davis KR (2000) Jasmonic acid signaling modulates ozone-induced hypersensitive cell death. Plant Cell 12:1633–1646
Rao MV, Lee HI, Davis KR (2002) Ozone-induced ethylene production is dependent on salicylic acid, and both salicylic acid and ethylene act in concert to regulate ozone-induced cell death. Plant J 32:447–456
Rao MV, Paliyath G, Ormrod DP, Murr DP, Watkins CB (1997) Influence of salicylic acid on H_2O_2 production, oxidative stress, and H_2O_2-metabolizing enzymes. Salicylic acid-mediated oxidative damage requires H_2O_2. Plant Physiol 115:137–149
Ryals J, Lawton KA, Delaney TP, Friedrich L, Kessmann H, Neuenschwander U, Uknes S, Vernooij B, Weymann K (1995) Signal transduction in systemic acquired resistance. Proc Natl Acad Sci USA 92:4202–4205
Saito N, Munemasa S, Nakamura Y, Shimoishi Y, Mori IC, Murata Y (2008) Roles of rcn1, regulatory a subunit of protein phosphatase2A, in methyl jasmonate signaling and signal crosstalk between methyl jasmonate and abscisic acid. Plant Cell Physiol 49:1396–1401
Savitsky PA, Gazaryan IG, Tishkov VI, Lagrimini LM, Ruzgas T, Gorton L (1999) Oxidation of indole-3-acetic acid by dioxygen catalysed by plant peroxidases: Specificity for the enzyme structure. Biochem J 340(Pt 3):579–583
Schopfer P (2001) Hydroxyl radical-induced cell-wall loosening in vitro and in vivo: Implications for the control of elongation growth. Plant J 28:679–688
Schopfer P, Liszkay A, Bechtold M, Frahry G, Wagner A (2002) Evidence that hydroxyl radicals mediate auxin-induced extension growth. Planta 214:821–828
Schroeder JI, Hagiwara S (1990) Repetitive increases in cytosolic Ca^{2+} of guard cells by abscisic acid activation of nonselective Ca^{2+} permeable channels. Proc Natl Acad Sci USA 87:9305–9309
Schweikert C, Liszkay A, Schopfer P (2000) Scission of polysaccharides by peroxidase-generated hydroxyl radicals. Phytochemistry 53:565–570
Smith AM, Morrison WL, Milham PJ (1982) Oxidation of indole-3-acetic acid by peroxidase: Involvement of reduced peroxidase and coumpound III with superoxide as a product. Biochemistry 21:4414–4419

Suhita D, Kolla VA, Vavasseur A, Raghavendra AS (2003) Different signaling pathways involved during the suppression of stomatal opening by methyl jasmonate or abscisic acid. Plant Sci 164:481–488

Suhita D, Raghavendra AS, Kwak JM, Vavasseur A (2004) Cytoplasmic alkalization precedes reactive oxygen species production during methyl jasmonate- and abscisic acid-induced stomatal closure. Plant Physiol 134:1536–1545

Sumimoto H (2008) Structure, regulation and evolution of nox-family nadph oxidases that produce reactive oxygen species. FEBS J 275:3249–3277

Summermatter K, Sticher L, Metraux JP (1995) Systemic responses in *Arabidopsis thaliana* infected and challenged with *Pseudomonas syringae* pv syringae. Plant Physiol 108:1379–1385

Tamura S (1990). Historical aspects of gibberelins. Gibberellins N, Takahashi BO, Phinney J (eds) Macmillan, New York, pp 1–8

Turner JG, Ellis C, Devoto A (2002) The jasmonate signal pathway. Plant Cell 14(Suppl):S153–S164

Zhang X, Dong FC, Gao JF, Song CP (2001) Hydrogen peroxide-induced changes in intracellular pH of guard cells precede stomatal closure. Cell Res 11:37–43

Reactive Oxygen Species in Growth and Development

Elizabeth Bell, Seiji Takeda, and Liam Dolan

Abstract The spatial control of cell growth is a central process in plant development. Reactive oxygen species (ROS) are important regulators of cell and organ growth and are thought to operate by controlling the extensibility of the cell wall and modulating intracellular signalling processes. By increasing elasticity of the wall they promote growth, and by cross-linking polymers they increase rigidity and repress growth. ROS produced by NADPH oxidase proteins are also important regulators of tip growth in root hairs and pollen tubes, where they not only control wall rigidity but also control cell signalling events involving calcium and MAP kinases cascades. Models for the roles of ROS in the control of cell growth during development have been proposed and are reviewed here.

1 Introduction

Reactive oxygen species (ROS) have a signalling role during plant development, and play an important function in the regulation growth through their various effects on cell wall elasticity. While their signalling roles in a variety of biological processes, such as the response to pathogen attack, have been well documented, their role in development is less well understood. In this chapter, we will describe the developmental processes in which ROS have been implicated. As the best characterised developmental role for ROS is in the regulation of polarised growth of root hairs and pollen tubes, the major focus of this review will be the role of ROS in tip growth.

E. Bell, S. Takeda, and L. Dolan (✉)
Cell and Developmental Biology, John Innes Centre, Norwich, NR4 7UH, UK
e-mail: liam.dolan@bbsrc.ac.uk

L.A. del Río and A. Puppo (eds.), *Reactive Oxygen Species in Plant Signaling*, Signaling and Communication in Plants, DOI 10.1007/978-3-642-00390-5_3,

2 ROS are Involved in Polarised Cell Growth

During the development of plants, the growth of cells is polarised; more growth occurs on some surfaces of the cell than others. For example, the expansion of an elongating cell in the cortex of a stem or root is said to be polar because more expansion occurs along its sides than at its ends. Furthermore, the polarity of a cell may change during development. For example, the early stages of *Arabidopsis* trichome (leaf hair) development involve the elongation of a spike-like cellular projection by an extremely polarised form of growth. Once this axis reaches a certain length, new sites of expansion are initiated which form branches. The initiation of these new axes of polarity in the trichome can be considered to be symmetry breaking events, where the original polarity of growth is changed to another. In this section, we will review the role of ROS in the control of polar growth in root hair cells and pollen tubes, in which growth takes place at a single site in the cell. While these are extreme examples of cell polarity, it is likely that the general principles controlling the polarity of growth are the same for all cells.

2.1 *Tip Growing Cells Accumulate ROS at the Tip*

Root hairs and pollen tubes grow by a highly polarised form of cell expansion called tip growth, where new growth occurs at a single site within the cell. Root hairs are long tubular projections that emerge from the surface epidermal cells of the root and increase the surface area of the root for nutrient and water uptake. They also help to anchor the plant in the soil (Fig. 1). Hair growth is sensitive to environmental conditions such as nutrient availability (reviewed in López-Bucio et al. 2003). Pollen tubes are also long tubular projections, but grow from germinated pollen grains and transport the male gametophyte from the surface of the stigma where the pollen germinates to the female gametophyte where fertilisation occurs.

Root hairs and pollen tubes produce ROS at the growing tip, and these ROS have been shown to be necessary for growth (Foreman et al. 2003; Potocky et al. 2007). The root hairs of the *root hair defective 2* (*rhd2*) mutant of *Arabidopsis* do not accumulate ROS at the hair tip and fail to develop past the initiation stage (Fig. 1). The small root hair bulges that do form burst more readily than wild-type root hairs, and interestingly are phenocopied by treatment of wild-type root hairs with ROS scavengers (Dunand et al. 2007; Jones et al. 2007; Macpherson et al. 2008). This suggests that the cell walls of the *rhd2* mutant are altered in structure and unable to resist turgor pressure once hair initiation has occurred. Furthermore, it has also been shown that root hair ROS accumulation oscillates during growth, and that this is accompanied by oscillations in pH and growth rate (Monshausen et al. 2007). Together, these observations indicate that ROS are required for root hair development.

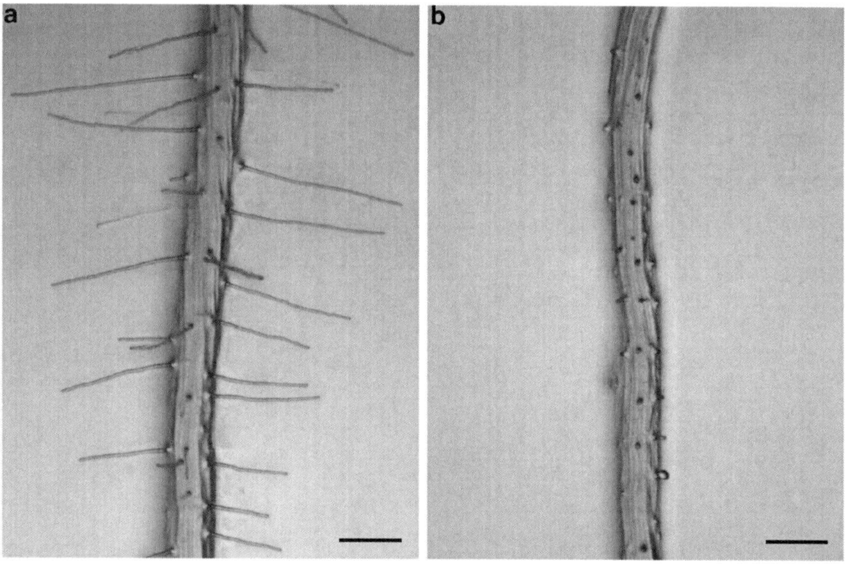

Fig. 1 *Arabidopsis* root hairs. (**a**) Wild-type and (**b**) *rhd2* mutant (Scale bars, 200 μm)

2.2 ROS are Produced by RBOH Proteins

RHD2 encodes an *Arabidopsis thaliana* respiratory burst oxidase homolog (AtrbohC) that is essential for root hair tip growth (Foreman et al. 2003). Members of this family of plasma membrane NADPH oxidases (NOX) (plant NADPH oxidases are called RBOH for respiratory burst oxidase homologues) are similar to gp91^{phox}, the glycosylated transmembrane subunit of the mammalian NOX cytochrome (Torres et al. 1998; see Chap. 1). NOX proteins oxidise NADPH, and transfer electrons to the apoplastic space where they are donated to oxygen molecules to form superoxide (O_2•$^-$). O_2•$^-$is the source from which other ROS such as H_2O_2 and OH• are derived (reviewed in Bedard et al. 2007; see Chap. 1).

Plant RBOH proteins consist of two cytosolic domains connected by six transmembrane domains. The amino-terminal domains contain two Ca^{2+}-binding EF hand motifs, and multiple putative phosphorylation sites (Nuhse et al. 2004; Kobayashi et al. 2007). The carboxy-terminal domain consists of FAD- and NADPH-binding catalytic domains (Fig. 2). The *Arabidopsis* genome encodes ten RBOH genes (*AtrbohA-J*). Unlike the mammalian homologues, plant RBOH contain a 300 amino acid N-terminal extension that includes two putative EF hands (Keller et al. 1998; Torres et al. 1998; see Chap. 1). This suggests that calcium may be an important regulator of RHD2/AtrbohC activity during root hair growth.

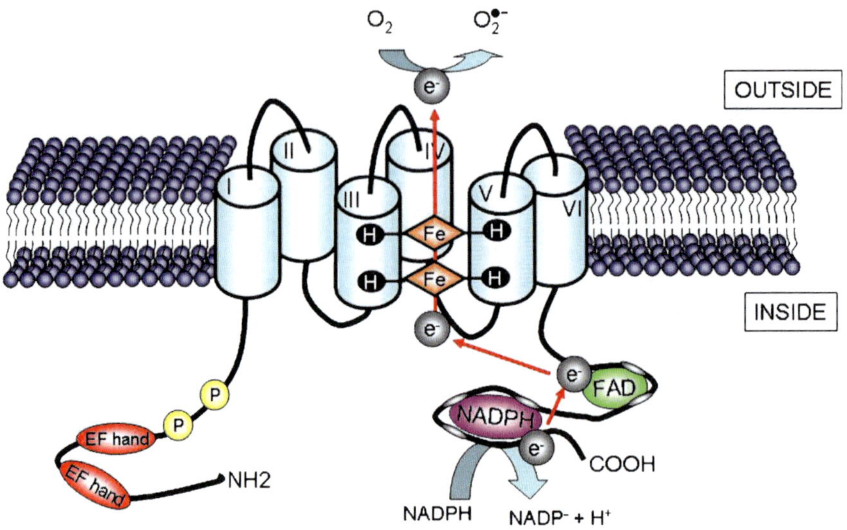

Fig. 2 Predicted structure of RBOH protein. RBOH proteins consist of a cytosolic amino-terminal region including Ca^{2+}-binding EF hand motifs and phosphorylation sites, six transmembrane including two heme (Fe) groups and catalytic carboxy-terminal domain containing FAD- and NADPH-binding domains. RBOH proteins generate superoxide outside the cell from molecular oxygen by transferring an electron from the oxidation of NADPH inside the cell (see Chap. 1).

2.3 RHD2/AtrbohC is Regulated by Calcium and Phosphorylation

Recent findings with RBOH proteins involved in the responses to plant pathogens have uncovered how RBOH proteins are activated to produce ROS. Potato StrbohB is directly phosphorylated in a Ca^{2+}-dependent manner by StCDPK4 and StCDPK5 calcium-dependent protein kinases (Kobayashi et al. 2007). AtrbohD, a RBOH protein in *Arabidopsis thaliana*, has been shown to undergo a conformational change upon Ca^{2+} binding and phosphorylation of its amino-terminal domains, triggering an increase in ROS production (Ogasawara et al. 2008). It has also been shown that the rice small GTPase OsRac1 directly binds to the amino-terminus of a rice RBOH protein in a calcium-dependent manner and triggers a stimulation of ROS production (Wong et al. 2007).

RHD2 has also been found to be regulated by calcium. Point mutations in the EF-hand motifs resulted in root hair growth defects, and phoshorylation sites within RHD2 have shown to be directly phosphorylated in a Ca^{2+} dependent manner. Point mutations within the phosphophorylation sites or the EF-hands also disrupted ROS production in vitro (Takeda et al. 2008). These data suggest that RHD2 ROS production is controlled positively by direct binding of Ca^{2+} to the EF-hand motifs and Ca^{2+}-dependent phosphorylation (which may be mediated by Ca^{2+}-dependent protein kinases). This dual regulation mechanism suggests that the control of RHD2 activity is a complex process and responds to a variety of internal and external stimuli.

ROS are known to stimulate the activity of the hyperpolarisation-activated calcium channels (HACCs) and induce the Ca^{2+} flux from outside the cell into the cytosol (Very and Davies 2000), generating the tip-focussed Ca^{2+} gradient that is essential for root hair tip growth (Wymer et al. 1997). Recently, depolarisation-activated calcium channels (DACC) have also been characterised at the apex of root hairs and are thought to increase the voltage range for Ca^{2+} uptake by root hairs (Miedema et al. 2008). The tip-focussed Ca^{2+} gradient is disrupted in the *rhd2* mutant and exogenous application of OH• has been found to increase cytoplasmic Ca^{2+} accumulation (Foreman et al. 2003), suggesting that RHD2-derived ROS may activate calcium channels at the hair tip causing influx of the Ca^{2+} into the cell and the generation of a tip-high Ca^{2+} gradient.

Based on these findings, it seems logical that RHD2, ROS, and Ca^{2+} work in concert to establish a positive feedback mechanism at the tip of growing hairs that serves to maintain tip growth (Fig. 3). At the tip, RHD2 enzymatic activity is regulated by Ca^{2+}-binding and direct phosphorylation to control $O_2•^-$ production. The resulting $O_2•^-$ can then be converted into other ROS that may loosen the wall

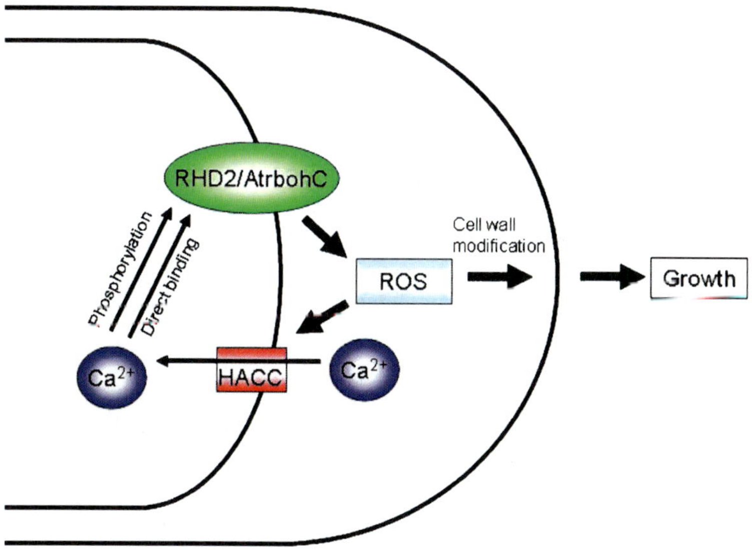

Fig. 3 A positive feedback mechanism at the root hair apex maintains tip growth. RHD2/AtrbohC, ROS and Ca^{2+} generate a positive feedback loop which maintains growth at the root hair tip. RHD2/AtrbohC produces ROS outside a cell, which in turn activates calcium channels on the plasma membrane to induce Ca^{2+} influx into the cell. Ca^{2+} directly binds the EF-hand motif of RHD2 to activate the ROS production. RHD2/AtrbohC is also activated by Ca^{2+}-dependent phosphorylation, which may be controlled by Ca^{2+}-dependent protein kinase (CDPK). This feedback loop is restricted to the hair tip by the control of RHD2 localisation and maintains the local growth to the hair tip.

at the tip (see below), allowing rapid growth, and activate the calcium channels at the tip. This influx of Ca^{2+} in turn may further activate RHD2 to produce more ROS at the tip. This positive feedback regulation together with control of RHD2 spatial localisation to the tip (see below) may be the central system to maintain the growth at the tip. It remains to be established how this positive feedback mechanism is negatively regulated.

Pollen specific NOX genes have been identified in *Nicotiana* (*NtNOX*) and *Arabidopsis* (*AtrbohJ* and *AtrbohH*), and phylogenetic analysis suggests that they form a distinct sub-family of NOX genes (Potocky et al. 2007). Transfection of *Nicotiana* pollen tubes with antisense *NtNOX* oligonucleotides inhibits pollen tube growth and decreases ROS accumulation at the pollen tube tip (Potocky et al. 2007). As pollen tubes and root hairs both develop tip-high Ca^{2+} gradients and accumulate ROS at their tips, it is likely that similar positive feedback mechanisms of NOX regulation exist both cell types.

2.4 RHD2 Protein Accumulates at the Site of ROS Production and Growth

Consistent with its role in promoting cell growth, the *RHD2* gene is expressed in all epidermal cells of the elongation zone of the root, but becomes restricted to the trichoblasts before hair outgrowth occurs. *RHD2* expression ceases around the time root hair growth stops; no transcripts (or protein) are found in non-growing older root hairs. It can be concluded therefore that RHD2 is expressed in growing cells, and its expression in hair cells is required for cell elongation. Expression of *RHD2* has also been shown to be upregulated by potassium, nitrogen and phosphorous deficiency (Shin et al. 2005). Since these environmental factors modulate root hair growth, it suggests that this environmental control may involve the regulation of *RHD2* transcription.

During wild-type root hair development, ROS accumulates at the site of growth. This suggests that the sites of ROS accumulation determine the places in the cell where growth occurs. Consistent with this view is the observation that exogenous application of OH• to *rhd2* mutant root hairs results in the formation of a spherical outgrowth which lacks polarity (Foreman et al. 2003). Restriction of ROS production to the tip of the root hair may, therefore, account for the spatial restriction of growth during wild-type root hair development. If localisation of ROS production determines where growth occurs in root hairs, we would predict that the ROS generating enzyme might be located at these growth sites. Transgenic plants expressing green fluorescent protein (GFP)-tagged RHD2 protein showed that RHD2 is indeed restricted to the tip of elongating hairs (Takeda et al. 2008). RHD2 protein remains at the hair tip until elongation ceases.

RHD2, like other plant RBOH proteins, has six predicted trans-membrane domains, and RHD2 membrane-localisation has also been shown to be dependent on endomembrane vesicle trafficking which is disrupted by the inhibitor brefeldin A (BFA) (Nebenfuhr et al. 2002). RHD2 is, therefore, located at the root hair tip, and this localisation is dependent on the activity of vesicle trafficking in the endomembrane system.

A further mechanism controlling spatial distribution of ROS within a root hair cell has been suggested from analysis of the *Arabidopsis suprecentipede1* (*scn1*) mutant. The *scn1* mutant root produces multiple bulges within one root hair cell. In these cells, $O_2^{\bullet-}$ was found to accumulate at the sites of these ectopic bulges, and the *rhd2* mutation was shown to suppress the formation of these multiple bulges (Carol et al. 2005). This suggests that ectopic RHD2 activity is required for the formation of these extra growth bulges in the *scn1* mutant. This is supported by the observation that GFP:RHD2 was found in these ectopic root hair bulges in the *scn1* mutant (Takeda et al. 2008). These data suggest that SCN1 controls the spatial localisation of RHD2, and thus the sites at which ROS accumulates during root hair tip growth. *SCN1* encodes a Rho GTPase dissociation inhibitor (RhoGDI). Rho GTPase dissociation inhibitors inactivate Rho GTPases by maintaining them in their GDP-bound inactivate state (Carol et al. 2005). The *Arabidopsis* genome encodes 11 such Rho GTPases which are called ROP (Rho in plants) GTPases in plants (Yang 2002). Over-expression of ROP2 results in the development of defective branching root hairs, confirming a role for a RhoGDI-ROP interaction in the control of hair growth (Jones et al. 2002). Furthermore, ROP2 has been found to localise to the tip of growing hairs and this localisation is aberrant in the *scn1* mutant background (Carol et al. 2005), suggesting that ROP2 localisation is dependent of SCN1 function. Taken together, the localisation of RHD2 to the growing tip of the root hair is dependent on Rho GDI–ROP signalling.

ROP proteins play a number of roles during cellular development including the regulation of microfilament organisation. Actin microfilament orientation is disrupted in the abnormal root hairs that result from ROP2 over-expression (Jones et al. 2002). This suggests that one function of ROP GTPases is to organise microfilaments during hair growth and that RHD2 localisation in the root hair cells is dependent on microfilaments. This view is supported by the observation that RHD2 localisation is defective in hairs in which microfilament organisation is disrupted (Takeda et al. 2008).

2.5 The Targets of ROS-Regulation in Polarised Growth Have Not Been Identified

It is likely that ROS regulates many processes in the cell, and we are only beginning to discover what these processes are. As mentioned above there is evidence that ROS activate calcium channels in the tip of the root hair and these allow the influx of calcium into the cytoplasm at the tip of the cell. This causes a tip-high calcium gradient to form that is required for root hair growth. There is also evidence that ROS controls other cellular processes.

RHD2-derived ROS may modify the structure or physical characteristics of the root hair cell wall. ROS can in some instances cause strengthening through cross-linking and in other cases causing weakening though the scission of cell wall hemicellulose polymers. For example OH• have been implicated in the cleavage of cell wall polysaccharides (Fry 1998) and cell wall loosening in living tissues (Schopfer 2001; Liszkay et al. 2004). It has also been suggested that the ROS involved in cell wall loosening are derived from NOX enzymes (Liszkay et al. 2003). In contrast, it is generally accepted that cell expansion by cell wall loosening and turgor pressure is followed by cell wall cross-linking and stiffening (Hohl et al. 1995). Cell wall stiffening is attributed to H_2O_2 in maize coleoptiles (Hohl et al. 1995) and in the lignification of xylem vessels (Ros Barceló 2005). Again, as different ROS can have distinct roles in cell growth, it is important to regulate the spatial production of ROS and to maintain a balance between different forms of ROS during cell growth.

It has been suggested that at least part of the *rhd2* root hair phenotype results from a cell wall defect (Monshausen et al. 2007; Macpherson et al. 2008). Around 30% of *rhd2* mutant root hairs rupture at the transition between initiation and tip growth. This lysis occurs rarely in wild-type. It is possible that a lack of ROS at the root hair tips of *rhd2* mutants causes a weakening of the cell wall. Similarly, it was found that by increasing the pH of the growth media, *rhd2-1* was able to form wild-type root hairs, suggesting that lack of cell wall rigidity in *rhd2* can be compensated for with higher pH (Monshausen et al. 2007). Alternatively, the ROS may be acting to increase the softening of the cell wall and protecting the root hairs against the hypo-osmotic shock that causes the root hairs to burst. As different ROS have conflicting functions in cell wall biochemistry, the balance between the different species may be critical. As the higher pH was shown to restore the tip-focussed calcium gradient, the phenotypic rescue could also be attributed to an activation of calcium channels.

It is likely that ROS produced at the tips of root hairs may control growth by modulating a variety of signal transduction pathways. For example, the *Arabidopsis OXIDATIVE SIGNAL INDUCIBLE 1* (*OXI1*) gene, which encodes a serine/threonine protein kinase, is induced in response to exogenous ROS treatment. The *oxi1* mutant has slightly shorter root hairs than wild-type, suggesting that OXI1 is a target of ROS that is important for root hair growth. OXI1 activates two *Arabidopsis* MAP kinases (MPK3 and MPK6) (Rentel et al. 2004), indicating that ROS modulates a MAP kinase cascade that controls root hair growth.

By comparing the root transcriptome of wild-type and *rhd2* mutants, a catalogue of genes that may act downstream of RHD2 during root hair growth has been identified (Jones et al. 2006). Although many of these genes may be involved in root hair development independently of ROS, others might be the direct targets of ROS-regulation during root hair growth. Detailed analysis of gene expression in response to ROS in root hairs will be helpful in our understanding of the ROS network during plant cell growth.

2.6 ROS in Diffuse Growth

Most plant cells expand by a form of polarised growth called diffuse or intercalary growth. In diffuse growth, cell expansion takes place at all cell surfaces and involves ROS production. Repression of RBOH enzymes in different plant species has suggested that the control of ROS production is important for diffuse growth. Antisense experiments with RBOH in tomato suggested ROS are involved in the growth of a variety of different cell types. Tomato lines with reduced RBOH expression displayed leaf and flower abnormalities suggesting that there is a general defect in the regulation of cell growth. Interestingly, the developmental defects were accompanied by ectopic expression of homeotic MADS box genes (Sagi et al. 2004). The distribution of ROS in growing maize has suggested that ROS may have functions in leaf growth and development. Higher concentrations of ROS were detected in the expansion zone of the leaf, and leaf expansion was inhibited by diphenyleneiodonium (DPI), suggesting a flavin-containing enzyme such as a NOX enzyme could be responsible for the ROS production and growth (Rodriguez et al. 2002). Maize coleoptiles treated with auxin display accelerated elongation, and this is accompanied by higher levels of ROS. Auxin-elicited elongation is reduced in the DPI-treated coleoptiles, again suggesting the involvement of NOX enzymes (Schopfer et al. 2002). These different observations together suggest that ROS is involved in cell and organ growth during development.

3 ROS in Developmentally Programmed Cell Death

There is evidence that ROS play a role in the control of programmed cell death (PCD) during development. PCD occurs at different times during the life of the plant and is exemplified by the death of xylem vessel element cells during their development. H_2O_2 is produced on the plasma membrane of differentiating xylem parenchyma cells where it diffuses to the lignifying xylem (Ros Barceló 2005) presumably before the final stages of PCD. The precise role of H_2O_2 in the PCD process remains to be determined. The aleurone layer is a secretory tissue in cereal grains that undergoes a hormonally-regulated PCD following germination. The PCD of barley aleurone cells has also been shown to involve ROS production (Bethke and Jones 2001). PCD has been shown to play a role in embryogenesis (Filonova et al. 2002), gametogenesis (Wang et al. 1999) and inhibition of incompatible pollen on the stigma of self-incompatible plant species (Thomas and Franklin-Tong 2004). While developmentally controlled PCD has been shown to involve ROS production, the precise role of these species remains to be determined. It is possible that they act in a similar fashion to those ROS that are involved in the PCD that occurs in response to pathogen infection.

4 Concluding Remarks

The tight regulation of ROS production both spatially and temporally contributes to the variety of cell and organ shapes in plants. NOX enzymes have been shown to control localised production of ROS and, therefore, play an important role in polarised cell growth. NOX activity is controlled by ROP GTPases, calcium ions and phosphorylation, and these are well-characterised signalling components in plant responses to exogenous stimuli. It seems likely, therefore, that NOX activity could be regulated in response to external signals from the environment. Understanding the role of ROS in the growth of different cells types and during developmentally controlled PCD will shed light on an important mechanism directing plant development.

References

Bedard, K., Lardy, B., and Krause, K.-H. (2007). NOX family NADPH oxidases: Not just in mammals. Biochimie **89,** 1107–1112.

Bethke, P.C., and Jones, R.L. (2001). Cell death of barley aleurone protoplasts is mediated by reactive oxygen species. Plant J **25,** 19–29.

Carol, R.J., Takeda, S., Linstead, P., Durrant, M.C., Kakesova, H., Derbyshire, P., Drea, S., Zarsky, V., and Dolan, L. (2005). A RhoGDP dissociation inhibitor spatially regulates growth in root hair cells. Nature **438,** 1013–1016.

Dunand, C., Crevecoeur, M., and Penel, C. (2007). Distribution of superoxide and hydrogen peroxide in Arabidopsis root and their influence on root development: possible interaction with peroxidases. New Phytologist **174,** 332–341.

Filonova, L.H., von Arnold, S., Daniel, G., and Bozhkov, P.V. (2002). Programmed cell death eliminates all but one embryo in a polyembryonic plant seed. Cell Death Differentiation **9,** 1057–1062.

Foreman, J., Demidchik, V., Bothwell, J.H., Mylona, P., Miedema, H., Torres, M.A., Linstead, P., Costa, S., Brownlee, C., Jones, J.D., Davies, J.M., and Dolan, L. (2003). Reactive oxygen species produced by NADPH oxidase regulate plant cell growth. Nature **422,** 442–446.

Fry, S.C. (1998). Oxidative scission of plant cell wall polysaccharides by ascorbate-induced hydroxyl radicals. Biochem J **332**, 507–515.

Hohl, M., Greiner, H., and Schopfer, P. (1995). The cryptic-growth response of maize coleoptiles and its relationship to H_2O_2-dependent cell wall stiffening. Physiologia Plantarum **94,** 491–498.

Jones, M.A., Raymond, M.J., and Smirnoff, N. (2006). Analysis of the root-hair morphogenesis transcriptome reveals the molecular identity of six genes with roles in root-hair development in Arabidopsis. Plant J **45,** 83–100.

Jones, M.A., Raymond, M.J., Yang, Z., and Smirnoff, N. (2007). NADPH oxidase-dependent reactive oxygen species formation required for root hair growth depends on ROP GTPase. J Exp Bot **58,** 1261–1270.

Jones, M.A., Shen, J.-J., Fu, Y., Li, H., Yang, Z., and Grierson, C.S. (2002). The Arabidopsis Rop2 GTPase is a positive regulator of both root hair initiation and tip growth. Plant Cell **14,** 763–776.

Keller, T., Damude, H.G., Werner, D., Doerner, P., Dixon, R.A., and Lamb, C. (1998). A plant homolog of the neutrophil NADPH oxidase gp91phox subunit gene encodes a plasma membrane protein with Ca2+ binding motifs. Plant Cell **10,** 255–266.

Kobayashi, M., Ohura, I., Kawakita, K., Yokota, N., Fujiwara, M., Shimamoto, K., Doke, N., and Yoshioka, H. (2007). Calcium-dependent protein kinases regulate the production of reactive oxygen species by potato NADPH oxidase. Plant Cell **19,** 1065–1080.

Liszkay, A., Kenk, B., and Schopfer, P. (2003). Evidence for the involvement of cell wall peroxidase in the generation of hydroxyl radicals mediating extension growth. Planta **217,** 658–667.

Liszkay, A., van der Zalm, E., and Schopfer, P. (2004). Production of reactive oxygen intermediates $O_2^{\bullet-}$, H_2O_2, and OH• by maize roots and their role in wall loosening and elongation growth. Plant Physiol **136,** 3114–3123.

López-Bucio, J., Cruz-Ramírez, A., and Herrera-Estrella, L. (2003). The role of nutrient availability in regulating root architecture. Curr Opin Plant Biol **6,** 280–287.

Macpherson, N., Takeda, S., Shang, Z., Dark, A., Mortimer, J., Brownlee, C., Dolan, L., and Davies, J. (2008). NADPH oxidase involvement in cellular integrity. Planta **227,** 1415–1418.

Miedema, H., Demidchik, V., Very, A.-A., Bothwell, J.H.F., Brownlee, C., and Davies, J.M. (2008) Two voltage-dependent calcium channels co-exist in the apical plasma membrane of *Arabidopsis thaliana* root hairs. New Phytol **179**, 378–385.

Monshausen, G.B., Bibikova, T.N., Messerli, M.A., Shi, C., and Gilroy, S. (2007). Oscillations in extracellular pH and reactive oxygen species modulate tip growth of Arabidopsis root hairs. Proc Natl Acad Sci U S A **104,** 20996–21001.

Nebenfuhr, A., Ritzenthaler, C., and Robinson, D.G. (2002). Brefeldin A: Deciphering an Enigmatic Inhibitor of Secretion. Plant Physiol **130,** 1102–1108.

Nuhse, T.S., Stensballe, A., Jensen, O.N., and Peck, S.C. (2004). Phosphoproteomics of the Arabidopsis plasma membrane and a new phosphorylation site database. Plant Cell **16,** 2394–2405.

Ogasawara, Y., Kaya, H., Hiraoka, G., Yumoto, F., Kimura, S., Kadota, Y., Hishinuma, H., Senzaki, E., Yamagoe, S., Nagata, K., Nara, M., Suzuki, K., Tanokura, M., and Kuchitsu, K. (2008). Synergistic activation of the arabidopsis NADPH oxidase AtrbohD by Ca2+ and Phosphorylation. J Biol Chem **283,** 8885–8892.

Potocky, M., Jones, M.A., Bezvoda, R., Smirnoff, N., and Zarsky, V. (2007). Reactive oxygen species produced by NADPH oxidase are involved in pollen tube growth. New Phytol **174,** 742–751.

Rentel, M.C., Lecourieux, D., Ouaked, F., Usher, S.L., Petersen, L., Okamoto, H., Knight, H., Peck, S.C., Grierson, C.S., Hirt, H., and Knight, M.R. (2004). OXI1 kinase is necessary for oxidative burst-mediated signalling in Arabidopsis. Nature **427,** 858–861.

Rodriguez, A.A., Grunberg, K.A., and Taleisnik, E.L. (2002). Reactive oxygen species in the elongation zone of maize leaves are necessary for leaf extension. Plant Physiol **129,** 1627–1632.

Ros Barceló, A. (2005). Xylem parenchyma cells deliver the H_2O_2 necessary for lignification in differentiating xylem vessels. Planta **220,** 747–756.

Sagi, M., Davydov, O., Orazova, S., Yesbergenova, Z., Ophir, R., Stratmann, J.W., and Fluhr, R. (2004). Plant respiratory burst oxidase homologs impinge on wound responsiveness and development in *Lycopersicon esculentum*. Plant Cell **16,** 616–628.

Schopfer, P. (2001). Hydroxyl radical-induced cell-wall loosening in vitro and in vivo: implications for the control of elongation growth. Plant J **28,** 679–688.

Schopfer, P., Liszkay, A., Bechtold, M., Frahry, G., and Wagner, A. (2002). Evidence that hydroxyl radicals mediate auxin-induced extension growth. Planta **214,** 821–828.

Shin, R., Berg, R.H., and Schachtman, D.P. (2005). Reactive oxygen speices and root hairs in Arabidopsis root response to nitrogen, phosphorus and potassium deficiency. Plant Cell Physiol **46**, 1350–1357.

Takeda, S., Gapper, C., Kaya, H., Bell, E., Kuchitsu, K., and Dolan, L. (2008). Local positive feedback regulation determines cell shape in root hair cells. Science **319,** 1241–1244.

Thomas, S.G., and Franklin-Tong, V.E. (2004). Self-incompatibility triggers programmed cell death in Papaver pollen. Nature **429,** 305–309.

Torres, M.A., Onouchi, H., Hamada, S., Machida, C., Hammond-Kosack, K.E., and Jones, J.D.G. (1998). Six Arabidopsis thaliana homologues of the human respiratory burst oxidase (gp91phox). Plant J **14,** 365–370.

Very, A.-A., and Davies, J.M. (2000). Hyperpolarization-activated calcium channels at the tip of Arabidopsis root hairs. Proc Natl Acad Sci USA **97,** 9801–9806.

Wang, M., Hoekstra, S., van Bergen, S., Lamers, G.E.M., Oppedijk, B.J., van der Heijden, M.W., de Priester, W., and Schilperoort, R.A. (1999). Apoptosis in developing anthers and the role of ABA in this process during androgenesis in *Hordeum vulgare L.* Plant Mol Biol **39,** 489–501.

Wong, H.L., Pinontoan, R., Hayashi, K., Tabata, R., Yaeno, T., Hasegawa, K., Kojima, C., Yoshioka, H., Iba, K., Kawasaki, T., and Shimamoto, K. (2007). Regulation of rice NADPH oxidase by binding of Rac GTPase to its N-terminal extension. Plant Cell **19,** 4022–4034.

Wymer, C.L., Bibikova, T.N., and Gilroy, S. (1997). Cytoplasmic free calcium distributions during the development of root hairs of *Arabidopsis thaliana*. Plant J **12,** 427–439.

Yang, Z. (2002). Small GTPases: versatile signaling switches in plants. Plant Cell **14,** S375–S388.

ROS Signalling in Stomata

Jasmine Pham and Radhika Desikan

Abstract ROS are now well established to act as signalling molecules in stomatal guard cells. Generation of ROS occurs in guard cells via specific enzymes; the most studied enzymatic source being the NADPH oxidase components RBOH, which are involved in both abiotic and biotic stresses. Hormones such as abscisic acid, salicylic acid, ethylene, auxin and cytokinin affect ROS-mediated responses in stomata. Biotic stresses such as virulent bacteria and fungi are also known to affect stomatal movements, and ROS are likely to play a role. Regulation of ROS production and action also occurs at various levels, including antioxidant activities, as well as ROS-sensitive proteins, such as kinases and phosphatases, which transduce ROS to effect stomatal closure. Data are slowly emerging on ROS regulation of ion channels in stomata. A complex integration of signalling pathways mediated via ROS makes guard cells an attractive tool to study ROS signal transduction.

1 Introduction

1.1 The Importance of Stomata

In order for photosynthesis to occur, access to CO_2 in the atmosphere is required. This access is provided through stomata, pores that are formed between two guard cells in the leaf epidermis, with opening and closure of the stomata controlled by the turgidity of these two cells (Blatt 2000). The need to balance water loss via transpiration with that of photosynthetic activity requires the ability to adjust stomatal apertures (i.e., open and close stomata) and to sense environmental changes. This ability is afforded by the two highly specialised guard cells surrounding the stomatal pore. The sensitivity of stomata to various stimuli, such as light, drought

J. Pham and R. Desikan (✉)
Division of Biology, Imperial College London, South Kensington, London, SW7 2AZ, UK
e-mail: r.desikan@imperial.ac.uk

L.A. del Río and A. Puppo (eds.), *Reactive Oxygen Species in Plant Signaling*, Signaling and Communication in Plants, DOI 10.1007/978-3-642-00390-5_4,

stress and pathogen attack, and the obvious detrimental effect that defects in stomata would have on plant health indicate the importance of stomata to plant physiology. The complexity of the signalling pathways/networks controlling stomatal movements reflects the diversity of stimuli to which guard cells respond. This, coupled with the easily quantifiable response of stomata to stimuli (i.e., changes in stomatal aperture), has resulted in the popularity of guard cells as systems for studying plant cell signalling (Schroeder et al. 2001).

Reactive oxygen species (ROS) are reactive molecules of oxygen resulting from the incomplete reduction of molecular oxygen. They include superoxide, hydrogen peroxide, singlet oxygen and hydroxyl radical; being the most stable, hydrogen peroxide (H_2O_2) is the most well-studied molecule. Although traditionally known as being deleterious to living cells at high concentrations, it is well established now that ROS at low levels within cells act as important second messengers mediating signalling and defence responses to a number of stimuli. This has been extensively studied in various plant systems over the last couple of decades (Apel and Hirt 2004). ROS are produced by plant cells in response to a number of stimuli, including hormones, reduced water availability, extreme temperatures, altered light quality, air pollutants and pathogens, as well as developmental signals such as gravitropism (Desikan et al. 2005b). Guard cells therefore offer an attractive system to study sensing and integration of various signals via ROS leading to a single response, stomatal closure. This focussed chapter covers various situations where guard cells produce ROS leading to regulation of stomatal movements, via multiple signalling pathways.

1.2 Mechanisms of Stomatal Opening and Closure

The mechanisms by which plants control stomatal apertures have been well studied, involving the control of guard cell turgor and cellular volume, with increased guard cell turgor leading to stomatal opening and decreased turgor leading to stomatal closure. This control of stomatal aperture occurs through regulation of ion movements across the guard cell plasma membrane. Stomatal opening is initiated by activation of plasma membrane proton (H^+)-pumps which extrude H^+, causing hyperpolarisation of the plasma membrane and extracellular acidification/ cytosolic alkalisation, leading to uptake of K^+ (through inward-rectifying K^+ channels, K^+_{in}) and Cl^- ions (Blatt 2000; Schroeder et al. 2001). This increase in ionic content consequently leads to increased guard cell turgor and stomatal opening. In the case of stomatal closure, plasma membrane H^+-pumps are inhibited, and efflux of Cl^- ions driven by slow and sustained (S)-type and rapid (R)-type anion channels activated by Ca^{2+} causes membrane depolarisation. This depolarisation inhibits K^+_{in} and activates outward rectifying K^+ channels (K^+_{out}) leading to loss of guard cell turgor and stomatal closure (Blatt 2000; Schroeder et al. 2001). Ca^{2+} also plays a role in stomatal movements, and its role in stomatal closure mediated by the plant hormone abscisic acid has been well studied (Blatt 2000; Schroeder et al. 2001). How ROS will act on these components to mediate opening and closing will be addressed later.

2 Sources of ROS

There are several sources of ROS in plant cells, with constitutive levels being produced through the normal metabolic processes occurring in chloroplasts, mitochondria and peroxisomes (Neill et al. 2002; Desikan et al. 2005b). In addition to constitutive levels of ROS, plants also possess enzymes which specifically produce ROS as signalling molecules in response to stimuli. Such enzymes include NADPH oxidase (NOX), a ROS generating enzyme which has been well studied.

Chloroplasts are known to produce ROS as a by-product of photosynthesis and recently, it has been suggested that chloroplasts in guard cells may act as a source of ROS for signalling purposes, since the photosynthetic capacity of guard cell chloroplasts is thought to be minimal (Wang and Song 2008). In support of this, ROS production in response to ozone (discussed in Sect. 4.3.2), is first seen in guard cell chloroplasts which then triggers ROS production in the guard cell plasma membrane by NOX (Joo et al. 2005).

In *Arabidopsis*, ten NOX subunit genes, *AtrbohA-J* (*Arabidopsis* respiratory burst oxidase homologs), were identified as homologs of the mammalian NOX gene $gp91^{phox}$, with each gene expressed in a tissue-specific manner (Torres et al. 1998; Kwak et al. 2003; Sagi and Fluhr 2006). Of these ten genes, *AtrbohD* and *AtrbohF* were found to be expressed in guard cells and have been shown to play key roles in guard cell signalling (Kwak et al. 2003). AtrbohF was shown by Keller et al. (1998) to be plasma membrane bound, and all the Atrboh sequences have plasma membrane localisation motifs (Torres et al. 1998). AtrbohD and AtrbohF have since been shown to be involved in plant defences against pathogen attack (Torres et al. 2002) and hormone-regulated stomatal closure discussed later (Kwak et al. 2003).

In addition to NOX, peroxidases, xanthine oxidase, oxalate oxidase and amine oxidase can also act as sources of ROS in plant cells (del Rio et al. 2002; Mittler 2002; Desikan et al. 2005b). Peroxidases are found in the cell wall and have been implicated in SA-mediated stomatal closure (Mori et al. 2001) and ROS production in response to pathogen challenge (Bolwell et al. 2002), although the identity of the peroxidase substrate(s) remains elusive (Bolwell et al. 2002). Xanthine oxidase is found in peroxisomes where they produce superoxide radicals and oxidise xanthine and hypoxanthine to uric acid (del Rio et al. 2002).

Examples where different stimuli utilise these sources to generate ROS in stomatal guard cells are detailed below.

3 ROS Production in Response to Stimuli

As ROS production is enhanced in response to many stresses, it is perhaps not surprising to find evidence of ROS acting as ‘cellular indicators of stress’ as integral signalling components in response to both abiotic and biotic stresses (Mittler 2002).

3.1 Involvement of ROS in Plant Hormone Signalling

McAinsh et al. (1996) were the first to show ROS cause stomatal closure and inhibits stomatal opening in *Commelina communis* through increases in cytosolic free calcium ($[Ca^{2+}]_{cyt}$) levels. Whilst low concentrations of ROS induced reversible effects in stomata via calcium, higher concentrations led to irreversible changes independent of calcium, resulting in loss of membrane integrity and viability. There have been no further reports on the effect of ROS on guard cell viability; however, concentrations of ROS are technically very difficult to estimate accurately and therefore a range of concentrations need to be used with every system testing exogenous ROS effects.

As the plant hormone abscisic acid (ABA) plays a key role in controlling stomatal movements during drought stress, much work has focused on ABA signalling in guard cells. Pei et al. (2000) were the first to demonstrate that ABA induces H_2O_2 production in guard cells leading to stomatal closure. Kwak et al. (2003) further showed that the two NOX subunit genes *AtrbohD* and *AtrbohF* are both involved in ABA mediated stomatal closure. Although the *atrbohD* single mutant was not affected, the *atrbohD/F* double mutant was more impaired in ABA-mediated stomatal closure than the *atrbohF* single mutant, suggesting partial redundancy in function. The *atrbohD/F* double mutant showed reduced stomatal closure and ROS production in response to ABA, compared to wild-type plants. Exogenous application of H_2O_2 restored stomatal closure, thus linking ROS production by NOX to ABA-mediated stomatal closure. H_2O_2 produced by NOX has also been implicated in ABA-mediated stomatal closure in *Vicia faba*, suggesting the ABA signalling pathway is similar between different plant species (Zhang et al. 2001).

In addition to H_2O_2, nitric oxide (NO), another plant signalling molecule which like H_2O_2 is involved in many signalling pathways in response to various stimuli, is also involved in ABA-mediated stomatal closure and acts downstream of H_2O_2 (Bright et al 2006). Here, ABA-induced H_2O_2 production via AtrbohD/F was required for NO synthesis leading to stomatal closure (Bright et al. 2006). As well as being involved in ABA-mediated stomatal closure, ROS also play a role in ABA inhibition of stomatal opening (a process distinct from stomatal closure) with exogenous application of H_2O_2 inhibiting stomatal opening (Mishra et al. 2006; Yan et al. 2007). However, the mechanisms of this process are not known.

Although ABA-induced stomatal closure is impaired in the *atrbohD/F* mutant, it is not abolished (Kwak et al. 2003), suggesting other sources of H_2O_2 may contribute to ABA signalling. Indeed, a recent study suggests the apoplast-located copper amine oxidase also contributes to ABA-induced H_2O_2 production in *Vicia faba* guard cells independent of and in addition to NOX (An et al. 2008). ABA treatment induced the activation of a copper amine oxidase which, via its degradation of putrescine to H_2O_2, caused increases in cytosolic calcium leading to stomatal closure.

Ethylene has also been shown to induce stomatal closure in *Arabidopsis* via H_2O_2 production by AtrbohF (Desikan et al. 2006). Until this publication, there was no evidence suggesting that ethylene causes stomatal closure in plants. Desikan et al. showed that ethylene, perceived by the ethylene receptor ETR1, induced H_2O_2 synthesis via AtrbohF (not AtrbohD), leading to stomatal closure. Intriguingly, it appears that ethylene, in the presence of ABA, does not cause stomatal closure in entire leaves (Desikan et al. 2006; Tanaka et al. 2005). Distinction between unique *Rboh* genes for specific stimulus-dependent ROS synthesis reflects specificity and tight regulation of ROS synthesis and action in guard cells.

In addition to ABA and ethylene-mediated stomatal closure, salicylic acid (SA), a plant hormone well known for its role in plant systemic acquired resistance (Feys and Parker 2000), also induces stomatal closure via ROS production in *V. faba* (Mori et al. 2001). In contrast to ABA-mediated stomatal closure, a cell wall-bound peroxidase, possibly guaiacol peroxidase (GPX), is the source of SA-induced ROS production and not plasma membrane-bound NOX (Mori et al. 2001). This suggests a division of labour between the different ROS generating enzymes with stimulus-specific sources of ROS allowing for fine-tuning of the plant responses and the ability to respond appropriately to distinct and multiple stimuli. However, SA-mediated stomatal closure was also found to require extracellular Ca^{2+} (although increases in $[Ca^{2+}]_{cyt}$ levels in response to SA was not shown), suggesting that the ABA and SA pathways may converge downstream of ROS production (Mori et al. 2001).

Methyl jasmonate (or jasmonic acid, JA) has also been shown to induce stomatal closure via ROS (Suhita et al. 2004). The *atrbohD/F* mutant did not display JA-induced ROS synthesis or stomatal closure, implying that RBOH is again involved in JA-induced ROS production. The JA-insensitive mutant *jar1-1* also did not show stomatal closure in response to JA, although they did respond to ABA. Convergence of JA and ABA signalling occurred at various intersecting nodes – calcium influx, pH changes, ROS production and K^+_{out} channels. These data provided strong evidence for JA as an important stress signal in stomatal guard cells.

As discussed so far, ABA, ethylene, JA and SA mediate stomatal closure via increased ROS production. It is perhaps not surprising, therefore, to find the process of stomatal opening by the plant hormones cytokinin and auxin involves the opposite. Cytokinin and auxin are involved in plant growth and development but can also stimulate stomatal opening, and recently the mechanism by which they do so has been elucidated. Although both hormones prevented dark-induced stomatal closure by inhibiting H_2O_2 production, only cytokinin could reopen dark-closed stomata, coinciding with reduction in the level of H_2O_2 which had accumulated under dark conditions (Song et al. 2006). In addition, stomatal closure and intracellular guard cell dichlorofluorescein fluorescence in response to exogenous H_2O_2 was inhibited by cytokinin, but not auxin, suggesting that cytokinin initiates scavenging of H_2O_2, acting after ROS production, whereas auxin may act to suppress the production of ROS (Song et al. 2006). Interestingly, Tanaka et al. (2006) reported cytokinin and auxin inhibited ABA-mediated stomatal closure via ethylene. It is apparent that much is still to be learnt about the role of H_2O_2 in hormone signalling cross-talk in stomata and the physiological relevance of this.

3.2 Abiotic Stimuli

Under field conditions, plants encounter many environmental stresses and stimuli such as light/dark, drought, UV radiation, ozone and elevated CO_2 levels. Such stresses and stimuli have been shown to induce ROS production and modify stomatal behaviour.

As mentioned earlier, stomata play a vital role in photosynthesis, allowing gaseous exchange to occur at the leaf surface. The need for CO_2 for photosynthesis must be balanced with the need to conserve water, which is lost by transpiration through the stomata. Stomata therefore are stimulated to open in the light (when photosynthesis can occur) and close in the dark. Under field conditions, conflicting stimuli often occur, such as drought stress under light conditions: how do stomata respond in this case? Blue light is a component of natural light and is known to stimulate stomatal opening via the action of a plasma membrane H^+-ATPase (which is activated by phosphorylation of its C-terminus, mediated by an unidentified protein kinase) that drives the accumulation of K^+ within the guard cell, leading to stomatal opening (Schroeder et al 2001). This blue light opening, however, can be inhibited by ABA, the levels of which increase during drought stress, reducing water loss by decreasing stomatal aperture (Leung and Giraudat 1998). Both H_2O_2 and NO have been shown to act downstream of ABA in inhibiting blue light stomatal opening by preventing the phosphorylation, and therefore activation, of the blue light dependent H^+-ATPase (Zhang et al. 2004, 2007). In this way, the ABA signal can override blue light stimulus to control stomatal behaviour. The production of ROS in ABA signalling was discussed earlier.

In the case of CO_2, the behaviour of stomata seemingly depends on the concentrations of CO_2 present, as low levels of CO_2 seem to increase stomatal opening whereas high levels result in stomatal closure, the reason for this being unclear (Kolla et al. 2007). In *Arabidopsis*, stomatal closure in response to high CO_2 levels (using bicarbonate as the source of CO_2) was found to require H_2O_2 generated by AtrbohD/F as with ABA-induced stomatal closure (Kolla et al. 2007). However, in the case of CO_2-induced stomatal closure, the level of H_2O_2 was lower and accumulation was more prolonged compared to that initiated by ABA, thereby leading to a lesser effect of CO_2 on stomatal closure (Kolla et al 2007). In addition, the mechanism by which H_2O_2 mediates CO_2-induced stomatal closure does not seem to involve cytoplasmic alkalisation as seen with ABA, but does trigger increases in cytosoplasmic free calcium, underlining the subtleties and versatility of ROS signalling in initiating specific downstream events in response to different stimuli (Vavasseur and Raghavendra 2005; Kolla et al. 2007).

The atmospheric pollutant ozone (O_3) is known to cause cellular damage to plants, linked to a biphasic oxidative burst, similar to that seen in response to pathogen challenge (Joo et al. 2005). In *Arabidopsis*, this ROS production was found to depend on chloroplastic generation of ROS in guard cells and propagation of the signal to adjacent epidermal cells through activation of AtrbohD and AtrbohF (Joo et al. 2005). O_3 also reduces CO_2 uptake by plants, as exposure to O_3 was found to inhibit K^+_{in} channels, leading to stomatal closure (Torsethaugen et al. 1999). The inhibitory affect of O_3 on stomatal aperture is likely to be via ROS signalling, given that ozone-sensitive anion channels are also sensitive to ROS (Vahisalu et al. 2008).

As well as blue light, UV-B radiation is also found in natural sunlight and can have damaging effects on plant metabolism and physiology (Jansen and van der Noort 2000). In addition, UV-B radiation affects stomatal behaviour, causing both opening and closing dependent on the metabolic state of the guard cells (Jansen and van der Noort 2000). In the case of stomatal closure to UV-B in *V. faba*, both H_2O_2 and NO were required for this response, suggesting cross-talk between H_2O_2 and NO signalling pathways and possibly a signalling amplification loop since the presence of one induces the production of the other (He et al. 2005). Although the exogenous application of catalase was shown to significantly reduce UV-B-induced H_2O_2 production (He et al. 2005), the source of ROS in UV-B-induced stomatal closure has not been identified.

H_2O_2 is also involved in stomatal closure in response to darkness. The early events leading to closure (inactivation of light-induced proton pumps, increased cytosolic pH and efflux of K^+) in response to H_2O_2 is similar to that seen with ABA-induced stomatal closure (Desikan et al 2004). Since H_2O_2 acts downstream of ABA in drought-induced stomatal closure, and treatment with the inhibitors diphenylene iodonium (DPI, an inhibitor of NOX), catalase and *N*-acetyl cysteine (NAC, an antioxidant) affected dark-induced H_2O_2 production and stomatal closure, it is possible that ABA-and dark-induced stomatal closure converge at the point of H_2O_2 (Desikan et al. 2004).

3.3 Biotic Stimuli

Plants are known to respond to pathogens and elicitors by production of an ‘oxidative burst’ early on in the interaction as part of the plant immune response. This oxidative burst not only acts as an antimicrobial agent against the invading pathogen, but also acts as a source of signalling molecule for initiation of plant defences. These defences include regulation of gene expression, leading to the hypersensitive response, whereby localised programmed cell death occurs, resulting in limitation of pathogen growth.

One of the earlier reports to show ROS are generated from distinct sources in response to different stimuli is that of Allan and Fluhr (1997). Using the fluorescent probe dichlorodihydrofluorescein, stimuli-specific sources of H_2O_2 in tobacco cells were identified: whereas cryptogein, a fungal elicitor, stimulated ROS production from an intracellular source (guard cell chloroplast), amines (such as putrescine) triggered ROS production in the apoplast. Further investigation by the authors using enzymatic and chemical inhibitors of ROS synthesis indicated that cryptogein induced ROS production by a flavin-containing oxidase and amines stimulated ROS production by a peroxidase or amine oxidase-type enzyme. It has since been shown that the ROS generating enzymes NOX (a flavin-containing enzyme) and copper amine oxidase are involved in ROS production in response to cryptogein and putrecine (in tobacco BY-2 cells) respectively (Kadota et al. 2004; An et al. 2008).

Previously, stomata were thought of as 'passive ports of entry' to the leaf interior for bacteria. However, a recent report by Melotto et al. (2006) suggests that this is not the case, but that *Arabidopsis* actively close stomata in response to the presence of bacteria and bacterial elicitors (flg22, the active peptide of bacterial flagellin, and lipopolysaccharide (LPS) found in outer cell walls of gram negative bacteria). Further investigation by Melotto et al. (2006) revealed a link between ABA, the downstream kinase OST1 and NO production by nitric oxide synthase. As H_2O_2 is also known to act downstream of ABA in mediating stomatal closure (Pei et al. 2000), it would be interesting to see whether H_2O_2 is also involved in stomatal defences against pathogens.

An earlier report by Lee et al. (1999) indicates this may be the case, at least in some plant–fungi interactions, where stomata are utilised as an entry point by the fungus. Here, they found stomata of tomato (*Lycopersicon esculentum* L.) and *Commelina communis* L. produced H_2O_2 and closed in response to oligogalacturonic acid (OGA), a breakdown product of the plant cell wall, and to chitosan, a component of the fungal cell wall. In addition to causing stomatal closure, OGA also inhibited stomatal opening. Further investigation revealed a requirement for H_2O_2, as addition of catalase or ascorbic acid coincided with suppression of stomatal responses to these elicitors. Although this study looked at the reaction of stomata to a molecule commonly produced during plant–fungi interactions (i.e. OGA) and to a common component of the fungal cell wall (i.e. chitosan), the response of stomata to actual fungal infection was not investigated here.

Manipulation of stomata by the necrotrophic fungus *Sclerotina sclerotium* was investigated by Guimarães and Stotz (2004), who found that the fungus induced stomatal opening, through which fungal hyphae could emerge onto the leaf surface. This ability to manipulate stomatal aperture was dependent on the virulence factor oxalic acid, which inhibited ABA-mediated stomatal closure, possibly through interference of H_2O_2 production downstream of ABA (Guimarães and Stotz 2004). In addition to ABA-induced ROS production, AtrbohD and AtrbohF are also required for the oxidative burst in response to pathogens (Torres et al. 2002). Since oxalic acid has previously been shown to inhibit elicitor-induced H_2O_2 production (Cessna et al. 2000), it is possible that it does so by inhibiting NOX, preventing ROS production and therefore stomatal closure during the plant-pathogen interaction.

Preliminary data in our laboratory indicate that ROS are produced in stomatal guard cells in response to the *Arabidopsis* virulent bacterium *Pseudomonas syringae* pv. *tomato* DC3000 (Desikan, personal communication). However, it is also possible that ROS generated in epidermal or mesophyll cells in response to pathogen attack acts as a mobile signal leading to the stomatal response (see Sect. 4.4.3). Nevertheless, the similarity in stomatal closure being overcome by both virulent bacteria (Melotto et al. 2006) and fungi (Guimarães and Stotz 2004) suggest that ROS will play a role in this dynamic process. It is important to note that ROS not only cause stomatal closure, but also inhibit stomatal opening (Desikan et al.2004; Mishra et al. 2006; Yan et al. 2007). This area of stomatal biology in response to pathogen attack remains an exciting area of further research, with many questions to be answered.

4 Regulators of ROS Synthesis and Action

4.1 Upstream Regulators of ROS Synthesis

In *Arabidopsis*, the OST1 (Open Stomata1) Ser/Thr protein kinase [previously named SRK2E, encoding for an SNF1-related protein kinase (SnRK)] was the first protein kinase identified as being involved in ABA-signalling during a screen for mutants defective in drought stress-induced stomatal closure (Mustilli et al. 2002; Yoshida et al. 2002). *OST1* is expressed in guard cells and the vascular system and its expression and kinase activity is upregulated and activated, respectively, by ABA (Mustilli et al. 2002). Stomata of the loss-of-function *ost1* mutants are specifically impaired in ABA-mediated stomatal closure and inhibition of stomatal opening, resulting in excess water loss through transpiration and a 'wilty' phenotype under mild drought stress, suggesting that OST1 acts as a positive regulator of ABA signalling (Mustilli et al. 2002; Yoshida et al. 2002). *ost1* mutants are defective in ABA-induced ROS production and application of exogenous H_2O_2 restored stomatal closure in response to ABA, suggesting that OST1 links ABA perception to ROS production. It is likely that OST1 regulates ATRBOH-mediated ROS production in response to ABA, although this is yet to be demonstrated (Mustilli et al. 2002). The C-terminal of OST1 was found to contain two functional domains, domains I and II (Yoshida et al. 2006). Whereas domain I is not involved in ABA activation of OST1, domain II is required for OST1 function in ABA signalling by interaction with ABI1 (ABA-Insensitive 1), a type-2C protein phosphatase (PP2C) known to negatively regulate ABA-mediated stomatal closure (Yoshida et al. 2006). ABI1 is also known to regulate ABA-induced ROS production (Murata et al. 2001). Therefore, it is possible that ABI1 regulation of OST1 function in turn mediates ROS production in response to ABA.

ROS production in response to both ABA and JA were shown to require cytoplasmic alkalinisation (Suhita et al. 2004). Kinetics of ABA and JA-induced pH changes and ROS accumulation in guard cells showed that pH changes (alkalinisation of the cytoplasm) precede that of ROS generation. Intriguingly, whilst ABA did not induce ROS or stomatal closure in *ost1* mutants, JA did; however, JA did not induce ROS or stomatal closure in the *jar1-1* mutant, whereas ABA did (Suhita et al. 2004). These data indicate a clear divergence in early upstream regulation of ROS production in response to either ABA or JA. Heterotrimeric G proteins also play an important role in regulating ROS generation in response to ozone (Joo et al. 2005). Whereas both the Gα and Gβ subunits are required for the first oxidative burst originating from a chloroplastic location, the second burst originating from multiple sources (including ATRBOH) requires only the Gα subunit (Joo et al. 2005).

4.2 Downstream Regulators of ROS Action

As mentioned earlier, one of the factors that determine stomatal aperture size is the concentration of ions in guard cells. There is evidence to suggest that ROS play a role in regulating ABA (or non-ABA)-induced ion changes in guard cells.

To further investigate the role of ROS in ABA signalling, Kwak et al. (2003) investigated activation of plasma membrane Ca^{2+}-permeable (I_{ca}) channels. H_2O_2 has been shown to mediate ABA signalling by activating I_{ca} channels, through which Ca^{2+} is released into the cytoplasm, triggering stomatal closure (Pei et al. 2000). Kwak et al. (2003) found a significantly smaller proportion of *atrbohD/F* guard cells displaying ABA-induced increases in $[Ca^{2+}]_{cyt}$ than wild-type guard cells. Moreover, activation of I_{ca} by exogenous H_2O_2 was not affected in *atrbohD/F* guard cells, suggesting that ATRBOH is the source of H_2O_2 activating calcium channels in ABA signalling (Kwak et al. 2003).

A study by Köhler et al. (2003) suggests that H_2O_2 may not act the same way as ABA during stomatal closure, but may operate via an alternative signalling pathway. Using voltage-clamp experiments, they found that K^+_{out} channels of *Vicia faba* guard cells responded differently to ABA and H_2O_2: whereas ABA depressed the activity of K^+_{in} channels in a reversible manner, H_2O_2 irreversibly depressed the activities of both K^+_{in} and K^+_{out} channels in guard cells bathed in H_2O_2 at concentration ranges of 1– 50 μM. It suggests that H_2O_2 induced from different stimuli could act on K^+_{out} channels, unlike ABA-induced H_2O_2. However, the possibility that exogenously applied H_2O_2 may not act in the same way as H_2O_2 produced/released within the guard cell cannot be ruled out. The authors suggest that, rather than acting as a critical component of the ABA pathway, H_2O_2 signalling converges with that of ABA at the point of Ca^{2+} activation. The identity of the K^+ channels regulated by H_2O_2 in response to ABA is not yet known, although it is known that JA-induced stomatal closure via ROS is affected in the K^+_{out} channel mutant *gork1* (Suhita et al. 2004). Supporting these observations, guard cells of the *Arabidopsis de-etiolated 3* (*det3*) mutant, a V-ATPase mutant, is disrupted in stomatal closure and $[Ca^{2+}]_{cyt}$ oscillation in response to H_2O_2 but not to ABA, again suggesting the existence of stimulus-specific signalling pathways (Allen et al. 2000).

As outlined above (Sect. 4.1.2), stomatal closure is driven by the efflux of Cl^- through S- and R-type anion channels, which subsequently activates the efflux of K^+ ions and results in loss of guard cell turgor. However, until recently, the membrane protein(s) controlling the activity of anion channels was unknown. It has since been shown that *SLAC1* (slow anion channel-associated *1*) encodes a putative membrane protein that modulates S-type anion channel activity (Negi et al. 2008; Vahisalu et al. 2008). Mutants in this gene are defective in S-type channel currents and in stomatal response to O_3, CO_2, light/dark transitions, H_2O_2, NO, Ca^{2+} and ABA (Vahisalu et al. 2008), making it an attractive target for manipulating water loss. However whether H_2O_2 acts directly on these ion channels or regulates their function via other signalling cascades is not known.

Along with ABI1, another PP2C, ABI2, is also a negative regulator of ABA signalling but acts downstream of ROS production. Whereas the *abi1-1* mutant is disrupted only in ABA-mediated ROS synthesis leading to stomatal closure, the *abi2-1* mutant was also impaired in H_2O_2-induced stomatal closure and activation of I_{ca} channels, but not H_2O_2 production in response to ABA (Murata et al. 2001).

In addition to the protein phosphatases ABI1 and ABI2, the MAP kinase MPK3 was found to be involved in ABA-inhibition of stomatal closure and required for

H_2O_2-induced stomatal closure as well as inhibition of opening (Gudesblat et al. 2006). MPK3 silenced plants showed wild-type levels of H_2O_2 synthesis in response to ABA, suggesting that MPK3 acts downstream of H_2O_2 (and ABA) leading to stomatal closure (Gudesblat et al. 2006).

The ethylene receptor ETR1 involved in H_2O_2 signalling is a two-component hybrid histidine kinase that has both sensing and transducing functions important in plant signalling and response to stimuli (Grefen and Harter 2004). The ETR1 receptor is expressed in guard cells and contains three domains: an N-terminal signal input domain, a transmitter domain and a C-terminal receiver domain. Perception of stimulus by the input domain initiates a set of phosphotransfer events that begins at a conserved His residue within the histidine kinase domain. The phosphoryl group is then transmitted to a conserved Asp residue within the receiver domain that is subsequently transferred to a conserved Asp residue of a response regulator that transduces the signal into a cellular response (Hwang et al. 2002; Grefen and Harter 2004; Desikan et al. 2005a). Work by Desikan et al. (2005a) demonstrated that the Cys-65 residue of the ETR1 N-terminal domain is essential for H_2O_2-induced stomatal closure in *Arabidopsis* as stomata of the *etr1-1* mutant (containing a Cys-65Tyr mutation) were insensitive to H_2O_2. The *etr1-1* mutant is also defective in ethylene-induced stomatal closure (Desikan et al. 2005a). *etr1-3*, on the other hand, which has a Ala-31Val point mutation, is not impaired in H_2O_2-induced responses implying that the Cys65 is required for H_2O_2 responses (Desikan et al. 2005a). Interestingly, the histidine kinase activity of ETR1 is not required for response to H_2O_2, although downstream response regulators such as ARR2 are required for H_2O_2-induced stomatal closure (Desikan et al. 2006). Mechanisms by which H_2O_2 is sensed and transduced via the two component system need to be studied in detail.

4.3 Compartmentalisation of ROS Production and Action

As mentioned earlier, H_2O_2 is produced by distinct enzymes in discrete locations within guard cells. For example, RBOH is plasma membrane localised, peroxidases are located in the cell wall/apoplast and xanthine oxidase in peroxisomes. Proteins acting downstream of ROS are located in the cytoplasm (ABI1/2), nucleus (MAPKs), endoplasmic reticulum (ETR1) and plasma membrane (SLAC1, K^+ channels, Ca^{2+} channels). Clearly, a careful co-ordination of ROS synthesis and action is required between these compartments to effect the required physiological response.

H_2O_2 may also act as an intercellular signalling molecule itself diffusing from cell to cell. Allan and Fluhr (1997) found that exposure of epidermal cells to white light not only increased ROS production in these cells, but also increased production in nearby stomatal guard cells which had not been exposed to the light. Since guard cells have no symplastic connection to surrounding epidermal cells, H_2O_2 may be acting as an intercellular signalling molecule by moving from cell to cell via the apoplast and by crossing the plasma membrane. However, it has also been shown that ozone-induced chloroplastic ROS is required for propagating the ROS signal in epidermal cells but not guard cells (Joo et al. 2005). Interestingly, a recent study provided evidence for diffusion of H_2O_2 across cellular membranes through two *Arabidopsis* aquaporins encoded for by the *TIP1;1* and *TIP1;2* genes (Biernet et al. 2007). However, these conclusions were derived from results obtained from assays performed in yeast cells; whether these aquaporins perform the same function in plant cells remains to be seen. It would also be interesting to see whether mutants in these genes are affected in ROS signalling.

4.4 ROS Scavenging

Thus far, we have discussed the importance of ROS production in guard cell signalling, but it is becoming apparent that ROS scavengers, in addition to their general function in detoxification of ROS, are also involved in fine-tuning and controlling ROS signalling (Mittler 2002). ROS signalling therefore involves a critical balance between production and compartmentalisation of ROS, ROS-scavenging activity, the type of scavenging molecules/antioxidants and enzymes present and the cellular redox state.

Superoxide dismutase (SOD) initiates ROS scavenging by dismutating superoxide (O_2•$^-$) to H_2O_2. H_2O_2 is then detoxified by the activities of catalase (CAT), ascorbate peroxidase (APX) and glutathione peroxidase (GPX), which requires reduced ascorbate (Asc) and glutathione (GSH) as the reducing agents (Mittler 2002; Apel and Hirt 2004). The pools of Asc and GSH required by APX and GPX are maintained by the actions of GSH reductase (GR), monodehydroascorbate reductase (MDAR) and dehydroascorbate reductase (DHAR) (Mittler 2002; Apel and Hirt 2004). Deficiency in ROS scavenging and disruption of cellular redox state affects both cellular levels of ROS and physiological responses controlled by ROS.

Asc is one of the most abundant antioxidants in plants, suggesting it may also play an important role in ROS signalling in guard cells in addition to its scavenging role. A study by Chen and Gallie (2004) found that both H_2O_2 and the redox state of Asc are diurnally regulated in tobacco: the highest levels of H_2O_2 were seen during mid-afternoon (correlating with stomatal closure) and were low during early morning, late afternoon and throughout the evening, inversely correlating with the Asc redox state, suggesting a link between the two. Stomata of knockout mutants of the *APX1* gene were unresponsive to shifts from dark to light periods, and leaves of these plants contained higher levels of H_2O_2 than wild-type plants, further supporting the link between APX and Asc with control of stomatal behaviour (Pnueli et al. 2003). Further investigation revealed that the Asc redox state controls stomatal responses: overexpression of DHAR (which causes an increase in the Asc redox state) results in reduced H_2O_2 production and stomatal closure in response to exogenous ABA and H_2O_2 (Chen and Gallie 2004). Since overexpression of DHAR increases the levels of DHA available for conversion to Asc rather than the levels of Asc per se, it suggests that it is the Asc redox state itself (i.e. the balance between Asc and DHA) that is important for stomatal responses (Chen and Gallie 2004).

In *Arabidopsis*, the glutathione peroxidase ATGPX3 is strongly expressed in guard cells and the *atgpx3* mutant is less sensitive to ABA and H_2O_2, suggesting that ATGPX3 may be the link between H_2O_2 and signalling events occurring downstream

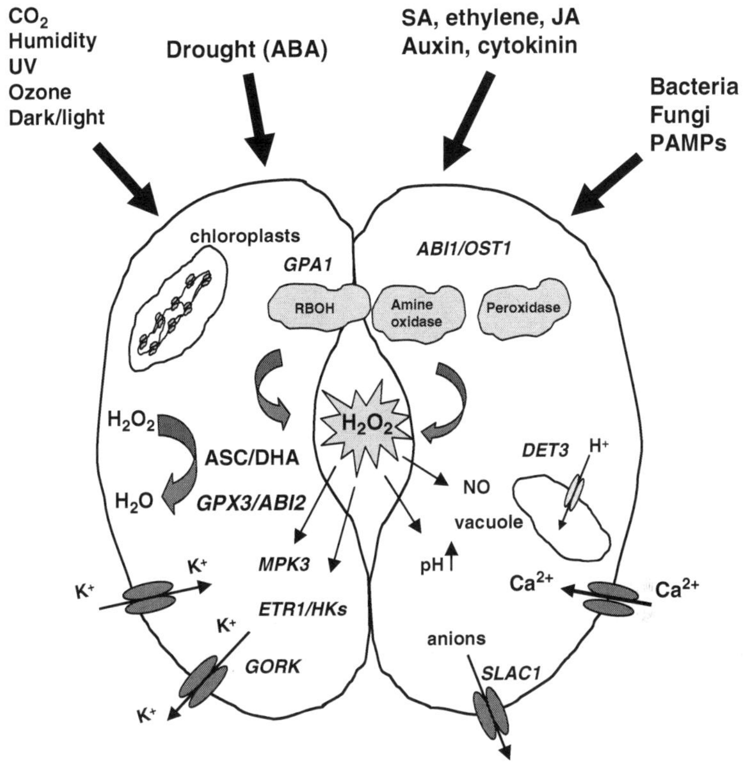

Fig. 1 ROS signalling in stomatal guard cells. Current understanding of the signalling mechanisms by which ROS are generated in response to multiple stimuli, leading to regulation of various cell signalling processes is shown. For details, see text

of ABA in the stomatal closure response (Miao et al. 2006). In yeast two-hybrid and GST pull-down assays, ATGPX3 was found to interact with and reduce the activity of both ABI1 and ABI2 thereby reducing negative regulation of ABA signalling, leading to stomatal closure (Miao et al. 2006). H_2O_2 was also found to modify the redox states of both ATGPX3 and ABI2 (to which ATGPX3 bound most strongly), and H_2O_2- and ABA-induced Ca^{2+} currents were disrupted in *atgpx3* mutants. This suggests that ATGPX3 may detect and transduce the H_2O_2 signal via interaction with ABI2 to bring about initiation of Ca^{2+} currents leading to stomatal closure (Miao et al. 2006).

5 Conclusions and Further Remarks

From 12 years ago, when ROS were first demonstrated to act in stomatal guard cells (McAinsh et al. 1996), we have come a long way in researching ROS signalling in stomata. Figure 1 summarises our current knowledge of ROS signalling in stomatal guard cells. Multiple stimuli have been shown to induce ROS generation in guard cells via more than one mechanism. In addition, the antioxidant activities in guard cells are also proving to be an important facet of redox signalling in stomata. A number of potential candidates have been identified that not only regulate ROS generation but also regulate ROS signal transduction and propagation, indicating that a tight control of production/removal and action of ROS in distinct locations is essential for a unique response, stomatal opening/closing. However, this field is not only just unmasking important phenomena but also posing further questions. For example, how stomatal guard cells are able to distinguish between different stimuli to activate unique ROS generating and responding pathways remains unknown. It is also not known whether there are guard cell-specific components that are involved in ROS signalling, or whether the components identified thus far are also involved in ROS-mediated responses in other cell types. Integrating multiple signals via ROS in stomata will be an exciting area to pursue. In particular, how stomatal guard cells help plants adapt towards stresses will be important to understand in the present global climate change scenario. Identification of key targets in ROS signalling that are conserved between plant species will be vital towards breeding and developing stress-tolerant plants. The use of modelling tools to map out and predict changes at the cellular level in response to the environment will no doubt make significant advances in the coming years.

References

Allan AC and Fluhr R (1997) Two distinct sources of elicited reactive oxygen species in tobacco epidermal cells. Plant Cell 9:1559–1572

Allen GJ, Chu SP, Schumacher K, Shimazaki CT, Vafeados D, Kemper A, Hawke SD, Tallman G, Tsien RY, Harper JF, Chory J and Schroeder JI (2000) Alteration of stimulus-specific guard cell calcium oscillations and stomatal closing in *Arabidopsis det3* mutant. Science 289:2338–2342

An Z, Jing W, Liu Y and Zhang W (2008) Hydrogen peroxide generated by copper amine oxidase is involved in abscisic acid-induced stomatal closure in *Vicia faba*. J Exp Botany 59:815–825

Apel K and Hirt H (2004) Reactive oxygen species: metabolism, oxidative stress, and signal transduction. Annu Rev Plant Biol 55:373–399

Biernet GP, Møller ALB, Kristiansen KA, Schulz A, Møller IM, Schjoerring JK and Jahn TP (2007) Specific aquaporins facilitate the diffusion of hydrogen peroxide across membranes. J Biol Chem 282:1183–1192

Blatt MR (2000) Cellular signaling and volume control in stomatal movements in plants. Annu Rev Cell Dev Biol 16:221–241

Bollwell GP, Bindschedler LV, Blee KA, Butt VS, Davies DR, Gardner SL, Gerrish C and Minibayeva F (2002) The apoplastic oxidative burst in response to biotic stress in plants: a three-component system. J Exp Bot 53:1367–1376

Bright J, Desikan R, Hancock JT, Weir IS and Neill SJ (2006) ABA-induced NO generation and stomatal closure in *Arabidopsis* are dependent on H_2O_2 synthesis. Plant J 45:113–122

Cessna SG, Sears VE, Dickman MB and Low PS (2000) Oxalic acid, a pathogenicity factor for *Sclerotina sclerotium*, suppresses the oxidative burst of the host plant. Plant Cell 12:2191–2199

Chen Z and Gallie DR (2004) The ascorbic acid redox state controls guard cell signalling and stomatal movement. Plant Cell 16:1143–1162

del Rio LA, Corpas FJ, Sandalio LM, Palma JM, Gómez M and Barroso JB (2002) Reactive oxygen species, antioxidant systems and nitric oxide in peroxisomes. J Exp Bot 53:1255–1272

Desikan R, Cheung M-K, Clarke A, Golding S, Sagi M, Fluhr R, Rock C, Hancock J and Neill S (2004) Hydrogen peroxide is a common signal for darkness- and ABA-induced stomatal closure in *Pisum sativum*. Funct Plant Biol 31:913–920
Desikan R, Hancock JT, Bright J, Harrison J, Weir I, Hooley R and Neill SJ (2005a) A role for ETR1 in hydrogen peroxide signaling in stomatal guard cells. Plant Physiol 137:831–834
Desikan R, Hancock J and Neill S (2005b) Reactive oxygen species as signalling molecules. In: Antioxidants and reactive oxygen species in plants. Blackwell, Oxford, pp 169–195
Desikan R, Last K, Harrett-Williams R, Tagliavia C, Harter K, Hooley R, Hancock JT and Neill SJ (2006) Ethylene-induced stomtal closure in Arabidopsis occurs via AtrbohF-mediated hydrogen peroxide synthesis. Plant J 47:907–916
Feys B and Parker J (2000) Interplay of signalling pathways in plant disease resistance. Trends Genet 16:449–455
Grefen C and Harter K (2004) Plant two-component systems: principles, functions, complexity and cross talk. Planta 219:733–742
Gudesblat GE, Iusem ND and Morris PC (2006) Guard cell-specific inhibition of *Arabidopsis MPK3* expression causes abnormal stomatal responses to abscisic acid and hydrogen peroxide. New Phytol 173:713–721
Guimarães RL and Stotz HU (2004) Oxalate production by *Sclerotinia sclerotiorum* deregulates guard cells during infection. Plant Physiol 136:3701–3711
He J-M, Xu H, She X-P, Song X-G and Zhao W-M (2005) The role and interrelationship of hydrogen peroxide and nitric oxide in the UV-B-induced stomatal closure in broad bean. Funct Plant Biol 32:237–247
Hwang I, Chen H-C and Sheen J (2002) Two-component signal transduction pathways in Arabidopsis. Plant Physiol 129:500–515
Lee S, Choi H, Suh S, Doo I-S, Oh K-Y, Choi EJ, Schroeder Taylor AT, Low PS and Lee Y (1999) Oligogalacturonic acid and chitosan reduce stomatal aperture by inducing the evolution of reactive oxygen species from guard cells of tomato and *Commelina communis*. Plant Physiol 121:147–152
Jansen MAK and van der Noort RE (2000) Ultraviolet-B radiation induces complex alterations in stomatal behaviour. Physiol Plant 110:189–194
Joo JH, Wang S, Chen JG, Jones AM and Fedoroff NV (2005) Different signalling and cell death roles of heterotrimetic G protein α and β subunits in the Arabidopsis oxidative stress response to ozone. Plant Cell 17:957–970
Kadota Y., Goh T, Tomatsu R, Higashi K, Muto S and Kuchitsu K (2004) Cryptogein-induced initial events in tobacco BY-2 cells: pharmacological characterisation of molecular relationship among cytosolic Ca^{2+} transients, anion efflux and production of reactive oxygen species. Plant Cell Physiol 45:160–170
Keller T, Damunde HG, Werner D, Doerner P, Dixon RA and Lamb C (1998) A plant homolog of the neutrophil NADPH oxidase gp91^{phox} subunit gene encodes a plasma membrane protein with Ca^{2+} binding motifs. Plant Cell 10:255–266
Köhler B, Hills A and Blatt MR (2003) Control of guard cell ion channels by hydrogen peroxide and abscisic acid indicates their action through alternate signalling pathways. Plant Physiol 131:385–388
Kolla VA, Vavasseur A and Raghavendra AS (2007) Hydrogen peroxide production is an early event during bicarbonate induced stomatal closure in abaxial epidermis of *Arabidopsis*. Planta 225:1421–1429
Kwak JM, Mori IC, Pei Z-M, Leonhardt N, Torres MA, Dangl JL, Bloom RE, Bodde S, Jones JDG and Schroeder JI (2003) NADPH oxidase *AtrbohD* and *AtrbohF* genes function in ROS-dependent ABA signalling in *Arabidopsis*. EMBO J 22:2623–2633
Leung J and Giraudat J (1998) Abscisic acid signal transduction. Annu Rev Plant Physiol Plant Mol Biol 49:199–222
McAinsh RM, Clayton H, Mansfield TA and Hetherington AM (1996) Changes in stomatal behaviour and guard cell cytosolic free calcium in response to oxidative stress. Plant Physiol 111:1031–1042

Melotto M, Underwood W, Koczan J, Nomura K and He SY (2006) Plant stomata function in innate immunity against bacterial invasion. Cell 126:969–980

Miao Y, Lv D, Wang P, Wang X-C, Chen J, Miao C and Song C-P (2006) An *Arabidopsis* glutathione peroxidise functions as both a redox transducer and a scavenger in abscisic acid and drought stress responses. Plant Cell 18:2749–2766

Mishra G, Zhang W, Deng F, Zhao J and Wang X (2006) A bifurcating pathway directs abscisic acid effects on stomatal closure and opening in Arabidopsis. Science 312:264–266

Mittler R (2002) Oxidative stress, antioxidants and stress tolerance. Trends Plant Sci 7:405–410

Mori IC, Pinontoan R, Kawano T and Muto S (2001) Involvement of superoxide generation in salicylic acid-induced stomatal closure in *Vicia faba*. Plant Cell Physiol 42:1383–1388

Murata Y, Pei ZM, Mori IC and Schroeder JI (2001) Abscisic acid activation of plasma membrane Ca^{2+} channels in guard cells requires cytosolic NAD(P)H and is differentially disrupted upstream and downstream of reactive oxygen species production in *abi1-1* and *abi2-1* protein phosphatase 2C mutants. Plant Cell 13:2513–2523

Mustilli A-C, Merlot S, Vavasseur A, Fenzi F and Giraudat J (2002) Arabidopsis OST1 protein kinase mediates the regulation of stomatal aperture by abscisic acid and acts upstream of reactive oxygen species production. Plant Cell 14:3089–3099

Negi J, Matsuda O, Nagasawa T, Oba Y, Takahashi H, Kawai-Yamada M, Uchimiya H, Hashimoto M, Iba K (2008) CO_2 regulator SLAC1 and its homologues are essential for anion homeostasis in plant cells. Nature 452:483–486

Neill SJ, Desikan R and Hancock JT (2002) Hydrogen peroxide signalling. Curr Opin Plant Biol 5:388–395

Pei ZM, Murata Y, Bening G, Thomine S, Klusener B, Allen GJ, Grill E and Schroeder JI (2000) Calcium channels activated by hydrogen peroxide mediate abscisic acid signalling in guard cells. Nature 406:731–734

Pnueli L, Liang H, Rozenberg M and Mittler R (2003) Growth suppression, altered stomatal responses and augmented induction of heat shock proteins in cytosolic ascorbate peroxidase (*APX1*)-deficient *Arabidopsis* plants. Plant J 34:187–203

Sagi M and Fluhr R (2006) Production of reactive oxygen species by plant NADPH oxidases. Plant Physiol 141:336–340

Schroeder JI, Allen GJ, Hugouvieux V, Kwak JM, Waner D (2001) Guard cell signal transduction. Annu Rev Plant Physiol Plant Mol Biol 52:627–658

Song X-G, She X-P, He J-M, Huang C, Song T-S (2006) Cytokinin- and auxin-induced stomatal opening involves a decrease in levels of hydrogen peroxide in guard cells of *Vicia faba*. Funct Plant Biol 33:573–583

Suhita D, Raghavendra AS, Kwak JM, Vavasseur A (2004) Cytoplasmic alkalinisation precedes reactive oxygen species production during methyl jasmonate- and abscisic acid-induced stomatal closure. Plant Physiol 134:1536–1545

Tanaka Y, Sano T, Tamaoki M, Nakajima N, Kondo N and Hasezawa S (2005) Ethylene inhibits abscisic acid-induced stomtal closure in *Arabidopsis*. Plant Physiol 138:2337–2343

Tanaka Y, Sano T, Tamaoki M, Nakajima N, Kondo N and Hasezawa S (2006) Cytokinin and auxin inhibit abscisic acid-induced stomatal closure by enhancing ethylene production in *Arabidopsis*. J Exp Bot 57:2259–2266

Torres MA, Dangl JL and Jones DG (2002) *Arabidopsis gp91phox* homologues *AtrbohD* and *AtrbohF* are required for accumulation of reactive oxygen intermediates in the plant defence response. Proc Nat Acad Sci USA 99:517–522

Torres MA, Onouchi H, Hamada S, Machida C, Hammond-Kosack KE and Jones JDG (1998) Six *Arabidopsis thaliana* homologues of the human respiratory burst oxidase (gp91phox). Plant J 14:365–370

Torsethaugen G, Pell EJ and Assman SM (1999) Ozone inhibits guard cell K^+ channels implicated in stomatal opening. Proc Natl Acad Sci USA 96:13577–13583

Vahisalu T, Kollist H, Wang Y-F, Nishimura N, Chan W-Y, Valerio G, Lamminmäki A, Brosché M, Moldau H, Desikan R, Schroeder JI and Kangasjärvi J (2008) SLAC1 is required for plant guard cell S-type anion channel function in stomatal signalling. Nature 452:487–492

Vavasseur A and Raghavendra AS (2005) Guard cell metabolism and CO^2 sensing. New Phytol 165: 665–682

Wang P and Song C-P (2008) Guard-cell signalling for hydrogen peroxide and abscisic acid. New Phytol 178:703–718

Yan J, Tsuichihara N, Etoh T and Iwai S (2007) Reactive oxygen species and nitric oxide are involved in ABA inhibition of stomatal opening. Plant Cell Environ 30:1320–1325

Yoshida R, Hobo T, Ichimura K, Mizoguchi T, Takahashi F, Aronso J, Ecker JR and Shinozaki K (2002) ABA-activated SnRK2 protein kinase is required for dehydration stress signalling in *Arabidopsis*. Plant Cell Physiol 43:1473–1483

Yoshida R, Umezawa T, Mizoguchi T, Takahashi S, Takahashi F and Shinozaki K (2006) The regulatory domain of SRK2E/OST1/SnRK2.6 interacts with ABI1 and integrates abscisic acid (ABA) and osmotic stress signals controlling stomatal closure in *Arabidopsis*. J Biol Chem 281:5310–5318

Zhang X, Takemiya A, Kinoshita T and Schimazaki K (2007) Nitric oxide inhibits blue light-specific stomatal opening via abscisic acid signalling pathways in *Vicia* guard cells. Plant Cell Physiol 48:715–723

Zhang X, Wang H, Takemiya A, Song C, Kinoshita T and Shimazaki K (2004) Inhibition of blue light-dependent H^+ pumping by abscisic acid through hydrogen peroxide-induced dephosphorylation of the plasma membrane H^+-ATPase in guard cell protoplast. Plant Physiol 136:4150–4158

Zhang X, Zhang L, Dong F, Gao J, Galbraith DW and Song C-P (2001) Hydrogen peroxide is involved in abscisic acid-induced stomatal closure in *Vicia faba*. Plant Physiol 126:1438–1448

Reactive Oxygen Species in Plant Cell Walls

Alfonso Ros Barceló and Laura V. Gómez Ros

Abstract Plant cell walls are dynamic structures composed of polysaccharides, phenolics, and proteins. The plant cell wall is important not only for maintaining cell shape, but it also responds to endogenous and environmental clues through the release of signaling molecules, such as H_2O_2, which may act following autocrine and paracrine pathways. However, the primary function of reactive oxygen species (ROS) production in the plant cell wall is to modify cell wall components by processes of cell wall stiffening/softening, which, in the last intance, control plant growth and morphogenesis. Four possible enzymatic machineries could explain how this H_2O_2 is produced: NADPH oxidases (NOX), peroxidases, poly(di)amine oxidases, and oxalate oxidases, but most of the molecular evidence, particularly in epidermal, vascular, and suberizing tissues, supports the exclusive participation of NOX in this process. Given the limited efficacy of the ROS-scavenging systems in plant cell walls, it may be concluded that ROS accumulation in the cell walls of these tissues is only the static image of their high redox state and large oxidative metabolism.

1 Introduction

Plant cell walls are dynamic structures composed of polysaccharides, phenolics, and proteins. The cell wall is not only important for maintaining cell shape and rigidity, but also responds to endogenous (hormonal) and environmental clues through the release of signaling molecules, such as H_2O_2 [the only biotic reactive oxygen species (ROS) that may be transported long distances such as the thickness of a plant cell], and oligosaccharins, which may act following autocrine and paracrine pathways (Brownlee 2002). In fact, extracellular H_2O_2 itself is capable of bursting transcriptional responses during various developmental processes (Neill et al. 2002), the possible proteins involved in H_2O_2 perception having been

A.R. Barceló (✉) and L.V.G. Ros
Department of Plant Biology (Plant Physiology), University of Murcia, E-30100, Murcia, Spain
e-mail: rosbarce@um.es

L.A. del Río and A. Puppo (eds.), *Reactive Oxygen Species in Plant Signaling*, Signaling and Communication in Plants, DOI 10.1007/978-3-642-00390-5_5,

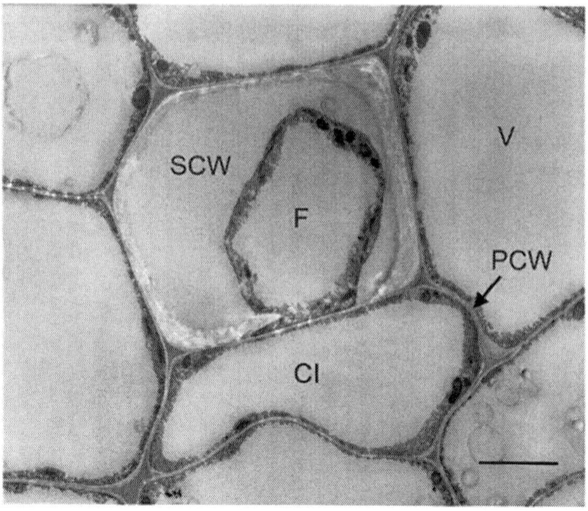

Fig. 1 Overview of primary and secondary cell walls in vascular tissues of *Lupinus albus*. *CI* Cambial initial cell, *F* fibre, *V* vessel, *PCW* primary cell wall, *SCW* secondary cell wall. *Bar* 3 μm

tentatively identified (Pitzschke and Hirt 2006; Hancock et al. 2006). Plants contain two types of cell wall: the primary wall, formed at the cell plate during division, controls and facilitates cell growth, while the secondary cell wall, produced during the later phase of development of vascular tissues (Fig. 1), confers mechanical strength to the aerial organs of the plant.

However, the primary function of ROS production in the plant cell wall is probably the modification of cell wall components, thereby making a major contribution to the mechanics of the plant cell and the control of plant growth and morphogenesis (Cosgrove 2005). Since ROS (Gómez Ros et al. 2006) and reactive nitrogen species (RNS) (Gabaldón et al. 2005) are simultaneously produced in the plant cell wall, it is then possible that NO•, a free radical itself, may react with $O_2^{\bullet-}$ to form the highly reactive peroxynitrite anion, $ONOO^-$, and subsequent cellular effects may then be induced by $ONOO^-$. In this context, $ONOO^-$ may be regarded as a substrate of plant peroxidases (Floris et al. 1993), supplanting in some aspects the action of ROS (Neill et al. 2002). The main themes of the present chapter are: (1) the ROS-dependent metabolic reactions that modify plant cell wall polymers, (2) how ROS are produced and how their levels are regulated at the plant cell wall, (3) the existence of enzymes that could feasibly catalyze such reactions, and (4) a critical evaluation of the evidence that such reactions actually happen in vivo. We focus our attention on ROS generation in epidermal and vascular cell walls during plant growth and development. ROS generation in cell walls of plant exposed to both biotic (see Chap. 9) and abiotic stresses (see Chaps. 11, 13 and 14) is treated elsewhere.

2 Plant Cell Wall Composition

Plant cell walls are mainly composed of polysaccharides, cationic ions, smaller proportions of glycoproteins and, in some specialized cell-types, several hydrophobic substances such as lignin, suberin, and cutin. Cell wall polysaccharides fall into three main categories (Fry 2004a): pectins, hemicelluloses, and cellulose. Pectins and hemicelluloses are components of the cell wall matrix, within which is embedded the cellulosic microfibril skeletal. Pectins are partly methyl-esterified galacturonate-rich acidic polysaccharides (galacturonan), which may be free or linked to rhamnogalacturonans (RG) I and II. Major hemicelluloses are the xylans (arabinoxylans, glucuronoarabinoxylans, etc.) and the xyloglucans, which interact through hydrogen bonds with cellulose. Cellulose is composed of neutral, β-(1→4)-linked D-glucose chains, hydrogen-bonded together to form microfibrils. The composition of secondary plant cell walls is, in outline, very similar to that of the primary plant cell wall, but greatly varies in the proportion of the individual components. Secondary plant cell walls are especially enriched in cellulosic materials and lack, or practically lack, pectins (Ros Barcelo 1997).

Plant cell wall proteins are encoded by multigene families and make up approximately 10% of the cell wall weight (Jamet et al. 2006). They are classified as structural or enzymatic, the latter group being responsible for the dynamic nature of the cell wall. Despite the central role of apoplastic proteins in most processes that control growth and development (Cosgrove 2005), the apoplast proteome is less characterized than that of other sub-cellular compartments, and so the understanding of cell wall physiology is still very limited. A priori, the annotation of the *Arabidopsis* genome shows that about 17% of the genome (i.e. 5,000 genes) encodes proteins with a predicted signal peptide that targets them to the secretory pathway (Jamet et al. 2006), although the function of most of them remains unknown.

Despite the gaps in the knowledge of most cell wall proteins, structural proteins are the major glycoproteins present in the primary plant cell wall, and these fall into two main categories (Jamet et al. 2006): hydroxyproline (Hyp)-rich glycoproteins (HRGP) and proline-rich proteins (PRP). HRGP may be classified into extensins and arabinogalactan proteins (AGP). Extensins are cell wall structural HRGP which showed numerous Ser-Hyp_n motifs, whereas AGP are cell wall highly glycosylated HRGP that contain repetitive motifs such as (Ser, Thr, Ala)-Hyp-(Ser, Thr, Ala) or (Ser, Thr, Ala)-Hyp-Hyp. PRP are ill-defined structural cell wall proteins rich in Pro residues.

Besides structural proteins, there is a vast array of cell wall enzymes involved in polysaccharide cell wall metabolism, and enzymes of an oxidoreductase nature, such as peroxidases, lacases, poly(di)amine oxidases (PDAOs), oxalate oxidases (OXOs), and superoxide dismutases (SODs), which are the main enzymes that consume or generate ROS in the plant cell wall.

3 Why Plant Cell Walls Need Generating ROS for Growth and Development?

Most of the polysaccharides and structural glycoproteins of the plant cell wall are released to the apoplast as water-soluble macromolecules. However, in the apoplastic space, these molecules are assembled by means of cross-links to form a three-dimensional and insoluble network (the plant cell wall), which is essential if a structurally competent material is to be formed that holds together in an aqueous environment. Some of the cross-links involved are noncovalent while others are covalent (Fry 2004a). Covalent cross-links are few in number but individually strong. For example, some polysaccharide chains may be cross-linked by the oxidative coupling of ferulic acid side-chains (Ralph et al. 2004a); some glycoproteins are cross-linked through the oxidative coupling of tyrosine residues (Fry 2004b), and lignin is covalently attached to certain polysaccharides through benzyl ether bridges to form lignin–carbohydrate complexes (Ralph et al. 2004b).

Our understanding of plant cell wall cross-links, and therefore of the wall architecture, is far from complete. However, the enzyme-catalyzed formation of the covalent cross-links described above requires ROS, and this constitutes the major fate for all the cell wall-located enzymes of an oxido-reductase nature, enzymes which are the responsible for all the irreversible cell wall stiffening reactions.

4 Cell Wall Stiffening

Irreversible process of cell wall stiffening may be dependent or independent from the cell type. Thus, extensin and ferulic acid cross-linking may constitute temporal and transitory events within the general program of development of any plant cell, while lignification and suberization are terminal processes of determinate, and highly differentiated, plant cells capable of forming secondary cell walls. In the case of lignification, this process is restricted to water-conducting vascular cells and their neighboring fibres (Ros Barceló 1997), while suberization is restricted to the cell wall of epidermal tissues of underground plant parts (roots, stolons, and tubers), the endodermis, and the cells of bark tissues (cork and periderm) (Bernards et al. 2004). In either case, lignification and suberization have the hallmark of being highly tissue/cell specific. Indeed, cells that ultimately become suberized or lignified either derive from specific progenitor cells in meristems or arise from the vascular cambium, respectively.

4.1 Extensin Cross-Linking

Extensins are the most studied family of HRGPs. The polypeptide backbone of extensins contains many repeats of the structural $\mathrm{Ser(Hyp)}_{4\text{–}6}$ motif. These structural motifs are often flanked by short sequences rich in Tyr, Lys, Val, and His, the Val–Tyr–Lys motifs being the sites for extensin cross-linking (Fry 2004b). Extensins are secreted into the apoplast as soluble monomers where the positively charged Lys and the protonated His residues interact ionically with the negatively charged uronic acids of pectins. The formation of the insoluble extensin network is a well-characterized peroxidase-mediated and H_2O_2-dependent process which, it has been proposed, involves the coupling of extensin Tyr residues to form isodityrosine linkages, and larger Tyr oligomers, such as di-isodityrosine or pulcherosine (Fry 2004b).

4.2 Ferulic Acid Cross-Linking

Ferulic acid is ester-bonded to cell wall polysaccharides (Ralph et al. 2004a). In dicots, nonreducing terminal arabinose (Ara) and galactose residues of pectic polysaccharides are feruloylated, while in monocots feruloylation is mainly restricted to the *O*-5 position of some Ara residues of arabinoxylans (Fry 2004a). Feruloyl residues can form covalent bonds with each other by oxidative coupling. This process only occurs in the presence of peroxidase and H_2O_2, and it has been speculated (Ralph et al. 2004a) that the resulting dehydrodiferuloyl residue (5,5′-dehydrodiferulate) forms a cross-link between the polysaccharides to which it is esterified.

In addition to the first-discovered dimer, 5,5′-dehydrodiferulate, several of its isomers have been obtained by alkaline hydrolysis of plant cell wall polysaccharides; and several specific trimers and tetramers have now been characterized (Ralph et al. 2004a). Depending on the linkage formed (intra- or inter-polysaccharide), up to four polysaccharide chains could potentially be cross-linked by a tetramer of ferulic acid, which suggests that these peroxidase-mediated oxidatively generated oligomers of ferulate may play an important role in determining cell wall assembly.

4.3 Lignification

Lignins are three-dimensional, amorphous, heteropolymers which result from the oxidative coupling of three *p*-hydroxycinnamyl alcohols (monolignols), *p*-coumaryl, coniferyl, and sinapyl alcohols, in a H_2O_2-dependent reaction mediated by peroxidases, enzymes which generate the corresponding free radicals from the three monolignols (Ros Barceló 1997). The regio- and stereospecificity of the cross-coupling reaction of monolignol radicals produces a hydrophobic heteropolymer composed of *p*-hydroxyphenyl, guaiacyl, and syringyl units, respectively. Both the chemistry and biochemistry of the lignification have been recently revised (Ralph et al. 2004b), as well as the nature of the peroxidase responsible for this process (Ros Barceló et al. 2007).

4.4 Suberization

Both lignification and suberization involve the formation of a three-dimensional poly(phenolic) matrix initially within the carbohydrate matrix of the primary cell wall. In addition to this phenolic component, suberized cells develop a distinct poly(aliphatic) layer described (Bernards et al. 2004) as a polyester connected through primary ester bonds between the major aliphatic components (α,ω-dioic acids and ω-hydroxyalkanoic acids), in close association with *p*-hydroxycinnamic acids, such as *p*-coumaric, caffeic, and ferulic acids, which constitute the cell wall-bound poly(phenolic) domain (PPD). The macromolecular assembly of potato PPD occurs via a H_2O_2-dependent peroxidase-mediated free radical coupling process (Bernards et al. 2004), analogous to the process of lignin formation from monolignols, but this enzyme shows a marked substrate preference for *p*-hydroxycinnamates (i.e., ferulic acid and derivatives).

5 Cell Wall Softening

Not all the H_2O_2-dependent oxidoreductase reactions that occur in the plant cell wall lead to processes of cell wall stiffening. It has been proposed (Fry 1998) that wall polysaccharides may be subjected in vivo to non-enzymatic scission mediated by hydroxyl radicals (OH•) resulting in cell wall softening. Wall polysaccharides are readily cleaved by OH• radicals in vitro (Fry 1998) and OH• can be generated from H_2O_2 in the cell walls by a Fenton reaction, which requires a reduced transition metal ion, Cu^+ being 60 times more effective than Fe^{2+}:

$$Cu^+ + H_2O_2 \rightarrow Cu^{2+} + OH\bullet + OH^-$$

Endogenous $OH^\bullet$ has also proposed to be produced by wall peroxidases (Liszkay et al. 2003) from $O_2\bullet^-$ and H_2O_2 through a Haber-Weiss-type reaction:

$$O_2\bullet^- + FeIII \rightarrow O_2 + FeII$$

$$2O_2\bullet^- + 2H^+ \rightarrow H_2O_2 + O_2$$

$$FeII + H_2O_2 \rightarrow FeIII + OH\bullet + OH^-$$

where FeIII is ferriperoxidase, the native form of the enzyme, and FeII is ferroperoxidase, the reduced form of the enzyme. Although OH• is exceedingly reactive, its production can be controlled, e.g., by the correct sitting of Cu^+ atoms relative to wall polysaccharide molecules. Once produced at a particular site within the cell wall, an OH• radical would rarely diffuse more than 1 nm before reacting with polysaccharides, which is about 1% of the thickness of a typical plant cell wall. Thus, if produced at the right time and right place within the wall, OH• could be very precisely targeted to cause xyloglucan scission (Schopfer 2001). How this might be achieved has yet to be determined. Indeed, evidence for OH•-induced polysaccharide splitting in growing cell walls might be obtained by finding the predicted products of OH• action on either pectins or xyloglucans. Evidence for such products has been found in softening fruit, but not in growing cells (Cosgrove 2005).

All these results support that, in growing cells, $O_2{\bullet}^-/H_2O_2$ directly or indirectly (as sources of OH•) may also promote cell wall softening reactions. In any case, the dichotomy of ROS production, especially as regards the fate of H_2O_2, $O_2{\bullet}^-$, and OH•, for cell wall stiffening/softening reactions, is apparently far from being completely understood (Passardi et al. 2004).

6 ROS Generation in Plant Cell Walls During Development is a Cell/Tissue-Specific Event with a Marked Topographic Localization

The monitoring of ROS generation/accumulation in plant cell walls during development by means of histochemical and cytochemical methods revealed that $H_2O_2/O_2{\bullet}^-$ production/accumulation during development is mainly restricted to epidermal and vascular tissues, constituting therefore a cell/tissue-specific event (Schopfer 1994; Liu et al. 1995; Ros Barcelo 1998a; Córdoba-Pedregosa et al. 2003). This observation agrees with the fact that in epidermal and vascular tissues most of the ROS-dependent reactions for cell wall stiffening/softening take place (see Sects. 4 and 5), and consequently ROS accumulation in cell walls of these tissues (Fig. 2) is only the static image of their high redox state and large oxidative metabolism.

The fine structural (topographic) localization of H_2O_2 production in epidermal and vascular cells may be followed using $CeCl_3$. Ce^{3+} penetrates the tissues, albeit slowly, and reacts directly with H_2O_2 to form insoluble deposits of Ce^{4+} perhydroxide ($CeO_2{\bullet}{\cdot}nH_2O$). The reflectance properties imparted to these deposits by the electron-dense Ce^{4+} makes them easily detectable in plant tissues by both electron and epi-polarization microscopy (Liu et al. 1995; Ros Barceló 2005).

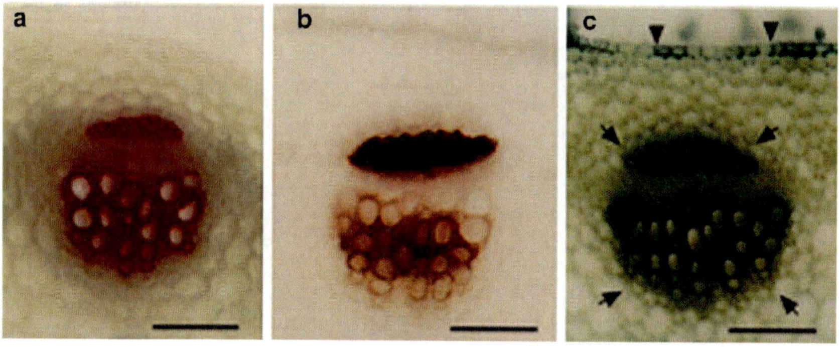

Fig. 2 Histochemical localization of lignins by means the phloroglucinol (**a**) and the Wiesner (**b**) reactions, and of H_2O_2 using 3,3′,5,5′-tramethylbenzidine (**c**) in stems of *Zinnia elegans*, showing the accumulation of H_2O_2 in non-lignifying epidermal (*arrowheads*) and lignifying vascular (*arrows*) tissues. *Bars* 150 μm

6.1 Epidermal Cells

The use of the $CeCl_3$ technique revealed that, at the level of epidermal cell walls, sites for H_2O_2 production are mainly located on the external face of the plasma membrane (Fig. 3), at the level of cell corners and the thin radial cell walls. This observation supports the role of NADPH oxidases (NOX) in this process, although the fate of this fine topographic localization is unknown. Besides, sites for H_2O_2 production may be seen in both growing and stop-growing epidermal cells (Fig. 3), which now emphasizes the poorly understood dichotomy of ROS production for cell wall stiffening/softening processes.

6.2 Vascular Cells

H_2O_2 production in vascular cells is a strongly regulated developmental process, which shows a burst at the precise moment at which xylem cell wall lignification begins (Ros Barceló et al. 2002). At this early developmental stage, when only differentiating xylem vessels have begun to lignify, electron dense deposits are mainly observed in the apoplastic space of differentiating (living) thin-walled xylem cells (Ros Barceló 2005). Such H_2O_2-dependent electron dense deposits are

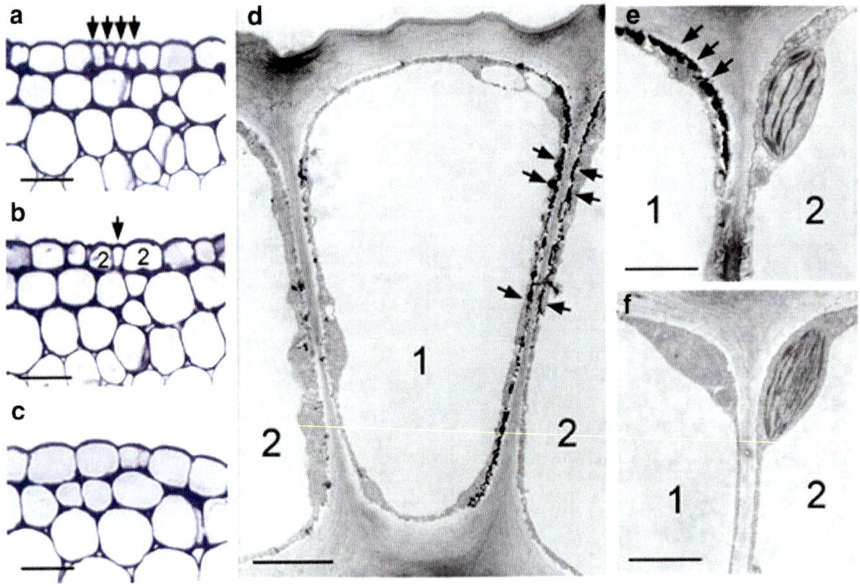

Fig. 3 Sequence (**a-c**) of local growth phenomena (*arrows*) in the epidermis of *Z. elegans* stems and cytochemical localization of H_2O_2 (**d, e**) using the $CeCl_3$ technique at the external face of the plasma membrane of the growing (1) and stop-growing (2) epidermal cells illustrated in (**b**). Control (**f**) in the presence of ascorbic acid. *Bars* 25 μm (**a-c**), 2 μm (**d**) and 1 μm (**e,f**)

also observed at the outer face at the plasma membrane of xylem parenchyma cells adjacent to differentiating thick-walled xylem cells (Ros Barceló 2005). These observations suggest that, in vascular tissues, non-lignifying xylem parenchyma cells may transitorily be the source of the H_2O_2 necessary for the polymerization of cinnamyl alcohols in the secondary cell wall of differentiating xylem vessels, suggesting a certain degree of cell-to-cell cooperation during lignin biosynthesis, which has been elegantly confirmed by confocal laser scanning microscopy using cultured tracheary elements (Gómez-Ros et al. 2006).

In the case of differentiating xylem vessels (Czaninski et al. 1993; Ros Barceló 2005), the reaction in the apoplast is generally so intense that it is impossible to determine whether the middle lamella, or the plasma membrane, is the precise site for H_2O_2 production. Caution should be exercised, however, in interpreting the presence of electron dense deposits in the middle lamella, since the ionic radius of Ce^{4+} (0.925 Å) is close to that of Ca^{2+} (0.940 Å), its charge/radius ratio being twofold higher. In fact, soluble pectins are precipitated by Ce^{3+}/Ce^{4+} ions (Fig. 4), and this observation would easily explain the presence of electron-dense Ce^{4+} deposits in the pectin-rich layer of the middle lamella.

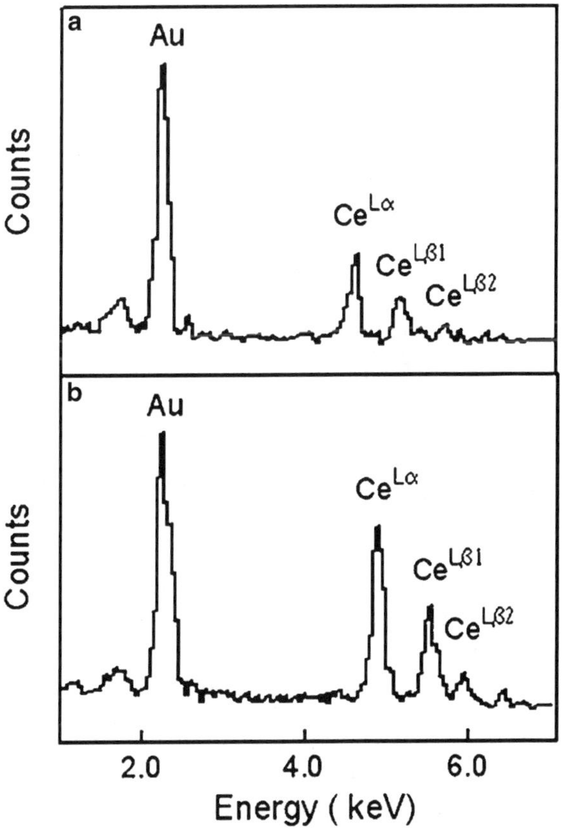

Fig. 4 Detection of Ce^{+3} (**a**) and Ce^{+4} (**b**) by coupled scanning electron microscopy-energy dispersive X-ray microanalysis (SEM-EDX) in insoluble pectin-cerium complexes using gold (*Au*) as internal standard

7 Sources of ROS in Cell Walls

As discussed in Sect. 6, H_2O_2 is constitutively produced by plant cells at the level of the cell wall and outer plasma membrane surface of epidermal and vascular tissues. However, the real molecular mechanism of H_2O_2 production at the plant cell surface is still unknown. Thus, although during recent years plant biologists have made great efforts to resolve the paradigm of the mechanism of H_2O_2 production at the plasma membrane/cell wall interface, the present knowledge of this enzymatic mechanism is fragmentary and not easily understood.

Four possible mechanisms have been proposed to explain how this H_2O_2 is produced: one is located at the level of the external face of the plasma membrane, and is mediated by NOX, and three are located at the level of the cell wall matrix, which would involve the action of peroxidases, PDAOs, and OXOs. Unlike PDAOs and OXOs, which directly generate H_2O_2, both NOX and peroxidases catalyze the initial formation of $O_2{\bullet}^-$, which later dismutates to H_2O_2. The peroxidase-mediated $O_2{\bullet}^-$ production may be distinguished from that catalyzed by NOX by the different k_M values for O_2 and the different sensitivities of the two enzymes to inhibitors such as cyanide, azide, and diphenylene iodonium (DPI) (Bolwell et al. 1998). However, caution should be exercised in the use of these inhibitors (Frahry and Schopfer 1998; Ros Barceló 1998b; Ros Barceló and Ferrer 1999).

7.1 *NADPH Oxidases*

As suggested from the topography of H_2O_2 production obtained from electron microscope studies (Fig. 3), epidermal and vascular cells very probably generate extracellular $O_2{\bullet}^-/H_2O_2$ via a plasma membrane-localized NOX. This enzyme catalyzes the formation of $O_2{\bullet}^-$ from O_2, using cytosolic NADPH as an electron donor (see Chap. 1), at the external face of the plasma membrane:

$$2O_2 + \text{NADPH} \rightarrow 2O_2{\bullet}^- + \text{NADP}^+ + \text{H}^+$$

In *Arabidopsis* (Foreman et al. 2003; see Chap. 5), cell wall softening (loosening) and plant cell growth stimulated by ROS is apparently mediated by a Rop GTPase-regulated NOX (encoded by *AtrbohC*), which results in $O_2{\bullet}^-$ production at active growing cells (Foreman et al. 2003). Likewise, suberization is an H_2O_2-dependent peroxidase-mediated process, in which the H_2O_2 necessary for the peroxidative oxidation of phenolics also comes from a DPI-sensitive NOX-like enzyme (Razem and Bernards 2003; Bernards et al. 2004), similarly to that occurring during lignification (Ros Barceló 1998c). In fact, a potato NOX homolog (*StrbohA*) has recently been implicated (Kumar et al. 2007) in the generation of H_2O_2 during suberization.

H_2O_2-generation is a cognate characteristic of all the lignifying xylem cells (Czaninski et al. 1993; Olson and Varner 1993; Ros Barceló 1998a; Weir et al. 2005), and may be over-stimulated by cadmium toxicity (Rodríguez-Serrano et al. 2006).

H_2O_2 production by the xylem is thought to be catalyzed by a DPI-sensitive NOX-like enzyme (Ogawa et al. 1997; Ros Barceló 1998c, 1999; Karlsson et al. 2005), which generates $O_2{\bullet}^-$. This $O_2{\bullet}^-$ in then dismutated to H_2O_2 by a tissue-specific CuZn-SOD (Ogawa et al. 1997; Karlsson et al. 2005; Srivastava et al. 2007) or, alternatively, by a Mn-SOD (Corpas et al. 2006).

Evidence for the presence of a NOX-like enzyme in the xylem also comes from the observation that a Rac small GTPase, identified as a component of the NOX complex (Wong et al. 2007), is preferentially and asymmetrically accumulated in xylem parenchyma cells and tracheary element precursor cells (Nakanomyo et al. 2002), which, as has been described in Sect. 6.2, are the main sites for $O_2{\bullet}^-/H_2O_2$ production (Ros Barceló 2005). Rac small GTPases are known as molecular switches that regulate cellular processes by transducing signals in the GTP-bound form (Wong et al. 2007). Their identification as putative components and regulatory proteins of the NOX complex in the lignifying xylem fit well with the observation that $O_2{\bullet}^-/H_2O_2$ production by the xylem is a highly regulated and orchestrated event (Fig. 5), in which calmodulin (sensitive to R-24571), phospholipase C (sensitive to neomycin sulfate) and protein kinase (sensitive to staurosporine) are key components in its regulation (Ros Barceló 1999). This strict metabolic regulation of $O_2{\bullet}^-/H_2O_2$ production suggests the participation of a metabolic cascade involving

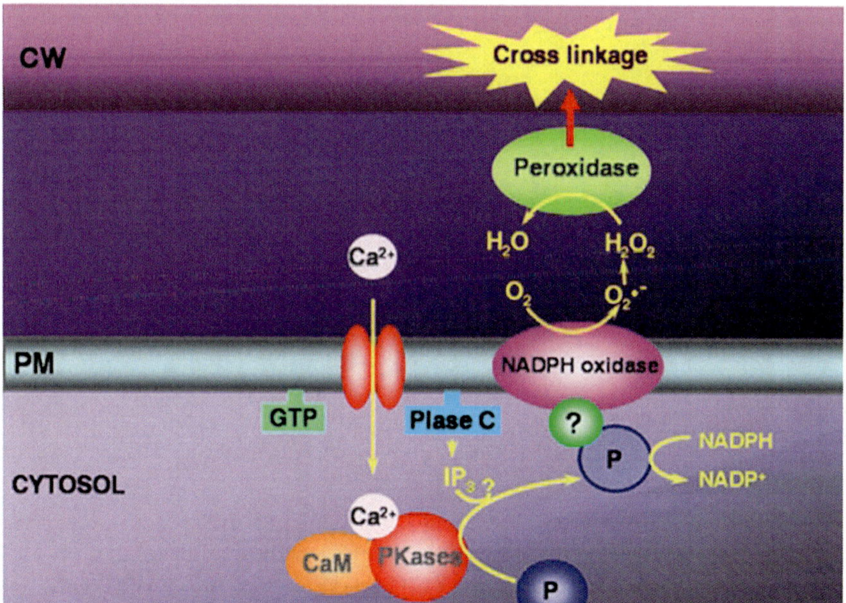

Fig. 5 Proposed scheme for the regulation of the NADPH-oxidase-like enzyme of the xylem by Ca^{2+}/calmodulin, phospholipase C, and protein kinases on the basis of the sensitivity of H_2O_2/$O_2{\bullet}^-$ production to R-24571, neomycin sulfate, and staurosporine, respectively

Ca^{2+}/calmodulin, inositol 1,4,5-triphosphate (IP_3) and protein phosphorylation. The metabolic cascade appears to act as the joint gateway for coordinating both the morphological and the biochemical processes which occur during xylem differentiation/lignification, since inhibitors of phospholipase C also inhibit tracheary element formation (Zhang et al. 2002).

Microarray analyses of gene expression during the *trans*-differentiation of *Zinnia* mesophyll cells into tracheary elements (Demura et al. 2002) have allowed the determination of a specific mRNA coding for a putative respiratory burst oxidase homolog (RBOH) (NOX). In fact, a differentially expressed cDNA (AU289770) showed a strong homology with the *AtrbohE* gene, which is also differentially expressed during the *trans*-differentiation of *Arabidopsis* mesophyll cells into tracheary elements (Kubo et al. 2005), making them the most obvious candidates for the generation of H_2O_2 in vascular tissues. The *AtrbohE* gene, however, clusters outside the two main groups in which all the known plant RBOH cluster (Yoshioka et al. 2001), including the *StrbohA* implicated in the generation of H_2O_2 during suberization.

7.2 Peroxidases

The key function of plant peroxidases is the single one-electron oxidation of several substrates of phenolic nature at the expense of H_2O_2 yielding diffusible phenoxy radicals (R•):

$$2RH + H_2O_2 \rightarrow 2R\bullet + 2H_2O$$

However, and paradoxically, peroxidases may transitorily generate $O_2\bullet^-/H_2O_2$ (Ros Barceló 2000; Passardi et al. 2004). This mechanism is restricted to the $O_2\bullet^-/H_2O_2$ generating step of peroxidases during their catalytic cycle, which is represented by the decay of Compound III (CIII) into FeIII, the native form of the enzyme. This cycle begins in the absence of H_2O_2 but in the presence of a reductant capable of reducing FeIII to FeII [E^o (FeIII/FeII) = −271 mV at pH 7.0] (Yamazaki and Yokota 1973). FeII has a high affinity for O_2 ($k = 5.8 \times 10^4\ M^{-1}\ s^{-1}$), and this reaction leads to the formation of CIII:

$$\text{FeII} + O_2 \rightarrow \text{CIII}.$$

Three different mechanisms for the decay of CIII to FeIII have been proposed (see Ros Barceló 2000, and references cited therein), but only one involves the dissociation of CIII ($k = 8.2 \times 10^{-3}\ s^{-1}$): CIII → FeIII + $O_2\bullet^-$. However, this reaction appears more properly to be displaced towards the formation of CIII ($k = 3.0 \times 10^7$ $M^{-1}\ s^{-1}$): FeIII + $O_2\bullet^-$ → CIII, and it is unlikely that it contributes noticeably to a net $O_2\bullet^-$ generation. Nevertheless, the slow decay of CIII may initiate a complex catalytic cycle (Yamazaki and Yokota 1973) which can be broken down into in the following Eqs.:

$$2\text{CIII} \rightarrow 2\text{FeIII} + 2\text{O}_2\bullet^- \quad (k = 8.2 \times 10^{-3}\,\text{s}^{-1})$$

$$2\text{O}_2\bullet^- + 2\text{H}^+ \rightarrow \text{H}_2\text{O}_2 + \text{O}_2 \quad (k = 1.8 \times 10^{7}\,\text{M}^{-1}\text{s}^{-1})$$

$$\text{FeIII} + \text{H}_2\text{O}_2 \rightarrow \text{CI} + \text{H}_2\text{O} \quad (k = 10^{7}\,\text{M}^{-1}\text{s}^{-1})$$

$$\text{CIII} + \text{CI} + \text{H}^+ \rightarrow \text{FeIII} + \text{CII} + \text{O}_2 \quad (k = 10^{4}\,\text{M}^{-1}\text{s}^{-1})$$

$$\text{CIII} + \text{CII} + \text{H}^+ \rightarrow 2\text{FeIII} + \text{O}_2 + \text{H}_2\text{O} \quad (k = 3.0 \times 10^{4}\,\text{M}^{-1}\text{s}^{-1})$$

where CI (compound I) and CII (compound II) constitute intermediates of the catalytic (peroxidative) cycle of the enzyme, and which leads to the complete CoIII auto-decomposition:

$$4\text{CIII} + 4\text{H}^+ \rightarrow 4\text{FeIII} + 3\text{O}_2 + 2\text{H}_2\text{O}$$

From these eqs., it is apparent that the decay of CIII would in no case contribute to a noticeable net $O_2\bullet^-/H_2O_2$ production. However, due to the fact that CI and CII are continuously generated and broken down during this oxidase/peroxidase cycle, the introduction in the bulk of the reaction of a phenolic compound (RH) acting as substrate of CI and CII:

$$\text{CII} + \text{RH} \rightarrow \text{FeIII} + \text{R}\bullet + \text{H}_2\text{O}$$

would result in the oxidation of the phenolic to its corresponding radical (R•), which would then undergo polymerization reactions. This is what occurs when a lignin precursor (e.g., coniferyl alcohol) (Ferrer et al. 1990) or soluble cell wall proteins (Wojtaszek et al. 1997) are introduced in the bulk of the oxidase/peroxidase reaction.

7.3 Poly(di)amine Oxidases

DAO catalyses the oxidative deamination of the diamine putrescine (Put), producing the corresponding 4-aminobutirylaldehyde, which spontaneously cyclizes to Δ^1-pyrroline, ammonia, and H_2O_2. PAO preferentially cleaves the aminopropyl side-chains at secondary amino groups of polyamine substrates, such as spermidine or spermine, producing H_2O_2, 1,3-diaminopropane, and Δ^1-pyrroline or Δ^1-(3-aminopropyl)-pyrroline, respectively (Smith 1985). Polyamines inhibit the NOX-mediated $O_2\bullet^-$ generation (Papadakis and Roubelakis-Angelakis 2005), and PDAOs are cell wall-located enzymes (Smith 1985), both histochemical and immunolocalization studies having demonstrated the presence of these enzymes in xylem tissues (Angelini and Federico 1989; Moller and McPherson 1998). However, since the carbon skeletons of the products of PDAO action [1,3-diaminopropane, Δ^1-pyrroline, and Δ^1-(3-aminopropyl)-pyrroline] cannot be recycled directly to the initial H_2O_2-generating polyamines (Sebela et al. 2001), these enzymes are generally considered to be involved in the terminal catabolism of polyamines. In fact, the biosynthesis of diamines (e.g., Put) is a highly expensive process:

$$2HNO_3 + 4CO_2 + 19NADPH + 19H^+ \rightarrow NH_2\text{–}CH_2\text{–}CH_2\text{–}CH_2\text{–}CH_2\text{–}NH_2 + 14H_2O + 19NADP^+$$

and therefore it makes no sense for plant cells to manufacture costly poly(di) amines to generate H_2O_2 in the cell wall, when they may directly and economically perform this task by means of NOX, through the direct and univalent reduction of O_2.

7.4 *Oxalate Oxidases*

OXOs are cell wall-located enzymes that can produce H_2O_2 in plants. They release CO_2 and H_2O_2 from O_2 and oxalic acid (OA), which is generally present in low levels in plant cell walls:

$$HOOC-COOH + O_2 \rightarrow 2CO_2 + H_2O_2$$

This enzyme was first isolated and characterized from monocots (Lane 1994). Wheat OXO, also known as germin, is the best-characterized member of the cupin family (Bernier and Berna 2001). Germin-like proteins (GLPs) have been isolated from both monocots and dicotyledonous, but only in monocots and pine have germins shown OXO activity (Bernier and Berna 2001). Various studies on the highly conserved family of GLPs revealed that GLPs may play certain roles in plant development (Bernier and Berna 2001), but the exact biological significance of the H_2O_2 production by OXOs during this process remains unknown.

8 H_2O_2 as Autocrine and Paracrine Signal

When relatively large amounts of H_2O_2 are generated in the apoplast, and it is exceeded the low cell wall antioxidant capacity, such as occurs in epidermal and vascular tissues (Sect. 6), H_2O_2 would rapidly move into the cell or through the apoplastic space of cells of the same tissue, and subsequently act as an intracellular and intercellular messenger (Neill et al. 2002; see also Chaps. 2–6), be sensed and metabolized. Thus, H_2O_2 can function as a mobile signal, but whether H_2O_2 is the sole signal or is part of a multiple systemic response remains to be determined.

Simulating ROS production at the plant cell wall by the exogenous supply of H_2O_2 influences the expression of a large number of genes (Kim et al. 2005), which suggests that plant cells have evolved strategies to utilize cell wall ROS as biological signals and that cell wall ROS control several genetic developmental programs

(Neill et al. 2002). Changes in transcriptional activity may be achieved through the oxidation of components of signalling pathways that subsequently activate transcription factors (TFs) or by modifying a redox-sensitive TF directly. The effects of ROS on components of the mitogen-activated protein kinase (MAPK) cascade may result in the indirect activation of TFs, and the activation of the MAPK cascade may be initiated by redox-controlled protein Tyr phosphatases (see Chap. 2).

H_2O_2 modulates the expression of a vast array of genes, including those encoding antioxidant enzymes and modulators of H_2O_2 production. A microarray study (Kim et al. 2005) showed that the expression of 9% of genes of *Arabidopsis* was altered in H_2O_2-treated seedlings. As would be expected, genes encoding antioxidant enzymes [ascorbate peroxidase (APX), dehydroascorbate reductase (DHAR), monodehydroascorbate radical reductase (MDAR)], oxidases such as ascorbate oxidases (AAO), enzymes of the lignin biosynthetic pathway [phenylalanine ammonia lyase, *p*-hydroxycinnamoyl CoA reductases (CCRs), and ferulate-5-hydroxylases], and RBOH proteins, were upregulated. One of the upregulated Atrboh genes was *AtrbohC* the RBOH involved in cell wall softening reactions (Foreman et al. 2003). As also expected, some of the downregulated genes corresponded to proteins involved in cell wall softening, such as expansins.

However, within the same multigenic family, members were found which were downregulated, while others were upregulated (Kim et al. 2005). This was the case for peroxidases, laccases, enzymes of the lignin biosynthetic pathway (caffeoyl-CoA *O*-methyl-transferases and caffeic acid *O*-methyl-transferases and CCRs), putative plasma membrane ABC monolignol transporters, extensins, and proteins involved in cell wall softening, such as xyloglucan-endo-trans-glycosylases, but also genes encoding proteins that are potentially involved in signalling, such as TFs of the MYB and WRKY family (Kim et al. 2005), all of which illustrates the complexity of the H_2O_2-regulated signalling pathways. Although H_2O_2-responsive promoters have been identified (Neill et al. 2002), specific H_2O_2-regulatory DNA sequences (*cis*-elements) and their cognate TFs have not yet been isolated and characterized.

9 ROS Homeostasis in Cell Walls

Non-enzymatic plant antioxidants include the major cellular redox buffers, ascorbic acid (AA) and glutathione (GSH), but only for AA does evidence exist concerning its presence in cell walls (Vanacker et al. 1998). In fact, although most of the cellular AA pool is localized within the cytoplasm, up to 5% is transported across the plasma membrane and into the apoplast, where it constitutes the major sink for ROS (Apel and Hirt 2004). AA is chemically oxidized to monodehydroascorbate radical (MDHAR), which dismutates into AA and dehydroascorbate (DHA). A high AA to DHA ratio ([AA]/[AA + DHA]) is essential for ROS scavenging in plant cell walls. In fact, AA interferes with the peroxidase-mediated oxidation of phenolics by reducing phenolic radicals back to the original phenolics, thereby scavenging H_2O_2 (Takahama 1993; de Gara 2004). It seems that, to allow cell wall stiffening reactions, apoplastic AA would first need to be completely oxidized to DHA, as has been confirmed in several studies (de Gara 2004; Ros Barceló et al. 2006).

Enzymatic ROS-scavenging mechanisms in plants include SOD, APX, glutathione peroxidase (GPX) and catalase, but evidence of their presence in cell walls exists only in the case of SOD and, to a lesser extent, of APX (Apel and Hirt 2004). SODs act as the first line of defense against ROS, dismutating $O_2{\bullet}^-$ to H_2O_2. APX can subsequently detoxify H_2O_2 at the expense of AA. To be effective in ROS scavenging, APXs require an AA and GSH regeneration system, the AA-GSH cycle (Foyer and Halliwell 1976). However, there is no conclusive evidence supporting the presence of the AA-GSH cycle in plant cell walls (Hernández et al. 2001; Apel and Hirt 2004).

In spite of this, different plasma membrane-associated carrier systems for AA and DHA translocation have been reported in several plant species (Horemans et al. 2000): DHA may be transported into the cytosol through the plasma membrane by a specific carrier that preferentially translocates DHA in exchange for the reduced form (AA), ensuring a continuous flux of reducing power to the cell wall (Horemans et al. 2000). Little is known about these plasma membrane AA transport systems, but it is known that they are too slow to maintain a highly reduced extracellular AA pool in situations of high ROS production (Horemans et al. 2000). It has also been speculated that AA may also be regenerated in the plant cell wall by the driving of a plasma membrane electron transport chain. In fact, plasma membranes contain an AA-dependent cytochrome b_{561}, which is reduced on one face by cytosolic AA and oxidized on the other by MDHAR, generating AA in the plant cell wall (Horemans et al. 2000).

However, due to the limited distribution of APXs in plant cell walls (de Gara 2004), there is increasing evidence that the levels of cell wall-located AA are mainly controlled by AAO (Pignocchi et al. 2003), a situation that is extrapolable to lignifying tissues (Ros Barceló et al. 2006). AAO is a widely distributed cell wall-localized enzyme, which regulates the redox status of AA in the apoplast since it oxidizes AA to MDHAR, with the concomitant reduction of O_2 to H_2O (Pignocchi and Foyer 2003). AAO belongs to the enzyme family of "the blue copper oxidases," which also includes plant laccases. AAO has a high degree of homology with laccase (Pignocchi and Foyer 2003), but although there seem to be several important differences between the two enzyme groups, no work has so far been published comparing the substrate specificities of laccases and AAO, which is especially important since both enzyme groups very probably exhibit overlapping substrate specificities.

If the redox status of the extracellular AA pool is regulated by AAOs (Pignocchi et al. 2003), this enzyme could be considered the first step in the AA degradation pathway in the apoplast (Green and Fry 2005), since DHA is unstable in aqueous solutions, and rapidly decays to 2,3-diketo-L-gulonate (DKG) (Green and Fry 2005). In fact, in the apoplast of cultured rose cells, DHA is degraded via several intermediates to OX and L-threonate, through a series of reactions that possibly generate H_2O_2 (Green and Fry 2005; Kärkonen and Fry 2006). Furthermore, OX, one of the end products in the AA degradation pathway of the apoplast, could generate H_2O_2 by the action of OXOs. Such an AA degradation pathway of the apoplast leading to L-threonate and CO_2 could confer some sense to lacasses/AAOs and OXOs, which are poorly understood enzymes in relation to their function (de Tullio et al. 2004;

Bernier and Berna 2001), and which are strongly upregulated during secondary cell wall formation and lignification (Demura et al. 2002; Kubo et al. 2005). In addition, regulation of the extracellular AA pool by AAOs would provoke an on/off transition of the redox state of the apoplast from a highly reducing state in undifferentiated cells to a highly oxidizing state in differentiating/differentiated cells, which is required not only for most of the terminal cell wall stiffening reactions, but also for vascular differentiation (de Pinto and de Gara 2004; Henmi et al. 2005).

10 Conclusion

The primary function of ROS production in the plant cell wall of epidermal, vascular, and suberizing tissues is modification of the cell wall components, through processes of cell wall softening/stiffening, which, in the last instance, trace the fine profiles of plant cell growth and morphogenesis. However, this produces a dichotomy in ROS production, especially as regards the fate of H_2O_2, $O_2^{\bullet -}$, and OH• in cell wall stiffening/softening reactions, which is far from being completely understood. Although four possible mechanisms have been proposed to explain how this H_2O_2 is produced in the plant cell wall, there is only certain creditable evidence (molecular and logistic) supporting the participation of NADPH oxidases in this process. There is also increasing substantiation that the redox buffering capacity of the apoplast is regulated by AAOs. Such regulation of the extracellular AA pool by AAOs would provoke an on/off transition of the redox state of the apoplast, from a highly reducing state in undifferentiated cells to a highly oxidizing state in differentiating/differentiated cells, which is required not only for most of the biochemical processes leading to cell wall stiffening reactions, such as suberization and lignification, but also for inducing cytological processes such as vascular differentiation.

Acknowledgement This work was supported by grants from the Fundación Séneca (project No. 08610/PI/08) and MCYT (BFU2006-11577)-FEDER.

References

Angelini R, Federico R (1989) Histochemical evidence of polyamine oxidation and hydrogen peroxide generation in the cell wall. J Plant Physiol 135:212–217

Apel K, Hirt H (2004) Reactive oxygen species: metabolism, oxidative stress, and signal transduction. Annu Rev Plant Biol 55:373–399

Bernards MA, Summerhurst DK, Razem FA (2004) Oxidases, peroxidases and hydrogen peroxide: the suberin connection. Phytochem Rev 3:113–126

Bernier F, Berna A (2001) Germins and germin-like proteins: plant do-all proteins. But what do they do exactly? Plant Physiol Bioch 39:545–554

Bolwell GP, Davies DR, Gerrish C, Auh C-K, Murphy TM (1998) Comparative biochemistry of the oxidative burst produced by rose and French bean cells reveals two distinct mechanisms. Plant Physiol 116:1379–1385

Brownlee C (2002) Role of the extracellular matrix in cell–cell signalling: paracrine paradigms. Curr Opin Plant Biol 5:396–401

Córdoba-Pedregosa MC, Córdoba F, Villalba JM, Gonzáles-Reyes JA (2003) Differential distribution of ascorbic acid, peroxidase activity, and hydrogen peroxide along the root axis in *Allium cepa* L. and its possible relationship with cell growth and differentiation. Protoplasma 221:57–65

Corpas FJ, Fernández-Ocaña A, Carreras A, Valderrama R, Luque F, Esteban FJ, Rodríguez-Serrano M, Chaki M, Pedrajas JR, Sandalio LM, del Río LA, Barroso JB (2006) The expression of different superoxide dismutase forms is cell-type dependent in olive (*Olea europaea* L.) leaves. Plant Cell Physiol 47:984–994

Cosgrove DJ (2005) Growth of the plant cell wall. Nature Rev Mol Cell Biol 6:850–861

Czaninski Y, Sachot RM, Catesson AM (1993) Cytochemical localization of hydrogen peroxide in lignifying cell walls. Ann Bot 72:547–550

de Gara L (2004) Class III peroxidases and ascorbate metabolism in plants. Phytochemistry Rev 3:195–205

de Pinto MC, de Gara L (2004) Changes in the ascorbate metabolism of apoplastic and symplastic spaces are associated with cell differentiation. J Exp Bot 55:2559–2569

de Tullio MC, Liso R, Arrigoni O (2004) Ascorbic acid oxidase: an enzyme in search of a role. Biol Plantarum 48:161–166

Demura T, Tashiro G, Horiguchi G, Kishimoto N, Kubo M, Matsuoka N, Minami A, Nagata-Hiwatashi M, Nakamura K, Okamura Y, Sassa N, Suzuki S, Yazaki J, Kikuchi S, Fukuda H (2002) Visualization by comprehensive microarray analysis of gene expression programs during transdifferentiation of mesophyll cells into xylem cells. Proc Natl Acad Sci USA 99:15794–15799

Ferrer MA, Pedreño MA, Muñoz R, Ros Barceló A (1990) Oxidation of coniferyl alcohol by cell wall peroxidases at the expense of indole-3-acetic acid and O_2. FEBS Lett 276:127–130

Floris R, Piersma, SR, Yang G, Jones P, Wever R (1993) Interaction of myeloperoxidase with peroxynitrite. A comparison with lactoperoxidase, horseradish peroxidase and catalase. Eur J Biochem 215:767–775

Foreman J, Demidchik V, Bothwell JHF, Mylona P, Miedema H, Torres MA, Linstead P, Costa S, Brownlee C, Jones JDG, Davies JM, Dolan L (2003) Reactive oxygen species produced by NADPH oxidase regulate plant cell growth. Nature 422:442–446

Foyer CH, Halliwell B (1976) The presence of glutathione and glutathione reductase in chloroplasts: a proposed role in ascorbic acid metabolism. Planta 133:21–25

Frahry G, Schopfer P (1998) Inhibition of O_2-reducing activity of horseradish peroxidase by diphenyleniodonium. Phytochemistry 48:223–227

Fry SC (1998) Oxidative scission of plant cell wall polysaccharides by ascorbate-induced hydroxyl radicals. Biochem J 332:507–515

Fry SC (2004a) Primary cell wall metabolism: tracking the careers of wall polymers in living plant cells. New Phytol 161:641–675

Fry SC (2004b) Oxidative coupling of tyrosine and ferulic acid residues: intra- and extra-protoplasmic occurrence, predominance of trimers and larger products, and possible role in inter-polymeric cross-linking. Phytochem Rev 3:97–111

Gabaldón C, Gómez Ros LV, Pedreño MA, Ros Barceló A (2005) Nitric oxide production by the differentiating xylem of *Zinnia elegans*. New Phytol 165:121–130

Gómez Ros LV, Paradiso A, Gabaldón C, Pedreño MA, de Gara L, Ros Barceló A (2006) Two distinct cell sources of H_2O_2 in the lignifying *Zinnia elegans* cell culture system. Protoplasma 227:175–183

Green MA, Fry SC (2005) Vitamin C degradation in plant cells via enzymatic hydrolysis of 4-*O*-oxalyl-L-threonate. Nature 433:83–87

Hancock J, Desikan R, Harrison J, Bright J, Hooley R, Neill S (2006) Doing the unexpected: proteins involved in hydrogen peroxide perception. J Exp Bot 57:1711–1718

Henmi K, Demura T, Tsuboi S, Fukuda H, Iwabuchi M, Ogawa K (2005) Change in the redox state of glutathione regulates differentiation of tracheary elements in *Zinnia* cells and *Arabidopsis* roots. Plant Cell Physiol 46:1757–1765

Hernández JA, Ferrer MA, Jiménez A, Ros Barceló A, Sevilla F (2001) Antioxidant systems and O_2^-/H_2O_2 production in the apoplast of pea leaves. Its relation with salt-induced necrotic lesion in minor veins. Plant Physiol 127:817–831

Horemans N, Foyer Ch, Asard H (2000) Transport and action of ascorbate at the plasma membrane. Trends Plant Sci 5:263–266

Jamet E, Canut H, Boudart G, Pont-Lezica RF (2006) Cell wall proteins: a new insight through proteomics. Trends Plant Sci 11:33–39

Kärkönen A, Fry SC (2006) Effect of ascorbate and its oxidation products on H_2O_2 production in cell-suspension cultures of *Picea abies* and in the absence of cells. J Exp Bot 57:1633–1644

Karlsson M, Melzer M, Prokhorenko I, Johansson T, Wingsle G (2005) Hydrogen peroxide and expression of hipI-superoxide dismutase are associated with the development of secondary cell walls in *Zinnia elegans*. J Exp Bot 56:2085–2093

Kim HS, Yu Y, Snesrud EC, Moy LP, Linford LD, Haas BJ, Nierman WC, Quackenbush J (2005) Transcriptional divergence of the duplicated oxidative stress-responsive genes in the *Arabidopsis* genome. Plant J 41:212–220

Kubo M, Udagawa M, Nishikubo N, Horiguchi G, Yamaguchi M, Ito J, Mimura T, Fukuda H, Demura T (2005) Transcription switches for protoxylem and metaxylem vessel formation. Gene Dev 19:1855–1860

Kumar GNM, Iyer S, Knowles NR (2007) *Strboh* A homologue of NADPH oxidase regulates wound-induced oxidative burst and facilitates wound-healing in potato tubers. Planta 227:25–36

Lane BG (1994) Oxalate, germin, and the extracellular matrix of higher plants. FASEB J 8:294–301

Liszkay A, Kenk B, Schopfer P (2003) Evidence for the involvement of cell wall peroxidase in the generation of hydroxyl radicals mediating extension growth. Planta 217:658–667

Liu L, Eriksson KEL, Dean JFD (1995) Localization of hydrogen peroxide production in *Pisum sativum* L. using epi-polarization microscopy to follow cerium perhydroxide deposition. Plant Physiol 107:501–506

Moller SG, McPherson MJ (1998) Developmental expression and biochemical analysis of the *Arabidopsis atao*1 gene encoding an H_2O_2-generating diamine oxidase. Plant J 13:781–791

Nakanomyo I, Kost B, Chua NH, Fukuda H (2002) Preferential and asymmetrical accumulation of a Rac small GTPase mRNA in differentiating xylem cells of *Zinnia elegans*. Plant Cell Physiol 43:1484–1492

Neill S, Desikan R, Hancock J (2002) Hydrogen peroxide signalling. Curr Opin Plant Biol 5:388–395

Ogawa K, Kanematsu S, Asada K (1997) Generation of superoxide anion and localization of CuZn-superoxide dismutase in the vascular tissue of spinach hypocotyls: their association with lignification. Plant Cell Physiol 38:1118–1126

Olson PD, Varner JE (1993) Hydrogen peroxide and lignification. Plant J 4:887–892

Papadakis AK, Roubelakis-Angelakis KA (2005) Polyamines inhibit NADPH oxidase-mediated superoxide generation and putrescine prevents programmed cell death induced by polyamine oxidase-generated hydrogen peroxide. Planta 220:826–837

Passardi F, Penel C, Dunand C (2004) Performing the paradoxical: how plant peroxidasas modify the cell wall. Trends Plant Sci 9:534–540

Pignocchi C, Foyer CH (2003) Apoplastic ascorbate metabolism and its role in the regulation of cell signalling. Curr Opinion Plant Biol 6:379–389

Pignocchi C, Fletcher JM, Wilkinson JE, Barnes JD, Foyer CH (2003) The function of ascorbate oxidase in tobacco. Plant Physiol 132:1631–1641

Pitzschke A, Hirt H (2006) Mitogen-activated protein kinases and reactive oxygen species signaling in plants. Plant Physiol 141:351–356

Ralph J, Bunzel M, Marita MJ, Hatfield RD, Fachuang Lu, Kim H, Schatz PF, Grabber JH, Steinhart H (2004a) Peroxidase-dependent cross-linking reactions of *p*-hydroxycinnamates in plant cell walls. Phytochem Rev 3:79–96

Ralph J, Lundquist K, Brunow G, Lu F, Kim H, Schatz PF, Marita JM, Hatfield RD, Ralph SA, Christensen JH, Boerjan W (2004b) Lignins: Natural polymers from oxidative coupling of 4-hydroxyphenylpropanoids. Phytochem Rev 3:29–60

Razem FA, Bernards MA (2003) Reactive oxygen species production in association with suberization: evidence for an NADPH-dependent oxidase. J Exp Bot 54:935–941

Rodríguez-Serrano M, Romero-Puertas MC, Zabalza A, Corpas FJ, Gómez M, del Río LA, Sandalio LM (2006) Cadmium effect on oxidative metabolism of pea (*Pisum sativum* L.) roots. Imaging of reactive oxygen species and nitric oxide accumulation in vivo. Plant Cell Environ 29:1532–1544

Ros Barceló A (1997) Lignification in plant cell walls. Int Rev Cytol 176:87–132

Ros Barceló A (1998a) Hydrogen peroxide production is a general property of the lignifying xylem from vascular plants. Ann Bot 82:97–103

Ros Barceló A (1998b) Use and misuse of peroxidase inhibitors. Trends Plant Sci 3:418

Ros Barceló A (1998c) The generation of H_2O_2 in the xylem of *Zinnia elegans* is mediated by an NADPH-oxidase-like enzyme. Planta 207:207–216

Ros Barceló A (1999) Some properties of the H_2O_2/O_2^- generating system from the lignifying xylem of *Zinnia elegans*. Free Rad Res 31:S147–S154

Ros Barceló A (2000) Peroxidase and H_2O_2 production by plant cells: truths and clues. Curr Top Phytochem 3:197–202

Ros Barceló A (2005) Xylem parenchyma cells deliver the H_2O_2 necessary for lignification in differentiating xylem vessels. Planta 220:747–756

Ros Barceló A, Ferrer MA (1999) Does diphenylene iodonium chloride have any effect on the O_2^--generating step of plant peroxidases? FEBS Lett 462:254–256

Ros Barceló A, Pomar F, López-Serrano M, Martínez P, Pedreño MA (2002) Developmental regulation of the H_2O_2-producing system and of a basic peroxidase isoenzyme in the *Zinnia elegans* lignifying xylem. Plant Physiol Biochem 40:325–332

Ros Barceló A, Gómez Ros LV, Ferrer MA, Hernández JA (2006) The apoplastic antioxidant enzymatic system in the wood-forming tissues of trees. Trees 20:145–156

Ros Barceló A, Gómez Ros LV, Esteban Carrasco A (2007) Looking for syringyl peroxidases. Trends in Plant Sci 12:486–491

Schopfer P (1994) Histochemical demonstration and localization of H_2O_2 in organs of higher plants by tissue printing on nitrocellulose paper. Plant Physiol 104:1269–1275

Schopfer P (2001) Hydroxyl radical-induced cell-wall loosening in vitro and in vivo: implications for the control of elongation growth. Plant J 28:679–688

Sebela M, Radova A, Angelini R, Tavladoraki P, Frebort I, Pec P (2001) FAD-containing polyamine oxidases: a timely challenge for researchers in biochemistry and physiology of plants. Plant Sci 160:197–207

Smith TA (1985) The di- and polyamine oxidase of higher plants. Biochem Soc Trans 13:319–322

Srivastava V, Schinkel H, Witzell J, Hertzberg M, Torp M, Srivastava MK, Karpinska B, Melzer M, Wingsle G (2007) Downregulation of high-isoelectric-point extracellular superoxide dismutase mediates alterations in the metabolism of reactive oxygen species and developmental disturbances in hybrid aspen. Plant J 49:135–148

Takahama U (1993) Regulation of peroxidase-dependent oxidation of phenolics by ascorbic acid: different effects of ascorbic acid on the oxidation of coniferyl alcohol by the apoplastic soluble and cell wall-bound peroxidases from epicotyls of *Vigna angularis*. Plant Cell Physiol 34:809–817

Vanacker H, Harbinson J, Ruisch J, Carver TLW, Foyer CH. (1998) Antioxidant defences of the apoplast. Protoplasma 205:129–140

Weir IE, Maddumage R, Allan AC, Ferguson IB (2005) Flow cytometric analysis of tracheary element differentiation in *Zinnia elegans* cells. Cytometry 68A:81–91

Wojtaszek P, Trethowan J, Bolwell P (1997) Reconstitution in vitro of the components and conditions required for the oxidative cross-linking of extracellular proteins in French bean (*Phaseolus vulgaris* L.). FEBS Lett 405:95–98

Wong HL, Pinontoan R, Hayashi K, Tabata R, Yaeno T, Hasegawa K, Kojima C, Yoshioka H, Iba K, Kawasaki T, Shimamoto K (2007) Regulation of rice NADPH oxidase by binding of Rac GTPase to its N-terminal extension. Plant Cell 19:4022–4034

Yamazaki I, Yokota KN (1973) Oxidation states of peroxidase. Mol Cell Biochem 2:39–52
Yoshioka H, Sugie K, Park H-J, Maeda H, Tsuda N, Kawakita K, Doke N (2001) Induction of plant gp91 *phox* homolog by fungal cell wall, arachidonic acid, and salicylic acid in potato. Mol Plant Microbe In 14:725–736
Zhang XG, Coté GG, Crain RC (2002) Involvement of phosphoinositide turnover in tracheary element differentiation in *Zinnia elegans* L. cells. Planta 215:312–318

Peroxisomes as a Cellular Source of ROS Signal Molecules

Luis A. del Río, Luisa M. Sandalio, Francisco J. Corpas, María C. Romero-Puertas, and José M. Palma

Abstract Peroxisomes are subcellular organelles with an essentially oxidative type of metabolism and devoid of DNA, and are probably the major sites of intracellular H_2O_2 production. Like mitochondria and chloroplasts, peroxisomes also produce superoxide radicals ($O_2\bullet^-$) and there are, at least, two sites of superoxide generation: one in the organelle matrix, the generating system being xanthine oxidase, and another site in the peroxisomal membranes dependent on NAD(P)H. In peroxisomal membranes, three integral polypeptides (PMPs) have been shown to generate $O_2\bullet^-$ radicals. Besides catalase, several antioxidative enzymes have been demonstrated in plant peroxisomes, including different superoxide dismutases, the enzymes of the ascorbate–glutathione cycle plus ascorbate and glutathione, and three NADP-dependent dehydrogenases. The presence of nitric oxide synthase (NOS) activity and its reaction product, nitric oxide (NO), has been demonstrated in plant peroxisomes. These organelles have a ROS-mediated cellular function in leaf senescence and in stress situations induced by xenobiotics and heavy metals, and can have an important role in plant cells as a source of signal molecules like $O_2\bullet^-$ radicals, H_2O_2, NO and *S*-nitrosoglutathione (GSNO).

1 Introduction

Peroxisomes are subcellular organelles bounded by a single membrane and devoid of DNA that contain as basic enzymatic constituents catalase and hydrogen peroxide (H_2O_2)-producing flavin oxidases, and occur in almost all eukaryotic cells (Huang et al. 1983; Fahimi and Sies 1987; Baker and Graham 2002). At the beginning of the 1960s, when these organelles were first isolated and characterized from

L.A. del Río (✉), L.M. Sandalio, F.J. Corpas, M.C. Romero-Puertas, and J.M. Palma
Departamento de Bioquímica, Biología Celular y Molecular de Plantas,
Estación Experimental del Zaidín, Consejo Superior de Investigaciones
Científicas (CSIC), Apartado 419, E-18080 Granada, Spain
e-mail: luisalfonso.delrio@eez.csic.es

L.A. del Río and A. Puppo (eds.), *Reactive Oxygen Species in Plant Signaling*,
Signaling and Communication in Plants, DOI 10.1007/978-3-642-00390-5_6,

mammalian tissues (De Duve and Baudhuin 1966), it was thought that their main function was the removal by catalase of toxic hydrogen peroxide generated by different oxidases. However, at present it is known that peroxisomes are involved in a range of important cellular functions in almost all eukaryotic cells (Minorsky 2002; Baker and Graham 2002; del Río et al. 2002a, 2003, 2006; Igamberdiev and Lea 2002; Titorenko and Rachubinski 2004; Corpas et al. 2009). Table 1 shows different functions that have been described so far for peroxisomes in plant cells. Apart from these functions, plant peroxisomes also play a role in the degradation of branched amino acids and the production of the compatible osmosolute Gly betaine (Reumann et al. 2004). Moreover, evidence is emerging for the existence of regulatory proteins in peroxisomes, like heat shock proteins, kinases, and phosphatases (Hayashi and Nishimura 2006; Reumann et al. 2007).

The peroxisome of plant cells is a highly dynamic compartment that is dependent upon the actin cytoskeleton, not microtubules, for its subcellular distribution and movements (Mano et al. 2002; Hashimoto et al. 2005). In plants, there are several types of peroxisomes which are specialized in certain metabolic functions. Glyoxysomes are specialized peroxisomes, occurring in the storage tissue of oilseeds, that contain the fatty acid β-oxidation and all the glyoxylate cycle enzymes to convert the seed reserve lipids into sugars which are used for germination and plant growth (Huang et al. 1983; Baker and Graham 2002). Leaf peroxisomes are specialized peroxisomes present in photosynthetic tissues that carry out the major reactions of photorespiration (Tolbert et al. 1987; Douce and Heldt 2000). Another type of specialized peroxisomes are root-nodule peroxisomes from certain tropical legumes, in which the synthesis of allantoin, the major metabolite for nitrogen transport within these plants, is carried out (Schubert 1986). In more recent years, by transcriptomic analysis of 286 peroxisomal genes present in the *Arabidopsis* genome, cotyledonary peroxisomes were proposed as a new class of peroxisomes (Arai et al. 2008).

A characteristic property of peroxisomes is their oxidative type of metabolism and metabolic plasticity, because their enzymatic content can vary depending on the organism, cell/tissue-type and environmental conditions (Fahimi and Sies 1987; Van den Bosch et al. 1992; Baker and Graham 2002). An example of the inducible

Table 1 Main functions described for peroxisomes in plant cells

Function	Reference
Photorespiration	Tolbert et al. (1987); Douce and Heldt (2000)
β-Oxidation of fatty acids	Huang et al. (1983); Baker and Graham (2002)
Glyoxylate cycle	Huang et al. (1983); Baker and Graham (2002)
Metabolism of ureides	Schubert (1986); Baker and Graham (2002)
Metabolism of ROS and RNS	del Río et al. (2002a, 2006)
Biosynthesis of phytohormones (auxin, jasmonic acid, salicylic acid)	Reumann et al. (2007); Nyathi and Baker (2006)
Oxidative reactions of senescence	del Río et al. (1998)
Photomorphogenesis	Desai and Hu (2008)
Sulfite detoxification	Hänsch et al. (2006)
Defense against pathogens	Koh et al. (2005); Kuźniak and Skłodowska (2005)

nature of peroxisomal metabolism is the light-induced transition of glyoxysomes, the specialized peroxisomes of oilseeds, to leaf-type peroxisomes (Huang et al. 1983; Mullen and Trelease 1996). During the senescence of leaves, the reverse process is observed, whereby leaf peroxisomes are converted into glyoxysomes (Vicentini and Matile 1993; Nishimura et al. 1996; del Río et al. 1998). These metabolic transitions are also observed in plants under conditions of abiotic stress by cadmium where this heavy metal causes leaf peroxisomes to adopt a glyoxysome-type metabolism (McCarthy et al. 2001).

Peroxisome proliferation and the induction of some peroxisomal β-oxidation enzymes was first documented in the livers of rats exposed to a variety of xenobiotics and subsequently found in other mammalian species (Reddy et al. 1982). In plants, the cellular population of peroxisomes can proliferate during senescence and under different stress conditions produced by xenobiotics, ozone, cadmium, H_2O_2 (del Río et al. 1998, 2002a, b; Romero-Puertas et al. 1999; Nila et al. 2006; Castillo et al. 2008), and by light (Desai and Hu 2008). Peroxisome proliferator-activated receptor (PPAR), the transcription factor involved in peroxisomal proliferation and induction of peroxisomal fatty acid β-oxidation in animal tissues, was recently expressed in tobacco plants. In the transgenic plants obtained, PPARα from frog (*Xenopus laevis*) was functional and its expression in tobacco led to changes in general lipid metabolism and induced the proliferation of peroxisomes, as reported in animal tissues (Nila et al. 2006).

In the last two decades, different experimental evidence has been obtained indicating the existence of cellular functions for plant peroxisomes related to reactive oxygen species. In this chapter, the production of ROS, the different antioxidant systems, the generation of nitric oxide and the function of peroxisomes in oxidative stress will be analyzed in the context of a new cellular function for peroxisomes as a source of ROS signal molecules.

2 Production of Reactive Oxygen Species in Peroxisomes

In plant cells, as in most eukaryotic organisms, peroxisomes are probably the major sites of intracellular H_2O_2 production, as a result of their essentially oxidative type of metabolism. The main metabolic processes responsible for the generation of H_2O_2 in different types of peroxisomes are the photorespiratory glycolate oxidase reaction, the β-oxidation of fatty acids, the enzymatic reaction of flavin oxidases, and the disproportionation or dismutation of superoxide radicals (Baker and Graham 2002; del Río et al. 2002a; Foyer and Noctor 2003).

In plants, the production of reactive oxygen species (ROS) has been demonstrated mainly in chloroplasts, mitochondria, plasma membrane, and the apoplastic space (Bolwell 1999; Mittler 2002; Halliwell and Gutteridge 2007). But during the last decade, it has been demonstrated that peroxisomes, like mitochondria and chloroplasts, also produce superoxide radicals ($O_2{\bullet}^-$) as a consequence of their normal metabolism. In peroxisomes from pea leaves and watermelon cotyledons, both biochemical and electron spin resonance spectroscopy (ESR) methods showed the existence of, at least, two sites of $O_2•^-$ generation: one in the organelle matrix, in which the generating systems was identified as xanthine oxidase (XOD), and another site in the peroxisomal membranes dependent on NAD(P)H (Sandalio et al. 1988; del Río et al. 1989, 2002a). Xanthine oxidase catalyses the oxidation of xanthine and hypoxanthine to uric acid and is a well-known producer of superoxide radicals (Fridovich 1995). Very recently, the peroxisomal xanthine oxidase from pea leaves has been characterized and the presence of this enzyme in peroxisomes has been confirmed by immunogold electron microscopy (Corpas et al. 2008). It is known that the conversion of xanthine dehydrogenase (XDH) into the $O_2•^-$ producing xanthine oxidase (XOD) can be carried out irreversibly by a proteolytic cleavage (Palma et al. 2002). Experiments of incubation of peroxisomal matrices from pea leaves with microbial xanthine oxidoreductase indicated that peroxisomal endoproteases (EPs) could carry out the irreversible conversion of XDH into the superoxide-generating XOD (Distefano et al. 1999). This result indicates that peroxisomal EPs could take part in a regulated modification of proteins which do not necessarily imply their complete degradation.

The other site of $O_2•^-$ production is the peroxisomal membrane, where a small electron transport chain appears to be involved. This is composed of a flavoprotein NADH:ferricyanide reductase of about 32 kDa and a cytochrome *b* (Bowditch and Donaldson 1990). The integral peroxisomal membrane polypeptides (PMPs) of pea leaf peroxisomes were identified by SDS-PAGE and three of these membrane polypeptides with molecular masses of 18, 29, and 32 kDa have been characterized and demonstrated to be responsible for $O_2•^-$ generation (López-Huertas et al. 1997, 1999). The main producer of superoxide radicals in the peroxisomal membrane was the 18-kDa PMP, which was proposed to be a cytochrome *b* (López-Huertas et al. 1997). While the 18- and 32-kDa PMPs use NADH as electron donor for $O_2•^-$ production, the 29-kDa PMP was clearly dependent on NADPH, and was able to reduce cytochrome *c* with NADPH as electron donor (López-Huertas et al. 1999; del Río et al. 2002a). On the basis of its biochemical and immunochemical properties, the PMP32 very probably corresponds to the monodehydroascorbate reductase (MDAR) (López-Huertas et al. 1999) whose activity was previously detected in pea leaf peroxisomal membranes (Jiménez et al. 1997). The third $O_2•^-$-generating polypeptide, PMP29, could be related to the peroxisomal NADPH:cytochrome P450 (López-Huertas et al. 1999).

3 Antioxidant Systems in Peroxisomes

3.1 Superoxide Dismutases

Besides catalase, a complex battery of antioxidative systems has been demonstrated in plant peroxisomes. These include different superoxide dismutases (SODs) a family of metalloenzymes that catalyze the disproportionation of $O_2•^-$ radicals into H_2O_2 and O_2, and protect cells against the toxic effects of superoxide radicals produced in different cellular loci (Fridovich 1995; Halliwell and Gutteridge 2007). SODs are distributed mainly in chloroplasts, cytoplasm, mitochondria, apoplast, and nuclei (Ogawa et al. 1996; Alscher et al. 2002; Corpas et al. 2006a; Rodríguez-Serrano et al. 2007), but the presence of SOD in peroxisomes was demonstrated for the first time in plant tissues, using immunoelectron microscopy and cell biology methods (del Río et al. 1983; Sandalio et al. 1987). Since then, the occurrence of SODs in plant peroxisomes has been reported in, at least, ten different plant species including pea, watermelon, carnation, castor bean, sunflower, cucumber, cotton, tomato, olive, and pepper (del Río et al. 2002a; Mateos et al. 2003). Results obtained concerning the presence of SOD in plant peroxisomes were extended years later to human and animal peroxisomes (see del Río et al. 2002a). Three SODs of peroxisomal origin have been purified and characterized, a CuZn-SOD and a Mn-SOD from watermelon cotyledons and a Mn-SOD from pea leaves (Bueno et al. 1995; Palma et al. 1998; Rodríguez-Serrano et al. 2007). However, the genes encoding the peroxisomal SODs have not yet been characterized.

3.2 Ascorbate–Glutathione Cycle

The ascorbate–glutathione cycle, also called Foyer–Halliwell–Asada cycle, is a way used by plant cells to dispose of H_2O_2 particularly in cellular compartments where no catalase is present (Halliwell and Gutteridge 2007; Noctor and Foyer 1998). This cycle occurs in chloroplasts, cytoplasm, and mitochondria (Noctor and Foyer 1998), but has also been demonstrated in peroxisomes. The four enzymes of the cycle, ascorbate peroxidase (APX), monodehydroascorbate reductase (MDAR), dehydroascorbate reductase (DAR), and glutathione reductase (GR), were present in peroxisomes purified from pea leaves (Jiménez et al. 1997) and years later were also found in peroxisomes from tomato leaves and roots (Mittova et al. 2004; Kużniak and Skłodowska 2005). Likewise, in intact peroxisomes, the presence of reduced ascorbate and glutathione and their oxidized forms was demonstrated by HPLC analysis (Jiménez et al. 1997). The intraperoxisomal distribution of the ascorbate–glutathione cycle was studied in pea leaves, and DHAR and GR were found in the soluble fraction of peroxisomes, whereas APX activity was bound to the cytosolic side of the peroxisomal membrane (Jiménez et al. 1997). The peroxisomal GR of pea leaves has been recently purified and characterized (Romero-Puertas et al. 2006).

MDAR was localized in both the membrane and matrix of these organelles (López-Huertas et al. 1999; Leterrier et al. 2005), and in pea and *Arabidopsis*, the peroxisomal MDAR was found to contain a PTS type 1 (Leterrier et al. 2005; Lisenbee et al. 2005). It has been proposed that the *trans*-membrane protein MDAR can oxidize NADH on the matrix side of the peroxisomal membrane and transfer the reducing equivalents as electrons to the acceptor MDA on the cytoplasmic side of the membrane (Bowditch and Donaldson 1990). In this process, O_2 could also act as an electron acceptor, with the concomitant formation of $O_2{\bullet}^-$ radicals (López-Huertas et al. 1999). The presence of APX and MDAR in leaf peroxisomal membranes could serve to reoxidize endogenous NADH to maintain a constant supply of NAD^+ for peroxisomal metabolism but could also have a protective role against H_2O_2 leaking from peroxisomes. H_2O_2 can easily permeate the peroxisomal membrane, but APX would degrade leaking H_2O_2, as well as the H_2O_2 that is being continuously formed by dismutation of the $O_2{\bullet}^-$ generated in the NAD(P)H-dependent electron transport system of the peroxisomal membrane, and MDAR would regenerate the reduced form of ascorbate to be used by APX (López-Huertas et al. 1999; del Río et al. 2002a). This membrane scavenging of H_2O_2 could prevent an increase in the cytoplasmic concentration of this metabolite, particularly under certain plant stress situations, when the level of H_2O_2 produced in peroxisomes can be substantially increased (del Río et al. 1996).

3.3 NADP–Dependent Dehydrogenases

Different studies carried out in recent years have shown that the NADPH-generating dehydrogenases have an important function in the mechanism of protection against oxidative stress of bacteria, yeast, and mammalian and plant cells (Martini and Ursini 1996; Lee et al. 2002; Valderrama et al. 2006). In plant cells, NADPH has an important role in the protection systems against oxidative stress due to its involvement in the ascorbate–glutathione cycle and water-water cycle (Noctor and Foyer 1998; Asada 2006). This evidence has supported the notion of NADP-dependent dehydrogenases as antioxidative enzymes which can be included in the group of catalase, SOD, APX and GR/peroxidase (Corpas et al. 1999).

In isolated plant peroxisomes, the presence of three NADP-dehydrogenases has been demonstrated, including glucose-6-phosphate dehydrogenase (G6PDH), 6-phosphogluconate dehydrogenase (6PGDH), and isocitrate dehydrogenase (ICDH) (Corpas et al. 1999; Valderrama et al. 2006). The presence in peroxisomes of these dehydrogenases implies that these organelles have the capacity to reduce $NADP^+$ to NADPH for its reutilization in their metabolism. NADPH is necessary for the function of the NADPH:cytochrome P450 reductase, whose presence has been detected in membranes of plant peroxisomes (Baker and Graham 2002). NADPH is also required for the GR activity of the ascorbate–glutathione cycle which recycles reduced glutathione (GSH) from its oxidized form (GSSG) to be used for the scavenging of H_2O_2 in the cycle (Noctor and Foyer 1998). NADPH has been reported to protect catalase from oxidative damage (Kirkman et al. 1999) and is also necessary for the peroxisomal NO-producing activity, nitric oxide synthase (NOS) (Corpas et al. 2004a), as well as for the reduction of double bonds of unsaturated fatty acids by 2,4-dienoyl-CoA reductase (Reumann et al. 2007).

3.4 Peroxiredoxins

Peroxiredoxins (Prxs) are a family of thiol-specific antioxidant enzymes, also known as thioredoxin-dependent peroxidases, which are present in bacteria, yeasts, plants, and mammals (Dietz 2003). The antioxidative functions of Prxs consists in the reduction and, therefore detoxification, of H_2O_2, peroxynitrite and different organic hydroperoxides (Dietz 2003; Wood et al. 2003). Although Prxs are mainly present in the cytoplasm, they also occur in mitochondria, chloroplasts, and nuclei (Dietz 2003). Using immunogold electron microscopy, a putative Prx with an atypical molecular mass of 60 kDa was localized in the matrix of pea leaf peroxisomes (Corpas et al. 2003), and in mammalian cells a Prx (Prx V) was also localized in these organelles (Seo et al. 2000). Further research is necessary to characterize this putative peroxisomal Prx. The localization of Prxs in peroxisomes would supply these organelles with an additional antioxidant enzyme system to control the peroxisomal level of H_2O_2 and also to remove peroxynitrite that could be formed in peroxisomes (see Sect. 4).

4 Nitric Oxide and Nitric Oxide Synthase Activity in Peroxisomes

In plants, nitric oxide (NO) can be produced by non-enzymatic systems and by enzymatic sources, mainly including nitrate reductase and L-arginine-dependent nitric oxide synthase-like activities (Rockel et al. 2002; del Río et al. 2004; Corpas et al. 2006b; Neill et al. 2008). The enzyme nitric oxide synthase (NOS; EC 1.14.13.39) catalyzes the oxygen- and NADPH-dependent oxidation of L-arginine to NO and citrulline in a complex reaction requiring FAD, FMN, tetrahydrobiopterin (BH_4), calmodulin, and calcium (Alderton et al. 2001). At present, there is abundant biochemical evidence showing the presence of L-arginine-dependent NOS activity in different plant species under physiological and stress conditions (Corpas et al. 2006b, 2007; Valderrama et al. 2006). The occurrence of this NO-producing activity in isolated peroxisomes was first demonstrated in plant tissues, in leaves of pea plants (Barroso et al. 1999; Corpas et al. 2004a). In olive leaves, pea leaves, and sunflower cotyledons, the occurrence of the enzyme in the matrix of peroxisomes was also demonstrated by immunocytochemistry (Barroso et al. 1999; Corpas et al. 2004b) and was ratified by laser confocal immunofluorescence microscopy (Corpas et al. 2004a). Following the localization of NOS activity in plant peroxisomes, the presence of NOS in animal peroxisomes (in hepatocytes) was also reported (Loughran et al. 2005).

The occurrence of nitric oxide as an endogenous metabolite in peroxisomes purified from pea leaves was demonstrated by fluorometric analysis, and by electron paramagnetic resonance (EPR) spectroscopy with the spin trap $Fe(MGD)_2$ (Corpas et al. 2004a). However, the identification of the gene(s) encoding the enzyme responsible for the NOS activity in plants remains to be achieved (Zemojtel et al. 2006; Neill et al. 2008).

The generation of NO and $O_2^{\bullet-}$ in peroxisomes suggests the possible formation of the powerful oxidant peroxynitrite ($ONOO^-$) in these organelles by reaction between both free radicals (Halliwell and Gutteridge 2007). This would be another reason to explain the presence of peroxiredoxin in leaf peroxisomes (see Sect. 3.3.4), since these thiol-dependent peroxidases can reduce and detoxify peroxynitrite, apart from H_2O_2 and different organic hydroperoxides (Wood et al. 2003).

5 Function of Peroxisomes in Oxidative Stress

In most biotic and abiotic stress conditions an overproduction of ROS has been demonstrated, and these species are thought to be responsible for the oxidative damage associated with plant stress (Dat et al. 2000; Mittler 2002; Møller et al. 2007). Conditions which induce the generation of ROS in plants mainly include pathogen infection, exposure to high light intensities, drought and salt stress, high- and low-temperature exposure, heavy metals, UV radiation, air pollutants, and physical and mechanical wounding (Dat et al. 2000; Mittler 2002; Desikan et al. 2001). In plants under normal physiological conditions, the production by peroxisomes of the ROS H_2O_2 and $O_2{\bullet}^-$ should be conveniently controlled by catalase and ascorbate peroxidase, and SOD, respectively, which are present in peroxisomes. But the risk of serious damage can arise when, under stress conditions, the peroxisomal generation of ROS is increased and the protective antioxidative systems of the organelle are depressed.

In leaf peroxisomes from pea plants subjected to stress conditions by xenobiotics, like the hypolipidemic drug clofibrate (ethyl-α-*p*-chlorophenoxyisobutyrate) and the herbicide 2,4-D (2,4-dichlorophenoxyacetic acid), an oxidative stress mechanism mediated by ROS was demonstrated to be involved (Palma et al. 1991; del Río et al. 2002b). Clofibrate also induced a proliferation of the peroxisomal population in leaves of pea, tobacco, and *Arabidopsis* plants (Palma et al. 1991; Nila et al. 2006; Castillo et al. 2008), a similar effect to that previously described in rodents by Reddy et al. (1982). In pea leaves sprayed with the herbicide 2,4-D, transcriptome analysis showed an induction of transcripts of the peroxisomal enzyme monodehydroascorbate reductase (*MDAR 1*) (Leterrier et al. 2005).

Peroxisomes appear to have a ROS-mediated function in the oxidative reactions characteristic of senescence. The senescence-induced changes in the reactive oxygen metabolism of peroxisomes are mainly characterized by the disappearance of catalase activity and an overproduction of $O_2{\bullet}^-$ and H_2O_2 and a strong decrease of APX and MDAR activities (del Río et al. 1998). Since $O_2{\bullet}^-$ radicals under physiological conditions quicky dismutate into H_2O_2 and O_2, the final result of senescence is a build-up in leaf peroxisomes of the more stable metabolite H_2O_2, which can diffuse into the cytoplasm. This represents a serious situation not only for peroxisomes but also for other cell organelles such as mitochondria, nuclei, and chloroplasts, due to the possible generation of the strongly oxidizing hydroxyl radicals (OH•) by the metal-catalyzed reaction of H_2O_2 with $O_2{\bullet}^-$ (Halliwell and Gutteridge 2007; Fridovich 1995). On the other hand, the endogenous proteases of peroxisomes are induced by senescence (Distefano et al. 1999) and they probably participate in the important metabolic changes that take place in these organelles as a result of senescence (McCarthy et al. 2001; Palma et al. 2002).

In the toxicity produced in pea plants by heavy metals, like cadmium and copper, leaf peroxisomes are also involved. In peroxisomes from Cu-tolerant plants, higher activities of the antioxidative enzymes Mn-SOD and catalase were found compared with Cu-sensitive plants (Palma et al. 1987). This suggests that Cu-tolerant plants could have evolved protection mechanisms against the peroxisomal generation of $O_2{\bullet}^-$-dependent toxic species. In leaf peroxisomes from pea plants treated with cadmium, an enhancement of the H_2O_2 concentration and the oxidative modification of some endogenous proteins was observed, which was parallel to a slight increase of the peroxisomal population of leaves (Romero-Puertas et al. 1999, 2002). Peroxisomes responded to cadmium toxicity by increasing the activity of antioxidative enzymes involved in the ascorbate–glutathione cycle and the NADP-dehydrogenases located in these organelles. Cadmium also increased the activity of the endogenous proteases and glyoxylate cycle enzymes, malate synthase and isocitrate lyase. The activity enhancement of these two enzymes suggests that Cd induces senescence symptoms in peroxisomes and, probably, a metabolic transition of leaf peroxisomes into glyoxysomes, with a participation of the peroxisomal proteases in all these Cd-induced metabolic changes (McCarthy et al. 2001; Palma et al. 2002).

In tobacco plants with 10% of wild-type catalase activity, stress by H_2O_2 showed that catalase was essential for maintaining the redox balance during oxidative stress (Willekens et al. 1997). In transformed *Arabidopsis* plants, a model was proposed whereby different stresses that generate H_2O_2 as a signaling molecule result in peroxisome proliferation by the upregulation of peroxisome biogenesis genes (*PEX*). This model establishes that the peroxisome proliferation by H_2O_2 might be a common mechanism of protection against oxidative stress, by making use of the antioxidants of peroxisomes (López-Huertas et al. 2000).

A ROS-dependent participation of plant peroxisomes in fungal infection has been proposed in tomato plants (Kuźniak and Skłodowska 2005). But peroxisomes could also have an important function in defense mechanisms conferring resistance to pathogen attack. It has been reported that the infection of *Arabidopsis* by oomycete pathogens promotes a redistribution of the Golgi bodies and peroxisomes that become preferentially localized at the infection site (Hardham 2007). In the *Arabidopsis* response to compatible fungal infections, epidermal peroxisomes appear to have a role in removing ROS generated at penetrating sites (Koh et al. 2005).

6 Peroxisomes as a Source of ROS and RNS Signal Molecules

Considering the superoxide radical generating systems and diverse antioxidants of peroxisomes, as well as the presence of L-arginine-dependent nitric oxide synthase (NOS) activity in these organelles, a model has been designed that shows that peroxisomes can release signal molecules, such as H_2O_2, $O_2{\bullet}^-$, NO and *S*-nitrosoglutathione (GSNO), to the cytoplasm (Fig. 1). Nitric oxide produced by the enzymatic reaction of NOS can react with superoxide radicals generated in the peroxisomal matrix by xanthine oxidase (XOD) to form the powerful oxidant peroxynitrite ($ONOO^-$), which can regulate the conversion of xanthine dehydrogenase into the superoxide-generating XOD (Sakuma et al. 1997).

Under normal physiological conditions, the production by peroxisomes of the ROS H_2O_2 and $O_2{\bullet}^-$ should be adequately controlled by catalase and ascorbate peroxidase, and SOD, respectively, which are present in peroxisomes. However, catalase is known to be inactivated by light and different stresses (Schäfer and

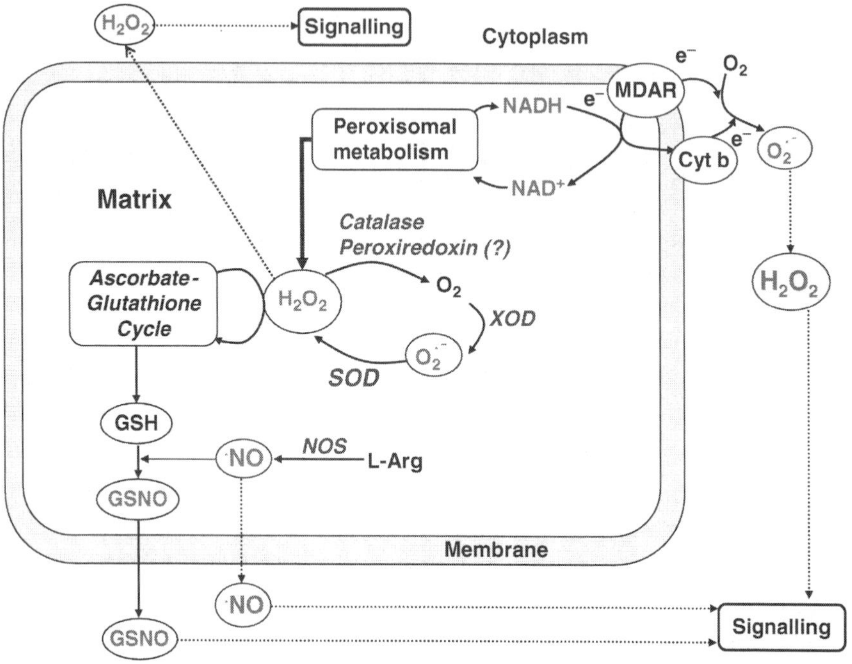

Fig. 1 Hypothetical model of the function of peroxisomes in the generation of the signal molecules H_2O_2, $O_2^{\bullet -}$, NO, and GSNO (modified from del Río et al. 2006). *MDAR* monodehydroascorbate reductase (PMP32), *Cyt b* a *b*-type cytochrome (PMP18), *GSH* reduced glutathione, *GSNO* *S*-nitrosoglutathione, *NOS* nitric oxide synthase, *PMP* peroxisomal membrane polypeptide, *SOD* superoxide dismutase, *XOD* xanthine oxidase, *L-Arg* L-arginine. *Broken arrows* indicate signaling

Feirabend 2000), and in these conditions an increase in the peroxisomal generation of H_2O_2 and $O_2^{\bullet -}$ can occur (del Río et al. 1996). An example of the accumulation of H_2O_2 in peroxisomes from plants treated with $CdCl_2$ is shown in Fig. 2. Narrow bands of cerium precipitates, due to H_2O_2, were observed in the membrane of peroxisomes in those sites in close contact with chloroplasts and tonoplast, suggesting a release of H_2O_2 into these cell compartments (Romero-Puertas et al. 2004).

Superoxide radicals are known to inhibit catalase activity (Fridovich 1995), and NO and peroxynitrite inhibit catalase and APX activity, the two major H_2O_2-scavenging enzymes of plant peroxisomes (Clark et al. 2000). In addition, in animal cells an enhanced synthesis of nitric oxide was found to increase the peroxisomal H_2O_2-producing β-oxidation. Taken together, these data indicate that if, under any type of plant biotic and/or abiotic stress an induction of the peroxisomal production of $O_2^{\bullet -}$ and NO radicals takes place, this can lead to the inhibition of catalase and APX activities, and possibly to an increase in the H_2O_2 level from the enhanced fatty acid β-oxidation. This breakdown of the peroxisomal antioxidant defenses would originate an overproduction of H_2O_2 in peroxisomes, leading to cellular oxidative damage and possibly cell death.

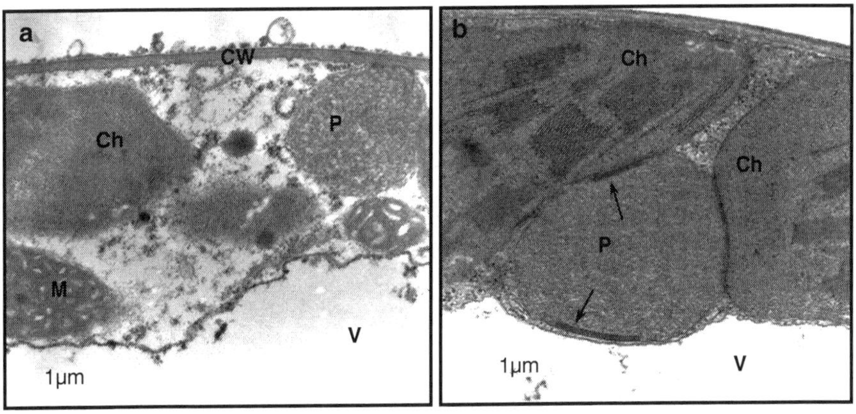

Fig. 2 Localization of H_2O_2 accumulation in leaf peroxisomes from Cd-treated plants. Pea plants were grown with 50 μM $CdCl_2$, and H_2O_2 was detected cytochemically by its reaction with $CeCl_3$ to produce electron-dense deposits of cerium perhydroxides. (**a**) Cell from control plants. (**b**) Cell from Cd-treated plants. In (**b**) the leaf tissue was treated with 1 mM aminotriazole, a catalase inhibitor, before $CeCl_3$ staining. *Ch* Chloroplast, *P* peroxisome, *M* mitochondrion, *V* vacuole. *Arrows* indicate $CeCl_3$ precipitates. *Bars* 1 μm (Romero-Puertas et al. 2004. Copyright 2004 Wiley-Blackwell Publishing)

Nevertheless, the rate of ROS and NO generation in plant cells can have an ambivalent effect. While a high cellular production of these active molecules can bring about extensive oxidative damage (Halliwell and Gutteridge 2007), low levels of NO and ROS, like H_2O_2 and $O_2{\bullet}^-$, are involved as signal molecules in many physiological processes, including the hypersensitive response to pathogens, growth, and development (Delledonne et al. 2001; Klessig et al. 2000; Desikan et al. 2001; del Río et al. 2002a; Neill et al. 2002; Gapper and Dolan 2006; see Chap. 4). Hydrogen peroxide has been described as a diffusible transduction signal in plant-pathogen interactions, response to wounding, stomatal closure, osmotic stress, and excess light stress, where H_2O_2 led to the induction of genes encoding different cellular protectants (Neill et al. 2002; del Río et al. 2002a; Mullineaux et al. 2006; Van Breusegem et al. 2008; see Chap. 4).

On the other hand, NO in the presence of O_2 can react with reduced glutathione (GSH), also present in peroxisomes (Jiménez et al. 1997), to form the reactive nitrogen species *S*-nitrosoglutathione (GSNO), a powerful inducer of defense genes that could function as a long-distance signal molecule, transporting glutathione-bound NO throughout the plant (Durner et al. 1998; Klessig et al. 2000). In this mechanism, leaf peroxisomes could participate through the endogenous production of GSNO which could diffuse to the cytoplasm. Accordingly, peroxisomes should be considered as cellular compartments with the capacity to generate and release into the cytoplasm important signal molecules such as $O_2{\bullet}^-$, H_2O_2, NO, and GSNO, which could contribute to a more integrated communication among cell compartments and tissues (Corpas et al. 2001). The idea that peroxisomes could be a cellular source of ROS (del Río and Donaldson 1995; del Río et al. 1996; Pastori and del Río 1997; Corpas et al. 2001) and RNS signal molecules (Corpas et al. 2001) was extended years later to other molecules synthesized in these organelles. In plants, peroxisomes are the only site for β-oxidation of fatty acids, and several signaling molecules derived from β-oxidation, which are generated in peroxisomes, have been suggested as possible signaling molecules. These mainly include jasmonic acid and its derivatives (methyl jasmonate, Z-jasmonate, and tuberonic acid), salicylic acid, and probably IAA (Nyathi and Baker 2006).

7 Conclusion

The existence of a reactive oxygen metabolism in plant peroxisomes and the presence in these organelles of a complex battery of antioxidative enzymes and the NO-generating activity nitric oxide synthase, demonstrate the importance of these organelles in cellular oxidative metabolism. The new enzymatic proteins recently found in plant peroxisomes show the role of these organelles as a pool of metabolites shared with other organelles, which supports the idea postulated two decades ago by Tolbert et al. (1987) on the effect of peroxisomal metabolism on metabolic pathways in other cell compartments.

Plant peroxisomes have a ROS- and RNS-mediated metabolic function in leaf senescence and in certain types of abiotic stress. For many years, mitochondria and chloroplasts have been considered almost exclusively responsible for the intracellular oxidative damage induced by different stresses, and the possible contribution of peroxisomes has been systematically neglected. However, peroxisomes can have a dual role, as oxidative stress-generators, and as indicators or sensors of plant stress and senescence by releasing ROS ($O_2^{\bullet -}$ and H_2O_2) and RNS (NO and GSNO) signaling molecules to the cytoplasm to participate in transduction pathways, which can trigger specific changes in defense gene expression. This signal molecule-producing function of plant peroxisomes is still more significant, from a physiological viewpoint, considering that the cellular population of these oxidative organelles can proliferate in plants during senescence and under different stress conditions. The cloning of the genes of peroxisomal ROS metabolism enzymes, and the characterization of mutants that are defective in stress responses, will throw more light on the cellular function of peroxisomes as a source of signaling molecules. It seems reasonable to think that a ROS and RNS signal molecule-producing function similar to that postulated for plant peroxisomes could also be performed by human, animal, and fungal peroxisomes. This would represent another example in modern biology where results initially obtained with plant cells were later found in animal cells.

Acknowledgments This work was supported by research projects from the Ministry of Education and Science (grants PB98-0493-01 and BFI 2002-04440-01), Spain, Research Training Network grants from the European Commission (contracts CHRX-CT94-0605 and HPRN-CT-2000-00094) and *Junta de Andalucía* (Group BIO-0192 and project P06-CVI-01820).

References

Alderton WK, Cooper CE, Knowles RG (2001) Nitric oxide synthases: structure, function and inhibition. Biochem J 357:593–615

Alscher RG, Erturk N, Heath LS (2002) Role of superoxide dismutases (SODs) in controlling oxidative stress in plants. J Exp Bot 53:1331–1341

Arai Y, Hayashi M, Nishimura M (2008) Proteomic analysis of highly purified peroxisomes from etiolated soybean cotyledons. Plant Cell Physiol 49:526–539

Asada K (2006) Production and scavenging of reactive oxygen species in chloroplasts and their functions. Plant Physiol 141:391–396

Baker A, Graham I (2002) Plant peroxisomes. Biochemistry, cell biology and biotechnological applications. Kluwer, Dordrecht, The Netherlands

Barroso JB, Corpas FJ, Carreras A, Sandalio LM, Valderrama R, Palma JM, Lupiáñez JA, del Río LA (1999) Localization of nitric-oxide synthase in plant peroxisomes. J Biol Chem 274:36729–36733

Bolwell GP (1999) Role of reactive oxygen species and NO in plant defence responses. Curr Opin Plant Biol 2:287–294

Bowditch MY, Donaldson RP (1990) Ascorbate free-radical reduction by glyoxysomal membranes. Plant Physiol 94:531–537

Bueno P, Varela J, Giménez-Gallego G, del Río LA (1995) Peroxisomal copper, zinc superoxide dismutase: characterization of the isoenzyme from watermelon cotyledons. Plant Physiol 108:1151–1160

Castillo MC, Sandalio LM, del Río LA, León J (2008) Peroxisome proliferation, wound-activated responses and expression of peroxisome-associated genes are cross-regulated but uncoupled in *Arabidopsis thaliana*. Plant Cell Environ 31:492–505

Clark D, Durner J, Navarre DA, Klessig DF (2000) Nitric oxide inhibition of tobacco catalase and ascorbate peroxidase. Mol Plant-Microbe Interact 13:1380–1384

Corpas FJ, Barroso JB, Sandalio LM, Palma JM, Lupiáñez JA, del Río LA (1999) Peroxisomal NADP-dependent isocitrate dehydrogenase. Characterization and activity regulation during natural senescence. Plant Physiol 121:921–928

Corpas FJ, Barroso JB, del Río LA (2001) Peroxisomes as a source of reactive oxygen species and nitric oxide signal molecules in plant cells. Trends Plant Sci 6:145–150

Corpas FJ, Pedrajas JR, Sandalio LM, León AM, Carreras A, Palma JM, Valderrama R, del Río LA, Barroso JB (2003) Localization of peroxiredoxin in peroxisomes from pea leaves. Free Radic Res 37(Suppl 2):19

Corpas FJ, Barroso JB, Carreras A, Quirós M, León AM, Romero-Puertas MC, Esteban FJ, Valderrama R, Palma JM, Sandalio LM, Gómez M, del Río LA (2004a) Cellular and subcellular localization of endogenous nitric oxide in senescent pea plants. Plant Physiol 136:2722–2733

Corpas FJ, Barroso JB, León AM, Carreras A, Quirós M, Palma JM, Sandalio LM, del Río LA (2004b) Peroxisomes as a source of nitric oxide. In: Magalhaes JR, Singh RP, Passos LP (eds) Nitric oxide signaling in higher plants. Studium Press, Houston, pp 111–129

Corpas FJ, Fernández-Ocaña A, Carreras A, Valderrama R, Luque F, Esteban FJ, Rodríguez-Serrano M, Chaki M, Pedrajas JR, Sandalio LM, del Río LA, Barroso JB (2006a) The expresión of different superoxide dismutase forms is cell-type dependent in olive (*Olea europaea* L.) leaves. Plant Cell Physiol 47:984–994

Corpas FJ, Barroso JB, Carreras A, Valderrama R, Palma JM, León AM, Sandalio LM, del Río LA (2006b) Constitutive arginine-dependent nitric oxide synthase activity in different organs of pea seedlings during plant development. Planta 224:246–254

Corpas FJ, Carreras A, Valderrama R, Chaki M, Palma JM, del Río LA, Barroso JB (2007a) Reactive nitrogen species and nitrosative stress in plants. Plant Stress 1:37–41

Corpas FJ, Palma JM, Sandalio LM, Valderrama R, Barroso JB, del Río LA (2008) Peroxisomal xanthine oxidoreductase: characterization of the enzyme from pea (*Pisum sativum* L.) leaves. J Plant Physiol 165:1319–1330

Corpas FJ, Barroso JB, Palma JM, del Río LA (2009) Peroxisomes as key organelles in the metabolism of reactive oxygen species, reactive nitrogen species and reactive sulphur species. In: Terlecky SR, Titorenko V (eds) Emergent functions of the peroxisome. Research Signpost, Kerala, India, pp 97–124

Dat J, Vandenabeele S, Vranová E, Van Montagu M, Inzé D, Van Breusegem F (2000) Dual action of the active oxygen species during plant stress response. Cell Mol Life Sci 57:779–795

De Duve C, Baudhuin P (1966) Peroxisomes (microbodies and related particles). Physiol Rev 46:323–357

Delledonne M, Zeier J, Marocco A, Lamb C (2001) Signal interactions between nitric oxide and reactive oxygen intermediates in the plant hypersensitive disease resistance response. Proc Natl Acad Sci USA 98:13454–13459

Desai M, Hu J (2008) Light induces peroxisome proliferation in *Arabidopsis* seedlings through the photoreceptor phytochrome A, the transcription factor HY5 HOMOLOG, and the peroxisomal protein PEROXIN11b. Plant Physiol 146:1117–1127

Desikan R, Mackerness SAH, Hancock JT, Neill SJ (2001) Regulation of the Arabidopsis transcriptome by oxidative stress. Plant Physiol 127:159–172

Dietz KJ (2003) Plant peroxiredoxins. Annu Rev Plant Biol 54:93–107

Distefano S, Palma JM, McCarthy I, del Río LA (1999) Proteolytic cleavage of plant proteins by peroxisomal endoproteases from senescent pea leaves. Planta 209:308–313

Douce R, Heldt HW (2000) Photorespiration. In: Leegood RC, Sharkey TD, von Cammerer S (eds) Photosynthesis: physiology and metabolism, Kluwer, Dordrecht, Holland, pp 115–136

Durner J, Wendehenne D, Klessig DF (1998) Defense gene induction in tobacco by nitric oxide, cyclic GMP, and cyclic ADP-ribose. Proc Natl Acad Sci USA 95:10328–10333

Fahimi HD, Sies H (eds) (1987) Peroxisomes in biology and medicine. Berlin, Springer-Verlag

Foyer CH, Noctor G (2003) Redox sensing and signaling associated with reactive oxygen in chloroplasts, peroxisomes and mitochondria. Physiol Plant 119:355–364

Fridovich I (1995) Superoxide radical and superoxide dismutases. Annu Rev Biochem 64:97–112

Gapper C, Dolan L (2006) Control of plant development by reactive oxygen species. Plant Physiol 141:341–345

Halliwell B, Gutteridge NM (2007) Free radicals in biology and medicine, 4th edition. Oxford University Press, Oxford

Hänsch R, Lang C, Riebeseel E, Lindigkeit R, Gessler A, Rennenberg H, Mendel RR (2006) Plant sulfite oxidase as novel producer of H_2O_2. J Biol Chem 281:6884–6888

Hardham AR (2007) Cell biology of plant-oomycete interactions. Cell Microbiol 9:31–39

Hashimoto K, Igarashi H, Mano S, Nishimura M, Shimmen T, Yokota E (2005) Peroxisomal localization of a myosin XI isoform in *Arabidopsis thaliana*. Plant Cell Physiol 46:782–789

Hayashi M, Nishimura M (2006) *Arabidopsis thaliana*–a model organism to study plant peroxisomes. Biochim Biophys Acta 1763:1382–1391

Huang AHC, Trelease RN, Moore Jr TS (1983) Plant peroxisomes. Academic, New York

Igamberdiev AU, Lea PJ (2002) The role of peroxisomes in the integration of metabolism and evolutionary diversity of photosynthetic organisms. Phytochemistry 60:651–674

Jiménez A, Hernández JA, del Río LA, Sevilla F (1997) Evidence for the presence of the ascorbate–glutathione cycle in mitochondria and peroxisomes of pea leaves. Plant Physiol 114:275–284

Kirkman HN, Rolfo M, Ferraris AM, Gaetani GF (1999) Mechanisms of protection of catalase by NADPH. Kinetics and stoichiometry. J Biol Chem 274:30451–30458

Klessig DF, Durner J, Noad R, Navarre DA, Wendehenne D, Kumar D, Zhou JM, Shah J, Zhang S, Kachroo P, Trifa Y, Pontier D, Lam E, Silva H (2000) Nitric oxide and salicylic acid signaling in plant defense. Proc Natl Acad Sci USA 97:8849–8855

Koh S, André A, Edwards H, Ehrhardt D, Somerville S (2005) *Arabidopsis thaliana* subcellular responses to compatible *Erysiphe cichoracearum* infections. Plant J 44:516–529

Kuźniak E, Skłodowska M (2005) Fungal pathogen-induced changes in the antioxidant systems of leaf peroxisomes from infected tomato plants. Planta 222:192–200

Lee SM, Koh HJ, Park DC, Song BJ, Huh TL, Park JW (2002) Cytosolic NADP-dependent isocitrate dehydrogenase status modulates oxidative damage to cells. Free Radic Biol Med 32:1185–1196
Leterrier M, Corpas FJ, Barroso JB, Sandalio LM, del Río LA (2005) Peroxisomal monodehydroascorbate reductase: genomic clone characterization and functional analysis under environmental stress conditions. Plant Physiol 138:2111–2123
Lisenbee CS, Lingard MJ, Trelease RN (2005) *Arabidopsis* peroxisomes possess functionally redundant membrane and matrix isoforms of monodehydroascorbate reductase. Plant J 43:900–914
López-Huertas E, Sandalio LM, Gómez M, del Río LA (1997) Superoxide radical generation in peroxisomal membranes: evidence for the participation of the 18 kDa integral membrane polypeptide. Free Radic Res 26:497–506
López-Huertas E, Corpas FJ, Sandalio LM, del Río LA (1999) Characterization of membrane polypeptides from pea leaf peroxisomes involved in superoxide radical generation. Biochem J 337:531–536
López-Huertas E, Charlton WL, Johnson B, Graham IA, Baker A (2000) Stress induces peroxisome biogenesis genes. EMBO J 19:6770–777
Loughran PA, Stolz DB, Vodovotz Y, Watkins SC, Simmons RL, Billiar TR (2005) Monomeric inducible nitric oxide synthase localizes to peroxisomes in hepatocytes. Proc Natl Acad Sci USA 102:13837–13842
Mano S, Nakamori C, Hayashi M, Kato A, Kondo M, Nishimura M (2002) Distribution and characterization of peroxisomes in *Arabidopsis* by visualization with GFP: dynamic morphology and actin-dependent movement. Plant Cell Physiol 43:331–341
Martini G, Ursini MV (1996) A new lease of life for an old enzyme. Bioessays 18:631–637
Mateos RM, León AM, Sandalio LM, Gómez M, del Río LA, Palma JM (2003) Peroxisomes from pepper fruits (*Capsicum annuum* L.): purification, characterisation and antioxidant activity. J Plant Physiol 160:1507–1516
McCarthy I, Romero-Puertas MC, Palma JM, Sandalio LM, Corpas FJ, Gómez M, del Río LA (2001) Cadmium induces senescence symptoms in leaf peroxisomes of pea plants. Plant Cell Environ 24:1065–1073
Minorsky PV (2002) Peroxisomes: organelles of diverse function. Plant Physiol 130:517–518
Mittler R (2002) Oxidative stress, antioxidants and stress tolerance. Trends Plant Sci 7:405–410
Mittova V, Guy M, Tal M, Volokita M (2004) Salinity up-regulates the antioxidative systems in root mitochondria and peroxisomes of the wild salt-tolerant tomato species *Lycopersicon Pennellii*. J Exp Bot 55:1105–1113
Møller IM, Jensen PE, Hansson A (2007) Oxidative modifications to cellular components in plants. Annu Rev Plant Biol 58:459–481
Mullen RT, Trelease RN (1996) Biogenesis and membrane properties of peroxisomes: does the boundary membrane serve and protect? Trends Plant Sci 1:389–394
Mullineaux PM, Karpinski S, Baker NR (2006) Spatial dependence for hydrogen peroxide-directed signaling in light-stressed plants. Plant Physiol 141:346–350
Neill SJ, Desikan R, Hancock JT (2002) Hydrogen peroxide signalling. Curr Opin Plant Biol 5:388–395
Neill S, Bright J, Desikan R, Hancock J, Harrison J, Wilson I (2008) Nitric oxide evolution and perception. J Exp Bot 59:25–35
Nila AG, Sandalio LM, López MG, Gómez M, del Río LA, Gómez-Lim MA (2006) Expression of a peroxisome proliferator-activated receptor gene (*xPPARα*) from *Xenopus laevis* in tobacco (*Nicotiana tabacum*) plants. Planta 224:569–581
Nishimura M, Hayashi M, Kato A, Yamaguchi K, Mano S (1996) Functional transformation of microbodies in higher plant cells. Cell Struct Funct 21:387–393
Noctor G, Foyer CH (1998) Ascorbate and glutathione: keeping active oxygen under control. Annu Rev Plant Physiol Plant Mol Biol 49:249–279
Nyathi Y, Baker A (2006) Plant peroxisomes as a source of signalling molecules. Biochim Biophys Acta 1763:1478–1495

Ogawa K, Kanematsu S, Asada K (1996) Intra- and extra-cellular localization of 'cytosolic' CuZn-superoxide dismutase in spinach leaf and hypocotyls. Plant Cell Physiol 37:790–799
Palma JM, Garrido M, Rodríguez-García MI, del Río LA (1991) Peroxisome proliferation and oxidative stress mediated by activated oxygen species in plant peroxisomes. Arch Biochem Biophys 287:68–74
Palma JM, Gómez M, Yáñez J, del Río LA (1987) Increased levels of peroxisomal active oxygen-related enzymes in copper-tolerant pea plants. Plant Physiol 85:570–574
Palma JM, López-Huertas E, Corpas FJ, Sandalio LM, Gómez M, del Río LA (1998) Peroxisomal manganese superoxide dismutase: purification and properties of the isozyme from pea leaves. Physiol Plant 104:720–726
Palma JM, Sandalio LM, Corpas FJ, Romero-Puertas MC, McCarthy I, del Río LA (2002) Plant proteases, protein degradation and oxidative stress: role of peroxisomes. Plant Physiol Biochem 40:521–530
Pastori GM, del Río LA (1997) Natural senescence of pea leaves: an activated oxygen-mediated function for peroxisomes. Plant Physiol 113:411–418
Reddy JK, Warren JR, Reddy MK, Lalwani ND (1982) Hepatic and renal effects of peroxisome proliferators: biological implications. Ann N Y Acad Sci 386:81–110
Reumann S, Ma C, Lemke S, Babujee L (2004) AraPerox. A database of putative *Arabidopsis* proteins from plant peroxisomes. Plant Physiol 136:2587–608
Reumann S, Babujee L, Ma C, Wienkoop S, Siemsen T, Antonicelli GE, Rasche N, Lüder F, Weckwerth W, Jahn O (2007) Proteome analysis of *Arabidopsis* leaf peroxisomes reveals novel targeting peptides, metabolic pathways, and defense mechanisms. Plant Cell 19:3170–3193
del Río LA, Lyon DS, Olah I, Glick B, Salin ML (1983) Immunocytochemical evidence for a peroxisomal localization of manganese superoxide dismutase in leaf protoplasts from a higher plant. Planta 158:216–224
del Río LA, Fernández VM, Rupérez FL, Sandalio LM, Palma JM (1989) NADH induces the generation of superoxide radicals in leaf peroxisomes. Plant Physiol 89:728–731
del Río LA, Donaldson RP (1995) Production of superoxide radicals in glyoxysomal membranes from castor bean endosperm. J Plant Physiol 146:283–287
del Río LA, Palma JM, Sandalio LM, Corpas FJ, Pastori GM, Bueno P, López-Huertas E (1996) Peroxisomes as a source of superoxide and hydrogen peroxide in stressed plants. Biochem Soc Trans 24:434–438
del Río LA, Pastori GM, Palma JM, Sandalio LM, Sevilla F, Corpas FJ, Jiménez A, López-Huertas E, Hernández JA (1998) The activated oxygen role of peroxisomes in senescence. Plant Physiol 116:1195–1200
del Río LA, Corpas FJ, Sandalio LM, Palma JM, Gómez M, Barroso JB (2002a) Reactive oxygen species, antioxidant systems and nitric oxide in peroxisomes. J Exp Bot 53:1255–1272
del Río LA, Sandalio LM, Palma JM, Corpas FJ, López-Huertas E, Romero-Puertas MC, McCarthy I (2002b) Peroxisomes, reactive oxygen metabolism, and stress-related enzyme activities. In: Baker A, Graham IA (eds) Plant peroxisomes. Biochemistry, cell biology and biotechnological applications. Kluwer, Dordrecht, The Netherlands, pp 221–258
del Río LA, Corpas FJ, Sandalio LM, Palma JM, Barroso JB (2003) Plant peroxisomes, reactive oxygen metabolism and nitric oxide. IUBMB Life 55:71–81
del Río LA, Corpas FJ, Barroso JB (2004) Nitric oxide and nitric oxide synthase activity in plants. Phytochemistry 65:783–792
del Río LA, Sandalio LM, Corpas FJ, Palma JM, Barroso JB (2006) Reactive oxygen species and reactive nitrogen species in peroxisomes. Production, scavenging, and role in cell signalling. Plant Physiol 141:330–335
Rockel P, Strube F, Rockel A, Wildt J, Kaiser WM (2002) Regulation of nitric oxide (NO) production by plant nitrate reductase in vivo and in vitro. J Exp Bot 53:103–110
Rodríguez-Serrano M, Romero-Puertas MC, Pastori GM, Corpas FJ, Sandalio LM, del Río LA, Palma JM (2007) Peroxisomal membrane manganese superoxide dismutase: characterization of the isozyme from watermelon (*Citrullus lanatus* Schrad.) cotyledons. J Exp Bot 58:2417–2427

Romero-Puertas MC, McCarthy I, Sandalio LM, Palma JM, Corpas FJ, Gómez M, del Río LA (1999) Cadmium toxicity and oxidative metabolism of pea leaf peroxisomes. Free Radic Res (Suppl) 31:S25–S31

Romero-Puertas MC, Palma JM, Gómez M, del Río LA, Sandalio LM (2002) Cadmium causes the oxidative modification of proteins in pea plants. Plant Cell Environ 25:677–686

Romero-Puertas MC, Rodríguez-Serrano M, Corpas FJ, Gómez M, del Río LA, Sandalio LM (2004) Cadmium-induced subcellular accumulation of $O_2^{\bullet -}$ and H_2O_2 in pea leaves. Plant Cell Environ 27:1122–1134

Romero-Puertas MC, Corpas FJ, Sandalio LM, Leterrier M, Rodríguez-Serrano M, del Río LA, Palma JM (2006) Glutathione reductase from pea leaves: response to abiotic stress and characterization of the peroxisomal isozyme. New Phytol 170:43–52

Sakuma S, Fujimoto Y, Sakamoto Y, Uchiyama T, Yoshioka K, Nishida H, Fujita T (1997) Peroxynitrite induces the conversion of xanthine dehydrogenase to oxidase in rabbit liver. Biochem Biophys Res Commun 230:476–479

Sandalio LM, Palma JM, del Río LA (1987) Localization of manganese superoxide dismutase in peroxisomes isolated from *Pisum sativum* L. Plant Sci 51:1–8

Sandalio LM, Fernández VM, Rupérez FL, del Río LA (1988) Superoxide free radicals are produced in glyoxysomes. Plant Physiol 87:1–4

Schäfer L, Feirabend J (2000) Photoinactivation and protection of glycolate oxidase in vitro and in leaves. Z. Naturfors 55c:361–372

Schubert KR (1986) Products of biological nitrogen fixation in higher plants: synthesis, transport, and metabolism. Annu Rev Plant Physiol 37:539–574

Seo MS, Kang SW, Kim K, Baines IC, Lee TH, Rhee SG (2000) Identification of a new type of mammalian peroxiredoxin that forms an intramolecular disulfide as a reaction intermediate. J Biol Chem 275:20346–20354

Titorenko VI, Rachubinski RA (2004) The peroxisome: orchestrating important developmental decisions from inside the cell. J Cell Biol 164:641–645

Tolbert NE, Gee R, Husic DW, Dietrich S (1987) Peroxisomal glycolate metabolism and the C2 oxidative photosynthetic carbon cycle. In: Fahimi HD, Sies H (eds) Peroxisomes in Biology and Medicine, Springer, Berlin, pp 213–222

Valderrama R, Corpas FJ, Carreras A, Gómez-Rodríguez MV, Chaki M, Pedrajas JR, Fernández-Ocaña A, del Río LA, Barroso JB (2006) The dehydrogenase-mediated recycling of NADPH is a key antioxidant system against salt-induced oxidative stress in olive plants. Plant Cell Environ 29:1449–1459

Van Breusegem F, Bailey-Serres J, Mittler R (2008) Unraveling the tapestry of networks involving reactive oxygen species in plants. Plant Physiol 147:978–984

Vicentini F, Matile P (1993) Gerontosomes, a multifunctional type of peroxisomes in senescent leaves. J Plant Physiol 142:50–56

Willekens H, Chamnongpol S, Davey M, Schraudner M, Langebartels C, Van Montagu M, Inzé D, Van Camp W (1997) Catalase is a sink for H2O2 and is indispensable for stress defense in C-3 plants. EMBO J 16:4806–4816

Wood ZA, Schröder E, Harris JR, Poole LB (2003) Structure, mechanism and regulation of peroxiredoxins. Trends Biochem Sci 28:32–40

Zemojtel T, Fröhlich A, Palmieri MC, Kolanczyk M, Mikula I, Wyrwicz LS, Wanker EE, Mundlos S, Vingron M, Martasek P, Durner J (2006) Plant nitric oxide synthase: a never-ending story? Trends Plant Sci 11:524–525

Reactive Oxygen Species in Plant–Pathogen Interactions

G. Paul Bolwell and Arsalan Daudi

Abstract Reactive oxygen species (ROS), superoxide, hydrogen peroxide and nitric oxide are produced at all levels of resistance reactions in plants. In basal resistance, they are linked to papilla formation and the assembly of barriers. In the hypersensitive response, they may be linked to programmed cell death, and in systemic acquired resistance, they interact with salicylate in signalling. Despite this importance, there is still a need to dissect the identities, activation and relative contributions of the ROS generating systems. Progress, however, is being made in identifying the molecular targets at the transcriptional, protein and cellular structure levels that are regulated by ROS in coordinating resistance responses.

1 Introduction

Plants have evolved sophisticated surveillance and defence systems triggered by recognition of pathogen- or microbial-associated molecular patterns (PAMPs or MAMPs). These are in combat with the microbial arsenal of effector proteins, which include suppressors of the defence response. Pathogens use diverse strategies to gain access to plant tissues. Bacteria commonly enter through stomata or wounds, whereas fungi extend their hyphae which either directly penetrate the epidermis or differentiate to form specialised nutrient exchange structures such as haustoria. The resultant cellular manifestations of resistant or susceptible interactions are the result of complex dynamic molecular and biochemical changes with checks and balances that determine the extent of ingress, which the plant attempts to prevent at many levels. The resistance response has now been dissected into several tiers or levels of pre-existing and/or activated defences all of which appear to involve the production of reactive oxygen species (ROS) (Chisholm et al. 2006;

G.P. Bolwell (✉) and A. Daudi
School of Biological Sciences, Royal Holloway, University of London,
Egham, Surrey, TW20 0EX, UK
e-mail: p.bolwell@rhul.ac.uk

L.A. del Río and A. Puppo (eds.), *Reactive Oxygen Species in Plant Signaling*,
Signaling and Communication in Plants, DOI 10.1007/978-3-642-00390-5_7,

Jones and Dangl 2006; He et al. 2007). Indeed, some of these responses are highly conserved across all eukaryotic phyla leading to the concept of innate immunity. However, in addition to these innate mechanisms, which form part of basal resistance, it is clear that reactive oxygen species are also involved at the later levels of the hypersensitive response and systemic acquired resistance (Torres et al. 2006). The main ROS that have received attention are superoxide, hydrogen peroxide and nitric oxide, and it is now realised they function not only in establishment of defence mechanisms, but also signalling by synergistic interaction between themselves and other signalling molecules such as salicylic acid. Counteracting the successful tiers of surveillance and response the successful pathogens have evolved suppression and detoxification strategies which include responses to ROS production.

2 Dissecting the Role of ROS in Molecular Plant–Pathogen Interactions

The plasma membranes of plant cells contain extracellular surface pattern-recognition receptors (PRRs), which are able to detect signals from invading pathogens (MAMPs) and elicit basal resistance. This leads to several early responses such as MAP kinase signalling cascades, transcriptional induction of defence genes, rapid microbursts of ROS and callose deposition to strengthen the cell wall at sites of infection as a result of complex cellular remodelling (Chisholm et al. 2006; Nurnberger et al. 2004). A well-studied example of such basal resistance is the response to the bacterial protein flagellin and its conserved 22 amino acid peptide flg22 (Gomez-Gomez and Boller 2002; Zipfel et al. 2004) mediated through the *Arabidopsis* receptor FLS2. FLS2 is a plasma membrane receptor-like kinase which consists of a leucine-rich-repeat (LRR) domain and a serine–threonine kinase domain. Mutant plants that lack the FLS2 protein have been shown to be insensitive to flagellin (Gomez-Gomez and Boller 2000; Zipfel et al. 2004). As a result of such molecular recognition and the resultant signal transduction, complex cellular and sub-cellular remodelling occurs at the interface of microbial penetration and host plant (Schulze-Lefert 2004). This includes the thickening of host cell walls and the deposition of polymers such as the β-1,3 glucan, callose, to form cell wall appositions (papillae) in the paramural space, the extracellular zone between the cell wall and the plasma membrane. The cellular remodelling described above is usually accompanied by the co-localised accumulation of proteins, phenolics and ROS such as hydrogen peroxide, superoxide and nitric oxide. Cell wall remodelling is linked with the rapid disassembly and subsequent reassembly of the cytoskeleton at fungal entry sites followed by dynamic secretory events (Bhat et al. 2005).

The hypersensitive response (HR) is also accompanied by a localised increase in the accumulation of ROS at sites of infection and is further characterised by rapid cell death at the site of infection (Hammond-Kosack and Jones 1996). The HR is dependent on the activation of *R* genes and numerous examples have been characterised in several different plant species against important microbial pathogens (Hammond-Kosack and Parker 2003; van't Slot and Knogge 2002). An abundant class of plant R genes encode intracellular nucleotide-binding/leucine-rich repeat (NB-LRR) proteins with variable N-terminal domains (Dangl and Jones 2001; Meyers et al. 2005). Plant NB-LRR proteins have evolved to recognise *avr* products of pathogens (He et al. 2007). As the receptors have structural homologues in animals, this is further evidence for the conservation of fundamental disease resistance mechanisms through the evolution of all species (Chisholm et al. 2006; Hammond-Kosack and Parker 2003; Jones and Dangl 2006; Nurnberger and Lipka 2005).

Systemic acquired resistance (SAR) is a mechanism of induced resistance in plants that provides them with extended temporal and spatial protection against a wide range of microbes. SAR can be achieved against pathogens, which need not be the same as those responsible for the initial challenge. SAR is associated with the increase in expression levels of several well characterised defence-related or pathogenesis-related (PR) genes (Durrant and Dong 2004). Reactive oxygen species appear to be involved in the establishment of systemic defences in conjunction with salicylic acid. It appears that ROS mediates systemic responses through interaction with NPR1, which regulates the expression of *PR* genes by interactions with leucine zipper protein transcription factors (Kanzaki et al. 2003).

The extent to which pathogen susceptibility is dependent upon the levels of ROS produced has long been debated, stimulated by the observation that many can still grow in high (mM) concentrations of H_2O_2. Nevertheless, pathogen strategies may include detoxification of ROS and signal transduction pathways for their detection appear conserved (Aguirre et al., 2005). This leads to activation of their antioxidant enzymes such as superoxide dismutase and catalase (Zhang et al. 2004; Unger et al. 2005). Most work, however, has been devoted to the suppression of various levels of the whole defence response including ROS production (Chisholm et al. 2006; Jones and Dangl 2006; He et al. 2007), and the complexities of these processes are beyond the present review. Broadly, however, the plant MAMPs recognition systems at the surface which are linked to the basal response can be overcome by successful pathogens delivering multiple virulence factors acting as suppressors. Some cultivars have then evolved internal surveillance resistance proteins to counteract these virulence factors leading to the HR. In turn, some bacteria have acquired virulence effectors to block this tier of immunity. At all levels, successful pathogens can potentially suppress ROS production as a component of basal resistance, HR or SAR.

3 Upstream Signalling Leading to ROS Production

Elicitors of the oxidative burst include bacterial peptides, such as flg22 derived from flagellin, other bacterial- and fungal-derived peptides, which are often cysteine–rich, or microbial enzymes such as xylanases. Predominantly carbohydrate elicitors of the burst include bacterial LPS, chitin oligomers, β-1-3 glucans and oligogalacturonides. Partially purified elicitor preparations from the microbial extracellular matrix are also conveniently used in these studies. Following such MAMPS interaction with the appropriate receptor, a number of well-defined and conserved events occur in all species examined, of which an increase in cytosolic Ca^{2+}, directly measured often within seconds of elicitation, is thought to be a primary signal essential for the subsequent downstream events (Chandra and Low 1997). This increase in cytoplasmic calcium was required for the production of ROS in *Arabidopsis* leaf cells (Grant et al. 2000a). Beside this early calcium influx into the cytosolic compartment, a rapid efflux of K^+ and Cl^- and extracellular alkalinisation of elicited cell cultures has also been observed (Fellbrich et al. 2000). Although these events can be detected in planta, detailed analysis is facilitated in elicitor-treated cell culture models. Both transcriptional analysis of flagellin- and chitin-treated suspension-cultured cells (Navarro et al. 2004; Ramonell et al. 2002) and proteomic analysis of cells treated with an elicitor preparation from the maize pathogen *F. moniliforme* (Chivasa et al. 2006) have shown the applicability of modelling responses in this way in *Arabidopsis*, where profiles are similar to those found in planta in response to pathogen attack. These studies with cell cultures have begun to link each of these early events with the activation of the individual ROS production systems. Amongst many such examples, work in two elicitation systems, French bean (*Phaseolus vulgaris*) cells treated with an elicitor from *Colletotrichum lindemuthianum* and *Arabidopsis* cells treated with an elicitor from *Fusarium oxysporum* showed, using modulators, the dependence of H_2O_2 generation upon cAMP, G proteins and Ca^{2+} and K^+ fluxes (Bindschedler et al. 2001; Bolwell et al. 1999, 2002; Davies et al. 2006). cAMP may directly act on cyclic-nucleotide gated channels (Clough et al. 2000) and induce an increase in $[Ca^{2+}]_{cyt}$ and alkalinisation of the medium. Also, in elicited *Arabidopsis* cells, NO has been shown to be produced much earlier than H_2O_2 and cGMP appears to be required for the induction of NO-dependent progranned cell death PCD (Clarke et al. 2000).

A secretory component may also be necessary for apoplastic generation of H_2O_2 (Bolwell et al. 1999; Davies et al. 2006). Use of inhibitors of the secretory system such as monensin, brefeldin A and NEM shows differential inhibition of ROS production (Davies et al. 2006). The potent inhibitory effect of monensin and NEM, together with the lack of effect of brefeldin A, can be interpreted as an involvement of post-Golgi vesicles only in the apoplastic oxidative burst. Notwithstanding, the secretory system has been shown to be responsible for many of the aspects of overall resistance (Schulze-Lefert 2004). In the context of the oxidative burst, this could involve unloading of substrates emanating from vesicles or larger organelles such as peroxisomes. There is both structural and genetic evidence for such involvement. Rearrangement and directed secretion of vesicles and organelles to sites of pathogen adhesion and/or attempted penetration has been observed in several systems leading to papilla formation (Bestwick et al. 1998; Brown et al. 1998; Freytag et al. 1994; Kobayashi et al. 1997; McLusky et al. 1999; Schmelzer 2002) including *Arabidopsis* (Soylu et al. 2005). Hydrogen peroxide is localised to these sites (Bestwick et al. 1998; Brown et al. 1998).

4 Sources of ROS

Following the recognition of an oxidative burst in the plant defence response against pathogen attack, the earliest studies concentrated upon the biochemistry of production of ROS including hydrogen peroxide and superoxide (Bolwell and Wojtaszek 1997; Lamb and Dixon 1997). Subsequently it was recognised that various plants, including the model plant *Arabidopsis*, exhibit different oxidative burst phases with an early production of ROS in both compatible and incompatible interactions and a later burst shown only in *R*-gene-mediated resistance responses (Apel and Hirt 2004; Grant and Loake 2000). It is also now established that the major sources of ROS can be plasma membrane-localised NADPH/NADH oxidase (NOX)-dependent which can generate superoxide or cell wall-localised peroxidase-dependent which can generate hydrogen peroxide directly or both sources (Bolwell 1999). Although these systems have been found in many species, it should not be forgotten that there may be other contributing sources of superoxide and hydrogen peroxide such as "leaky" oxidative and photosynthetic electron transport, discussed elsewhere in this volume, which could occur in defence. There are also specialised substrate-dependent sources such as oxalate oxidase confined to particular species. Depending upon the host and pathogen species involved, it is important to dissect the relative contribution of these sources especially plasma membrane NOX or cell wall peroxidases in relation to the production and role of superoxide and H_2O_2 in basal resistance and in the hypersensitive response (HR). Several lines of evidence from various plant species now suggest that the sources of ROS are different during non-host (basal) resistance and during the HR, but also that these sources may interact with each other. The apoplastic oxidative burst during basal resistance is likely to be peroxidase-dependent at least in *Arabidopsis* (Bindschedler et al. 2006; Soylu et al. 2005). It now appears that an earlier source of ROS may serve to activate NOX during the HR (Torres et al. 2005). By deduction, this source could be from the basal resistance response. Genetic evidence and use of inhibitors in *Arabidopsis thaliana* plants have provided data that would comply with such a model but there is no direct evidence as yet, since compromising the peroxidase-dependent apoplastic oxidative burst also affects the HR and it may be different in other species such as *Nicotiana* spp. The other major reactive oxygen species, nitric oxide, has a major role in the HR but may also participate in papilla formation and possibly systemic acquired resistance (Mur et al. 2006; Neill et al. 2008).

4.1 NADPH Oxidase and Superoxide

NADPH oxidases (NOX), also referred to as respiratory burst oxidases (RBOH), have been implicated in biotic interactions in a number of plants especially *Arabidopsis* and solanaceous species (Torres and Dangl 2005). They were first identified by the susceptibility of the ROS production in plants to inhibition by diphenylene iodonium (DPI) and confirmed in those cases where its Ki for ROS production was below 2 μM DPI. Detailed discussion of their biochemistry, domain structure and activation is presented in Chapter 1 of this volume. Activation is especially relevant here and is probably dependent upon the upstream Ca^{2+} influx. In *Arabidopsis*, this leads to phosphorylation on elicitation (Benschop et al. 2007; Nuhse et al. 2007) synergistically with Ca^{2+} binding to EF hands inducing a conformational change and activation of the AtrBOHD (Ogasawara et al. 2008). Recent work in potato suggests CDPKs are responsible for the phosphorylation at Ser82 in StrBOHD (Kobayashi et al. 2007). An additional level of regulation is interaction with plant Rac2 homologues which may be positive or negative (Torres et al. 2006).

Genetic approaches have revealed considerable insight into their role in biotic interactions. For example, *AtrbohD* and *AtrbohF* genes were identified as required for the production of a full oxidative burst in response to avirulent strains of the bacterial and oomycete pathogens *Pseudomona syringae* and *Hyaloperonospora parasitica*, respectively (Torres et al. 2002). This was also shown for *NtrbohD* in elicited tobacco cells (Simon-Plas et al. 2002). However, neither the *atrbohD* or *atbrohF* mutants, either singly or doubly, were more susceptible to either *P. syringae* or *H.parasitica*. More recently, an *lsd1-atrbohD-atrbohF* triple mutant has been constructed in which the *lsd1* (lesions simulating disease) mutation exhibits spreading lesions. Since this phenotype was found to be enhanced in the triple mutant, it was concluded that the role of NOX is to limit the spread of a salicylic acid-elicited cell death programme in cells surrounding an infection site (Torres et al. 2005). It was also concluded that the NOX need activation by an independent source of ROS to generate their own oxidative burst. The source of that ROS is likely to be generated during the basal response and this interpretation placed the NOX as having a function mainly in the HR. While a major role of NOX in resistance is therefore questionable in *Arabidopsis*, resistance of *Nicotiana benthamiana* to *Phytophthora infestans* was compromised by VIGS of *NbrbohA* and *NbrbohB* (Yoshioka et al. 2003). Therefore, the contribution of the various ROS producing systems to resistance has a different balance between species and also depends upon their synergistic action and signalling functions at the three tiers of plant–pathogen interaction.

4.2 Peroxidase and Hydrogen Peroxide

There are many cellular superoxide-generating systems besides NOX such as leakage from electron transport systems and xanthine oxidase, all of which form superoxide by the one electron reduction of molecular oxygen. However, there are also several ways in which haem proteins and even free haem can form hydrogen peroxide directly from iron-superoxide adducts without the intervention of superoxide dismutase. This reaction is inhibited strongly by the cytochrome inhibitors potassium cyanide and sodium azide, but is also sensitive to diphenylene iodonium (DPI) although not to the same extent as the NOX. Although DPI has been claimed to be a specific inhibitor of NOX, a concentration higher than 7 μM would also inhibit haem-generated ROS. However, this differential sensitivity allows distinguishing peroxidase-dependent processes since NOX, being a flavoprotein, is insensitive to azide and cyanide.

With respect to the biochemistry of H_2O_2 generation by peroxidase, the reaction is favoured at high pH in the presence of a number of reductants and the reaction has been most studied on the model peroxidase horse radish peroxidase (Bolwell et al. 1995, 1999; Frahry and Schopfer 1998; Pichorner et al. 1992). When extended to the apoplastic oxidative burst in elicitor-treated French bean cells, this model reaction has been placed into the context of a three component system involving a cationic cell wall peroxidase, the release of a reductant and/or a substrate, and pH change which gives rise to relative extracellular alkalinisation (Bolwell et al. 2002). The pH change is dependent upon K^+ efflux. The search for a substrate has remained elusive although there are many that are active with horse radish peroxidase. In addition to the model substrate cysteine (Bolwell et al. 2002; Pichorner et al. 1992), some C_2compounds are active including glycine, glyoxylate and glycollate. Others such as fatty acids above C_{12} in length promote dismutation of superoxide from peroxidase without their oxidation (Bolwell et al. 1999). Appearance of metabolites in the cell wall following elicitation include the fatty acids palmitate and stearate and the tri- and di-carboxylic acids, malate, citrate and succinate in French bean cells (Bolwell et al. 1999, 2002), but the apparent absence of C_2 compounds may be due to technical limitations as these are difficult to derivatise and detect. The same carboxylic acids also appear during the oxidative burst in *Arabidopsis* together with malonate and fumarate (Butt, Finch, Davies and Bolwell, unpublished data). Since these are indicators of peroxisomal as well as mitochondrial metabolism it may be that peroxisomes are unloaded into the wall as well as secretory vesicles (Lipka et al. 2005). These compounds may also indicate metabolism related to the provision of substrates for H_2O_2 generation, although the precise identity of these is still unknown.

Arabidopsis thaliana suspension-cultured cells also show an azide-sensitive but diphenylene iodonium-insensitive apoplastic oxidative burst that generates H_2O_2 in response to a *Fusarium oxysporum* cell wall preparation (Bindschedler et al. 2006; Bolwell et al. 2002; Davies et al. 2006). In addition, use of inhibitors has indicated the existence of the peroxidase-dependent burst in planta (Grant et al. 2000a), and this needs to be reconciled with previous observations that indicate that NOX play a key role in the oxidative burst (Grant et al. 2000b; Torres et al. 2002). Subsequently, a structural study investigated the response of *Arabidopsis* to either mutant or avirulent *P. syringae phaseolicola* using cerium chloride to study the accumulation of H_2O_2 at reaction sites (Soylu et al. 2005). The strongest H_2O_2 response was found during the HR activated by avirulent strains and other mutant strains. Accumulation of H_2O_2 during the HR, but not during cell wall alterations, was strongly suppressed by inhibition of NOX with DPI. However, the differential effect of DPI suggests an alternative source of H_2O_2 to modify the plant cell wall in basal resistance. In addition, a peroxidase rather than a NOX has been identified as the generator in suspension-cultured cells of *Arabidopsis* in response to hypo-osmotic stress (Rouet et al. 2006). While such studies establish the biochemical mechanism and potential for a peroxidase-dependent system in *Arabidopsis* as well, rigorous analysis of its importance requires genetic proof.

Transgenic *Arabidopsis* plants expressing an anti-sense cDNA encoding a French bean type III peroxidase (FBP1) (Blee et al. 2001) exhibited an impaired oxidative burst in response to avirulent strains of *P. syringae* as shown by a lack of diaminobenzidine detectable ROS at the cellular level in the leaf and cerium hydroperoxide staining in cell wall appositions at the subcellular level. Moreover *FBP1* antisense plants were more susceptible than wild-type plants to both fungal and bacterial pathogens (Bindschedler et al. 2006; Bolwell et al. 2002). Transcriptional profiling and RT-PCR analysis showed that the antisense *FBP1* transgenic plants had reduced levels of specific peroxidase-encoding mRNAs that encode two class III peroxidases, At3g39110 (AtPrx33) and At3g49120 (AtPrx34), with a high degree of homology to FBP1. These data indicate that peroxidases play a significant role in generating H_2O_2 during the *Arabidopsis* defence response and in conferring resistance to a wide range of pathogens (Bindschedler et al. 2006). An alternative strategy lending support for the importance of hydrogen peroxide generation by peroxidase in resistant responses has come from over-expression of an extracellular peroxidase from Capsicum in *Arabidopsis* leading to enhanced disease resistance accompanied by cell death, cyanide-sensitive but DPI-insensitive H_2O_2 accumulation and PR gene induction (Choi et al. 2007). Therefore an alternative source of ROS to that from NOX is involved in basal (non-host) resistance as suggested from knockout and inhibitor data in *Arabidopsis*.

4.3 Nitric Oxide

NO has been implicated in contributing to increased disease resistance against pathogenic microbes (Neill et al. 2003). A majority of detection studies have identified NO accumulation primarily in gene-for-gene interactions leading to the HR and the localised accumulation of ROS with the consequent increased expression of several defence-related genes (Delledonne et al. 1998; Romero-Puertas et al. 2004). However strong bursts of NO are also observed in the response to general MAMPs during basal resistance upstream or simultaneously to H_2O_2 production in a number of systems (Foissner et al. 2000; Lamotte et al. 2004). Studies of NO accumulation under these processes induced by general MAMPs such as LPS from bacterial cell walls led to the discovery of a putative nitric oxide synthase (NOS) gene in plants (Guo et al. 2003; Zeidler et al. 2004) which was not homologous to the mammalian NOS genes. However, further analysis revealed that *AtNOS1* did not exhibit direct NOS activity and was renamed *Arabidopsis* NOS Associated1 (*AtNOA1*) (Zemojtel et al. 2006). The NOS (or NOA) gene, however, does not appear to be the only source of NO in plants. Two lines of evidence suggested that NO could be generated in planta by reduction of nitrite mediated by NADPH-dependent nitrate reductase (NR). However, this source has so far been implicated in developmental processes rather than the defence response (Desikan et al. 2002; Bethke et al. 2007).

4.4 Oxalate Oxidase and Other Specialised Sources

Other ROS generating sources such as oxalate oxidase and amine oxidase have also received some attention. Oxalate is produced as a toxin by some pathogens, most conspicuously by *Sclerotinia sclerotiorum*, during the infection process. Oxalate oxidases or germins are members of the cupin superfamily and are involved in cereal defence responses to invasion by fungal pathogens (Dunwell et al. 2004; Svedruzic et al. 2005). This is an example of specialisation as they are not found in dicotyledonous plants. However, they have been shown to be effective in conferring resistance even when transferred into dicots. In addition to direct action of hydrogen peroxide on the pathogen and in the construction of cell wall appositions, they also metabolise oxalate which can act as an inhibitor of the HR. Apoplastic copper- or flavin-dependent amine oxidases also have the potential to produce H_2O_2 directly from polyamines, possibly contributing to ROS production in *Cicer* and *Nicotiana* interactions with some pathogens (Cona et al. 2006).

4.5 Lipid Peroxides and Oxylipin Production

The production of oxylipins are intimately related to ROS production particularly during herbivory or wounding (Farmer et al. 2003; Balbi and Devoto 2008). Technically, these are oxygenated fatty acids and, of course, some of the intermediates are very reactive and have been classified as reactive electrophile species (RES) rather than ROS. However, the extent to which they are directly involved in basal resistance is not clear, and most attention has been paid to their role in late resistance events. Together with jasmonates, these are potential signalling molecules interacting with ROS, SA and/or ethylene (Balbi and Devoto 2008). In this context, following elicitation of French bean cells with an elicitor from *Colletotrichum*, there was no production of oxylipins during the period of the oxidative burst (Bolwell et al. 2002) although, in planta, *cis*-3-hexenol and *trans*-2-hexanal were produced commencing around 12–15 h following inoculation of bean leaves with *Pseudomonas syringae* (Croft et al. 1993). Recently, over-expression of hydroperoxide lyase in *Arabidopsis*, which produces hexenols, was shown to improve resistance against *Botrytis cinerea*, possibly through direct fungicidal action (Kishimoto et al. 2008).

5 ROS and Molecular Targets

Considerable effort has gone into identifying the immediate consequences of ROS production in response to elicitor treatment or pathogen interaction. Experimentally, however, it is not always easy to determine whether ROS involvement occurs as a cause or consequence and, furthermore, exogenous application of ROS cannot duplicate subtle aspects of temporal and spatial localisation seen in the defence response. Some attempts to locally direct artificial ROS generating systems has allowed better control and, combined with the data from knocking out and over-expressing established endogenous generators, has allowed progress in understanding targets for ROS action.

5.1 Papilla Formation

The formation of cell wall appositions, papillae requiring reorganisation of the cytoskeleton and delivery of vesicles and other cell organelles such as peroxisomes, is a characteristic feature of non-host or basal resistance. Papillae consist of callose, cross-linked proteins and phenolics (Brown et al. 1998; McLusky et al. 1999) and hydrogen peroxide is detectable by cerium chloride during their formation (Brown et al. 1998; Soylu et al. 2005). Aspects of this, such as protein immobilisation, are H_2O_2-dependent (Bradley et al. 1995; Wojtaszek et al. 1995, 1997). Some of the proteins immobilised have anti-microbial function (Bindschedler et al. 2006). The role of callose, however, is complex since depletion of callose from papillae in plants lacking the callose synthase gene *gsl5* marginally enhanced the penetration of the grass powdery mildew fungus *Blumeria graminis* on the non-host *Arabidopsis*. Paradoxically, the absence of callose in papillae or haustorial complexes correlated with the effective growth cessation of several normally virulent powdery mildew species and of *Peronospora parasitica* (Jacobs et al. 2003).

Studies of a series of penetration mutants compromised in papilla formation allowing inappropriate ingress by barley powdery mildew in *Arabidopsis* have aided identification of genes and organelles required. These, based around *mlo* resistance initially in barley and subsequently in *Arabidopsis*, show that it does not involve the signalling molecules ethylene, jasmonic acid or salicylic acid, but requires a syntaxin, glycosyl hydrolase and ABC transporter which code for *pen1*, *pen 2* and *pen3* mutants, respectively (Consonni et al. 2006). PEN1 was found to encode plasma membrane syntaxin-121 involved in vesicle trafficking (Assaad et al. 2004; Collins et al. 2003). In a non-host interaction between *Arabidopsis* and *Colletotrichum*, the plant cells responded by the formation of PMR4/GSL5-dependent papillary callose (Shimada et al. 2006). In this interaction, *Arabidopsis* PEN1 syntaxin controls timely accumulation of papillary callose, but is functionally dispensable for effective pre-invasion (penetration) resistance in non-host interactions. The identity of PEN2, a peroxisomal-localised glycosidase and PEN3, a membrane-bound ABC transporter, also indirectly implicate the participation of endomembrane vesicles and peroxisomes in papilla formation (Lipka et al. 2005). The indication that peroxisomes act as a component of an inducible pre-invasion resistance mechanism, and that a glycosyl hydrolase is a crucial component, could also mean they are involved in provision of metabolites (Lipka et al. 2005). Inhibitor studies in elicitor-treated plant cells also implicate such events in the peroxidase-dependent apoplastic burst. The PEN1 syntaxin represents one of 24 sequence-related family members in the *Arabidopsis* genome which code for soluble *N*-ethylmaleimide (NEM)-sensitive factor adaptor protein receptors (Collins et al. 2003), and NEM is an effective blocker of the oxidative burst in elicitor-treated *Arabidopsis* cells (Davies et al. 2006).

Although required for protein cross-linking and phenolic polymerisation in the formation of barriers, the level of dependence of papilla formation on ROS signalling is obscure. Preliminary data do, however, suggest callose deposition is additionally dependent upon ROS. Callose production in the response of *Arabidopsis* seedlings to flg22 is compromised in the presence of inhibitors of the peroxidase-dependent burst and also in the peroxidase-downregulated lines (Bolwell and Ausubel, unpublished data). Peroxidase-dependent hydrogen peroxide production has clear potential to be a central component in orchestrating papilla formation.

5.2 *Transcription*

There have been a large number of studies of gene expression in response to pathogen attack, and the accumulated array data can be inspected in the publicly accessible databases such as Genevestigator. Upstream signalling requires some involvement of ROS. Exogenous application of H_2O_2 to *Arabidopsis* leaf cells shows that it signals through AtMPK3 and AtMPK6 (Kovtun et al. 2000). The same MAPKs are required for innate immunity signalling as part of a completely identified signalling pathway in response to flagellin with MEKK1, MKK4/MKK5 upstream and WRKY22/WRKY29 transcription factors downstream and leading to transcriptional activation (Asai et al. 2002). Others have also attempted to dissect out the contributions of ROS to regulation of gene expression by direct H_2O_2 application exogenously (Desikan et al. 2001). Approximately 1% of the *Arabidopsis* transcriptome was found to be regulated by H_2O_2. Upregulated genes included *pal1*, *gst6*, a heat shock protein, a zinc finger protein, calmodulin and an ethylene-responsive transcription factor. However, of particular interest was the syntaxin that was upregulated in terms of its importance in papilla formation. These types of experiments, however, may be limited by the lack of localised application and the high levels of hydrogen peroxide needed to elicit a response. However, specific differences in the response of wild-type and lines deficient in individual antioxidant enzymes to exogenously applied oxidative stress-causing agents were also observed, and could be analysed to determine transcription dependent upon hydrogen peroxide and superoxide (Gadjev et al. 2006). Marker genes upregulated relatively in all treatments included a defensin, a TIR-class protein and three proteins of unknown function. Those that could be specifically attributed to H_2O_2 induction were L-Asp oxidase family protein, xyloglucan endotransglycosylase, a Proline-rich family protein, a 2-oxoglutarate-dependant dioxygenase and two unknown proteins. Yet more alternative methods of inducing oxidative stress have been used, and detailed analysis of their effects on the transcriptome and metabolome (Baxter et al. 2007) has drawn analogies to the superoxide and hydrogen peroxide stimulon of yeast. Responses include the suppression of the TCA cycle and an upregulation of oxidative pentose phosphate pathway and increased synthesis of NADPH. General amino acid metabolism and polyamine synthesis were also suppressed. Similar approaches have been used to dissect differentially expressed NO-responsive genes and promoters in plants and cell cultures, and were somewhat dependent upon the mode of NO application. However, a core subset of 28 genes was upregulated. These include protein kinases and transcription factors as well as GSTs and peroxidases (Palmieri et al. 2008).

Another approach is to analyse the knockout lines for ROS generating systems such as NOX and apoplastic peroxidases, since there may also be pleiotropic effects on gene expression in uninfected plants that reduce pre-existing mechanisms of resistance and compromise their ability to recognise and respond to pathogen attack. There has been some analysis of changes in transcription in an AtrbohD mutant background (Davletova et al. 2005) which shows modification of catalase and ascorbate peroxidase expression. Analysis of the array data from the peroxidase downregulated lines (Bindschedler et al. 2006) indicates how changes in background gene expression could explain the increased susceptibility of the *fbp1* antisense lines. This analysis identified 25 upregulated and 68 downregulated genes besides the endogenous peroxidase, At3g49120, in uninfected tissue. Notable downregulated genes include those in signal transduction such as a TIR1, those in papilla formation, including syntaxin and the PR proteins, PR4, PR5, PDF1. Clearly, the heterologous expression of FBP1, known to be present in papillae in French bean (Brown et al. 1998), in *Arabidopsis* compromises the normal expression of genes involved in basal immunity and possibly the ability to mount an HR. One hypothesis to understand this hyper-susceptibility is that a low level of apoplastic ROS production is required to pre-prime basal resistance gene expression.

5.3 Proteins and Proteomes

In addition to immobilisation of the proline- and hyroxyproline-rich proteins during the apoplastic oxidative burst (Bradley et al. 1992; Wojtaszek et al. 1995, 1997), the induction of a number of extracellular defence proteins has long been documented before the advent of the ability to analyse global proteomes. These include the extracellular PR proteins such as chitinases, glucanases and thaumatin-like proteins (van Loon et al. 2006). More recently, proteomics has been applied to plant–pathogen interactions (Jones et al. 2006a), elicited cell cultures (Chivasa et al. 2006), to document global changes and to post-translational modifications (PTM) such as phosphorylation (Jones et al. 2006b; Benschop et al. 2007; Nuhse et al. 2007; Sugiyama et al. 2008) or during oxidative and nitrositive stress in plants (Lindermayr et al. 2005). Phosphorylation is linked to some of the most immediate downstream events following PAMPs recognition. Significant targets for phosphorylation in terms of the oxidative burst(s) and cell wall modifications are AtrBOHD, syntaxin, kinesins involved in vesicle transport, AtCDPK1, PEN3 and several HRGPs (Benschop et al. 2007). Recent technological advances in global/large-scale proteome and phospho-proteome analyses have significantly increased the coverage of identified proteins and shown that approximately 10% of phosphorylated proteins are implicated in signal transduction and responses to biotic stress (Sugiyama et al. 2008).

There is reasonable correspondence between elicitor-treated cell cultures (Chivasa et al. 2006) and leaf tissue responding to MAMPs inoculation in *Arabidopsis* (Jones et al. 2006a, b) with respect to notable global changes in the proteome. These include increases in antioxidant enzymes and decreases in primary metabolism enzymes and molecular chaperones in both cases. Changes in the extracellular matrix in response to elicitor treatment, other than immobilisation have also been documented in maize where the appearance of cytosolic heat shock proteins and glyceraldehyde 3-phosphate dehydrogenase (GAPDH) in the wall may be related to exocytotic events (Chivasa et al. 2006). GAPDH has been shown to be a specific target for inactivation by hydrogen peroxide (Hancock et al. 2005). Alcohol dehydrogenase (ADH), *S*-adenosyl-methione (SAM) synthase and glutamine synthase are also direct targets for protein oxidation by H_2O_2. More recently, quantitative proteomics has been applied to analyse responses to ROS application in whole *Arabidopsis* plants (Bindschedler et al. 2008). Changes in the protein profile of intercellular washing fluid were observed, and two glyceraldehyde-3-*P*-dehydrogenase isoforms, a glucan-β-1,3-glucosidase (PR2), a secretory peroxidase (P71), two glutathione-S-transferases and a putative chitinase (PR3) were significantly upregulated after the H_2O_2 treatment.

Other protein modifications arise from the nitration of tyrosine by peroxinitrate ($ONOO^-$) arising from the reaction of superoxide (O_2^-) and NO (Marla et al. 1997). This process is cytotoxic (Szabo et al. 1996) and can regulate NO signalling during plant immune reactions (Delledonne et al. 2001). NO is also highly reactive itself in its several different states such as NO, NO^- (nitroxyl ion) and NO^+ (nitrosonium ion) and can react with cysteine residues on proteins (Mur et al. 2006). NO/cGMP signalling was shown to lead to this S-nitrosylation of more than 100 candidate proteins of which the targets, GAPDH, SAM-synthase, metacaspase and possible K^+ channels may be of significance to the defence response (Palmieri et al. 2008). S-nitrosylation was also shown activate the GTP binding protein p21ras or inactivate the protein kinase JNK1 (Mur et al. 2006). The correspondence of many of these targets to more than one ROS suggests major roles in regulating the gross responses such as PCD (Hong et al. 2008)

5.4 Programmed Cell Death (PCD)

PCD is a major component of the HR and has been intricately linked with micro-bursts of ROS particularly NO (Mur et al. 2008). Understanding of PCD in plants has benefited from comparison with mammalian apoptosis where it is characterised by DNA cleavage along with nuclear and cytoplasmic shrinking resulting in overall loss of cell integrity (Mur et al. 2006, 2008). NO is implicated through binding to the haem group of cytochrome causing the release of superoxide which in turn causes the production of $ONOO^-$. Peroxynitrite induces disruption of the mitochondria, which in turn promotes cytochrome c release, caspase activation and, consequently, apoptosis (Mur et al. 2006). PCD in plants is similarly characterised by the disruption of the mitochondria, consequent calcium influxes, the formation of permeability transition pores and the release of cytochrome c along with oxidative stress (Lam et al. 2001; Tiwari et al. 2002). However, analogous caspases are not present in the known plant genomes and proteomes, but there is evidence for caspase-like activity in plants, mediated potentially by paracaspases/metacaspases (Uren et al. 2000) or other proteases with caspase-like activity, such as serine-dependent instead of cysteine-dependent proteases, referred to as saspases (Chichkova et al. 2004).

6 Downstream Signalling

It is apparent that the accumulation of ROS in response to pathogen attack via localised micro-bursts or sustained release contributes to a complex regulatory network of multiple signalling cascades which prime the plant against disease (Mur et al. 2006; Torres et al. 2006). Salicylic acid (SA) is a well-established player in disease resistance signalling in both local and systemic responses and was shown to accumulate locally around pathogen breach sites (Enyedi et al. 1992). Futhermore, the accumulation of H_2O_2 after the application of SA was demonstrated, which showed that SA itself was not a strong enough signal to account for enhanced defence responses, but instead interacted with other downstream components including ROS (Shirasu et al. 1997). It has also been shown that not only does SA accumulation lead to a burst of ROS production, but also a reduction in ROS-scavenging enzymes which in turn leads to the same effect of increased ROS in response to pathogen invasion (Torres et al. 2006). Additional complexities arise through other studies which show that in lesion mimic mutants (*lsd1*), which exhibit sustained spreading of HR, SA and ROS, appear to antagonise each other during cell death spread (Torres et al. 2005). These often contrasting examples of crosstalk between SA and ROS highlight the different avenues of complex signalling that appear to be occurring downstream of ROS production and merit careful dissection in the future. More recently, applications of SA together with jasmonic acid were shown to elicit the generation of apoplastic ROS when these signals have been considered to be antagonists (Mur et al. 2006). Other emerging predominantly transcriptional data suggest that SAR is also mediated by jasmonates (Truman et al. 2007).

There is also evidence for the interaction of ROS and ethylene. Large-scale gene expression analysis revealed that ethylene-responsive elements and other genes involved in ethylene signalling were upregulated by the exogenous application of H_2O_2 (Desikan et al. 2001). Ethylene is known to induce programmed cell death and fruit or flower senescence, and there is also evidence for the accumulation of H_2O_2 in response to ethylene in tomatoes (de Jong et al. 2002). More recent genetic work revealed that the ethylene receptor ETR1 was implicated in H_2O_2 signalling during the regulation of stomatal closure (Desikan et al. 2005). The *etr1* loss of function mutants were shown to be disrupted in H_2O_2 signalling in both plants and yeast. Additional experiments also highlighted the interplay between ROS and ethylene signalling in *Arabidopsis* resistance against cauliflower mosaic virus. This work interestingly showed that both ethylene and ROS were implicated in a SAR response against the virus, but SA levels had little role to play in this particular system (Love et al. 2005).

Calcium signalling is also known to occur downstream of ROS production besides the essential role of Ca^{2+} influx for initial ROS accumulation (Blume et al. 2000; Grant et al. 2000b). ROS production is also required for later additional calcium fluxes in the cell following elicitation. Rapid calcium influx occurs in soybean cells when treated with hydrogen peroxide which leads to programmed cell death (Levine et al. 1996). The fact that ROS production can induce downstream calcium signalling further indicates the complex interplay between ROS signalling and other second messengers.

References

Aguirre J, Rios-Momberg M, Hewitt D, Hansberg W (2005) Reactive oxygen species and development in microbial eukaryotes. Trends Microbiol 13:111–118

Apel K, Hirt H (2004) Reactive oxygen species: Metabolism, oxidative stress, and signal transduction. Annu Rev Plant Biol 55:373–399

Asai T, Tena G, Plotnikova J, Willmann MR, Chiu WL, Gomez-Gomez L, Boller T, Ausubel FM, Sheen J (2002) MAP kinase signalling cascade in *Arabidopsis* innate immunity. Nature 415:977–983

Assaad FF, Qiu JL, Youngs H, Ehrhardt D, Zimmerli L, Kalde M, Wanner G, Peck SC, Edwards H, Ramonell K, Somerville CR, Thordal-Christensen H (2004) The PEN1 syntaxin defines a novel cellular compartment upon fungal attack and is required for the timely assembly of papillae. Mol Biol Cell 15:5118–5129

Balbi V, Devoto A (2008) Jasmonate signalling network in *Arabidopsis thaliana*: Crucial regulatory nodes and new physiological scenarios. New Phytol 177:301–318

Baxter CJ, Redestig H, Schauer N, Repsilber D, Patil KR, Nielsen J, Selbig J, Liu JL, Fernie AR, Sweetlove LJ (2007) The metabolic response of heterotrophic *Arabidopsis* cells to oxidative stress. Plant Physiol 143:312–325

Benschop JJ, Mohammed S, O'Flaherty M, Heck AJR, Slijper M, Menke FLH (2007) Quantitative phosphoproteomics of early elicitor signaling in *Arabidopsis*. Mol Cell Proteom 6:1198–1214

Bestwick CS, Brown IR, Mansfield JW (1998) Localized changes in peroxidase activity accompany hydrogen peroxide generation during the development of a nonhost hypersensitive reaction in lettuce. Plant Physiol 118:1067–1078

Bethke PC, Libourel IGL, Aoyama N, Chung YY, Still DW, Jones RL (2007) The *Arabidopsis* aleurone layer responds to nitric oxide, gibberellin, and abscisic acid and is sufficient and necessary for seed dormancy. Plant Physiol 143:1173–1188

Bhat RA, Miklis M, Schmelzer E, Schulze-Lefert P, Panstruga R (2005) Recruitment and interaction dynamics of plant penetration resistance components in a plasma membrane microdomain. Proc Natl Acad Sci USA 102:3135–3140

Bindschedler L, Palmblad M, Cramer R (2008) Hydroponic isotope labelling of entire plants (HILEP) for quantitative plant proteomics:an oxidative stress case study. Phytochemistry 69:1962–1972

Bindschedler LV, Dewdney J, Blee KA, Stone JM, Asai T, Plotnikov J, Denoux C, Hayes T, Gerrish C, Davies DR, Ausubel FM, Bolwell GP (2006) Peroxidase-dependent apoplastic oxidative burst in *Arabidopsis* required for pathogen resistance. Plant J 47:851–863

Bindschedler LV, Whitelegge JP, Millar DJ, Bolwell GP (2006) A two component chitin-binding protein from French bean – Association of a proline-rich protein with a cysteine-rich polypeptide. FEBS Lett 580:1541–1546

Bindschedler LV, Minibayeva F, Gardner SL, Gerrish C, Davies DR, Bolwell GP (2001) Early signalling events in the apoplastic oxidative burst in suspension cultured French bean cells involve cAMP and Ca^{2+}. New Phytol 151:185–194

Blee KA, Jupe SC, Richard G, Zimmerlin A, Davies DR, Bolwell GP (2001) Molecular identification and expression of the peroxidase responsible for the oxidative burst in French bean (*Phaseolus vulgaris* L.) and related members of the gene family. Plant Mol Biol 47:607–620

Blume B, Nurnberger T, Nass N, Scheel D (2000) Receptor-mediated increase in cytoplasmic free calcium required for activation of pathogen defense in parsley. Plant Cell 12:1425–1440

Bolwell GP (1999) Role of active oxygen species and NO in plant defence responses. Curr Opin Plant Biol 2:287–294

Bolwell GP, Bindschedler LV, Blee KA, Butt VS, Davies DR, Gardner SL, Gerrish C, Minibayeva F (2002) The apoplastic oxidative burst in response to biotic stress in plants: A three-component system. J Exp Bot 53:1367–1376

Bolwell GP, Blee KA, Butt VS, Davies DR, Gardner SL, Gerrish C, Minibayeva F, Rowntree EG, Wojtaszek P (1999) Recent advances in understanding the origin of the apoplastic oxidative burst in plant cells. Free Rad Res 31:S137–S145

Bolwell GP, Butt VS, Davies DR, Zimmerlin A (1995) The origin of the oxidative burst in plants. Free Rad Res 23:517–532

Bolwell GP, Wojtaszek P (1997) Mechanisms for the generation of reactive oxygen species in plant defence – a broad perspective. Physiol Mol Plant Pathol 51:347–366

Bradley DJ, Kjellbom P, Lamb CJ (1992) Elicitor- and wound-induced oxidative cross-linking of a proline-rich plant cell wall protein: a novel, rapid defense response. Cell 70:21–30

Brown I, Trethowan J, Kerry M, Mansfield J, Bolwell GP (1998) Localization of components of the oxidative cross-linking of glycoproteins and of callose synthesis in papillae formed during the interaction between non-pathogenic strains of *Xanthomonas campestris* and French bean mesophyll cells. Plant J 15:333–343

Chandra S, Low PS (1997) Measurement of Ca^{2+} fluxes during elicitation of the oxidative burst in aequorin-transformed tobacco cells. J Biol Chem 272:28274–28280

Chichkova NV, Kim SH, Titova ES, Kalkum M, Morozov VS, Rubtsov YP, Kalinina NO, Taliansky ME, Vartapetian AB (2004) A plant caspase-like protease activated during the hypersensitive response. Plant Cell 16:157–171

Chisholm ST, Coaker G, Day B, Staskawicz BJ (2006) Host-microbe interactions: Shaping the evolution of the plant immune response. Cell 124:803–814

Chivasa S, Hamilton JM, Pringle RS, Ndimba BK, Simon WJ, Lindsey K, Slabas AR (2006) Proteomic analysis of differentially expressed proteins in fungal elicitor-treated *Arabidopsis* cell cultures. J Exp Bot 57:1553–1562

Choi HW, Kim YJ, Lee SC, Hong JK, Hwang BK (2007) Hydrogen peroxide generation by the pepper extracellular peroxidase CaPO2 activates local and systemic cell death and defense response to bacterial pathogens. Plant Physiol 145:890–904

Clarke A, Desikan R, Hurst RD, Hancock JT, Neill SJ (2000) NO way back: Nitric oxide and programmed cell death in *Arabidopsis thaliana* suspension cultures. Plant J 24:667–677

Clough SJ, Fengler KA, Yu IC, Lippok B, Smith RK, Bent AF (2000) The *Arabidopsis dnd1* "defense, no death" gene encodes a mutated cyclic nucleotide-gated ion channel. Proc Natl Acad Sci USA 97:9323–9328

Collins NC, Thordal-Christensen H, Lipka V, Bau S, Kombrink E, Qiu JL, Huckelhoven R, Stein M, Freialdenhoven A, Somerville SC, Schulze-Lefert P (2003) SNARE-protein-mediated disease resistance at the plant cell wall. Nature 425:973–977

Cona A, Rea G, Angelini R, Frederico R, Tavladoraki P (2006) Functions of amine oxidases in plant development and defence. Trends Plant Sci 11:80–88

Consonni C, Humphry ME, Hartmann HA, Livaja M, Durner J, Westphal L, Vogel J, Lipka V, Kemmerling B, Schulze-Lefert P, Somerville SC, Panstruga R (2006) Conserved requirement for a plant host cell protein in powdery mildew pathogenesis. Nature Genet 38:716–720

Croft KPC, Juttner F, Sluzarenko AJ (1993) Volatile products of the lipoxygenase pathway evolved from *Phaseolus vulgaris* (L) leaves inoculated with *Pseudomonas syringae* pv. *Phaseolicola*. Plant Physiol 101:13–24

Dangl JL, Jones JDG (2001) Plant pathogens and integrated defence responses to infection. Nature 411:826–833

Davletova S, Rizhsky L, Liang H, Shengqiang Z, Oliver DJ, Coutou J, Shulaev V, Schlauch K, Mittler R (2005) Cytosolic ascorbate peroxidase 1 is a central component of the reactive oxygen gene network of *Arabidopsis*. Plant Cell 17:268–281

Davies DR, Bindschedler LV, Strickland TS, Bolwell GP (2006) Production of reactive oxygen species in *Arabidopsis thaliana* cell suspension cultures in response to an elicitor from *Fusarium oxysporum*: Implications for basal resistance. J Exp Bot 57:1817–1827

de Jong AJ, Yakimova ET, Kapchina VM, Woltering EJ (2002) A critical role for ethylene in hydrogen peroxide release during programmed cell death in tomato suspension cells. Planta 214:537–545

Delledonne M, Xia YJ, Dixon RA, Lamb C (1998) Nitric oxide functions as a signal in plant disease resistance. Nature 394:585–588

Delledonne M, Zeier J, Marocco A, Lamb C (2001) Signal interactions between nitric oxide and reactive oxygen intermediates in the plant hypersensitive disease resistance response. Proc Natl Acad Sci USA 98:13454–13459

Desikan R, Griffiths R, Hancock J, Neill S (2002) A new role for an old enzyme: Nitrate reductase-mediated nitric oxide generation is required for abscisic acid-induced stomatal closure in *Arabidopsis thaliana*. Proc Natl Acad Sci USA 99:16314–16318

Desikan R, Hancock JT, Bright J, Harrison J, Weir L, Hooley R, Neill SJ (2005) A role for ETR1 in hydrogen peroxide signaling in stomatal guard cells. Plant Physiol 137:831–834

Desikan R, Mackerness SAH, Hancock JT, Neill SJ (2001) Regulation of the *Arabidopsis* transcriptome by oxidative stress. Plant Physiol 127:159–172

Dunwell JM, Purvis A, Khuri S (2004) Cupins: The most functionally diverse protein superfamily. Phytochemistry 65:7–17

Durrant WE, Dong X (2004) Systemic acquired resistance. Annu Rev Phytopathol 42:185–209

Enyedi AJ, Yalpani N, Silverman P, Raskin I (1992) Signal molecules in systemic plant-resistance to pathogens and pests. Cell 70:879–886

Farmer EE, Almeras E, Krishnamurthy V (2003) Jasmonates and related oxylipins in plant responses to pathogenesis and herbivory. Curr Opp Plant Biol 6:372–378

Fellbrich G, Blume B, Brunner F, Hirt H, Kroj T, Ligterink W, Romanski A, Nurnberger T (2000) *Phytophthora parasitica* elicitor induced reactions in cells of *Petroselinum crispum*. Plant Cell Physiol 41:692–701

Foissner I, Wendehenne D, Langebartels C, Durner J (2000) In vivo imaging of an elicitor-induced nitric oxide burst in tobacco. Plant J 23:817–824

Frahry G, Schopfer P (1998) Inhibition of O2-reducing activity of horseradish peroxidase by diphenyleneiodonium. Phytochemistry 48:223–227

Freytag S, Arabatzis N, Hahlbrock K, Schmelzer E (1994) Reversible cytoplasmic rearrangements precede wall apposition, hypersensitive cell-death and defense-related gene activation in potato phytophthora-infestans interactions. Planta 194:123–135

Gadjev I, Vanderauwera S, Gechev TS, Laloi C, Minkov IN, Shulaev V, Apel K, Inze D, Mittler R, Van Breusegem F (2006) Transcriptomic footprints disclose specificity of reactive oxygen species signaling in *Arabidopsis*. Plant Physiol 141:436–445

Gomez-Gomez L, Boller T (2002) Flagellin perception: a paradigm for innate immunity. Trends Plant Sci 7:251–256

Gomez-Gomez L, Boller T (2000) FLS2: An LRR receptor-like kinase involved in the perception of the bacterial elicitor flagellin in *Arabidopsis*. Mol Cell 5:1003–1011

Grant JJ, Loake GJ (2000) Role of reactive oxygen intermediates and cognate redox signalling in disease resistance. Plant Physiol 124:21–29

Grant JJ, Yun BW, Loake GJ (2000a) Oxidative burst and cognate redox signalling reported by luciferase imaging: identification of a signal network that functions independently of ethylene, SA and Me-JA but is dependent on MAPKK activity. Plant J 24:569–582

Grant M, Brown I, Adams S, Knight M, Ainslie A, Mansfield J (2000b) The RPM1 plant disease resistance gene facilitates a rapid and sustained increase in cytosolic calcium that is necessary for the oxidative burst and hypersensitive cell death. Plant J 23:441–450

Guo FQ, Okamoto M, Crawford NM (2003) Identification of a plant nitric oxide synthase gene involved in hormonal signaling. Science 302:100–103

Hammond-Kosack KE, Jones JDG (1996) Resistance gene-dependent plant defense responses. Plant Cell 8:1773–1791

Hammond-Kosack KE, Parker JE (2003) Deciphering plant-pathogen communication: fresh perspectives for molecular resistance breeding. Curr Opin Biotechnol 14:177–193

Hancock JT, Henson D, Nyirenda M, Desikan R, Harrison J, Lewis M, Hughes J, Neill SJ (2005) Proteomic identification of glyceraldehydes 3-phosphate dehydrogenase as an inhibitory target of hydrogen peroxide in *Arabidopsis*. Plant Physiol Biochem 43:828–835

He P, Shan L, Sheen J (2007) Elicitation and suppression of microbe-associated molecular pattern-triggered immunity in plant–microbe interactions. Cell Microbiol 9:1385–1396

Hong KH, Yun BW, Kang JG, Raja MU, Kwon E, Sorhagen K, Chu C, Wang Y, Loake G (2008) Nitric oxide function and signalling in plant disease resistance. J Exp Bot 59:147–154

Jacobs AK, Lipka V, Burton RA, Panstruga R, Strizhov N, Schulze-Lefert P, Fincher GB (2003) An *Arabidopsis* callose synthase, GSL5, is required for wound and papillary callose formation. Plant Cell 15:2503–2513

Jones AME, Thomas V, Bennett MH, Mansfield J, Grant M (2006a) Modifications to the *Arabidopsis* defence proteome occur prior to significant transcriptional change in response to inoculation with *Pseudomonas syringae*. Plant Physiol 142: 1603–1620

Jones AME, Bennett MH, Mansfield JW, Grant M (2006b) Analysis of the defence phosphoproteome of *Arabidopsis thaliana* using differential mass tagging. Proteomics 6:4155–4165

Jones JDG, Dangl JL (2006) The plant immune system. Nature 444:323–329

Kanzaki H, Saitoh H, Ito A, Fujisawa S, Kamoun S, Katou S, Yoshioka H, Terauchi R (2003) Cytosolic HSP90 and HSP70 are essential components of INF1-mediated hypersensitive response and non-host resistance to *Pseudomonas cichorii* in *Nicotiana benthamiana*. Mol Plant Pathol 4:383–391

Kishimoto K, Matsui K, Ozawa R, Takabayashi J (2008) Direct fungicidal activities of C6-aldehydes are important constituents of defense responses in *Arabidopsis* against *Botrytis cinerea*. Phytochemistry 69:2127–2132

Kobayashi M, Ohura I, Kawakita K, Yokota N, Fujiwara M, Shimamoto K, Doke N, Yoshioka H (2007) Calcium-dependent protein kinases regulate the production of reactive oxygen species by potato NADPH oxidase. Plant Cell 19:1065–1080

Kobayashi Y, Kobayashi I, Funaki Y, Fujimoto S, Takemoto T, Kunoh H (1997) Dynamic reorganization of microfilaments and microtubules is necessary for the expression of non-host resistance in barley coleoptile cells. Plant J 11:525–537

Kovtun Y, Chiu WL, Tena G, Sheen J (2000) Functional analysis of oxidative stress-activated mitogen-activated protein kinase cascade in plants. Proc Natl Acad Sci USA 97:2940–2945

Lam E, Kato N, Lawton M (2001) Programmed cell death, mitochondria and the plant hypersensitive response. Nature 411:848–853

Lamb C, Dixon RA (1997) The oxidative burst in plant disease resistance. Annu Rev Plant Physiol Plant Mol Biol 48:251–275

Lamotte O, Gould K, Lecourieux D, Sequeira-Legrand A, Lebrun-Garcia A, Durner J, Pugin A, Wendehenne D (2004) Analysis of nitric oxide signaling functions in tobacco cells challenged by the elicitor cryptogein. Plant Physiol 135:516–529

Levine A, Pennell RI, Alvarez ME, Palmer R, Lamb C (1996) Calcium-mediated apoptosis in a plant hypersensitive disease resistance response. Curr Biol 6:427–437

Lindermayr C, Saalbach G, Durner J (2005) Proteomic identification of S-nitrosylated proteins in *Arabidopsis*. Plant Physiol 137:921–930

Lipka V, Dittgen J, Bednarek P, Bhat R, Wiermer M, Stein M, Landtag J, Brandt W, Rosahl S, Scheel D, Llorente F, Molina A, Parker J, Somerville S, Schulze-Lefert P (2005) Pre- and post-invasion defenses both contribute to nonhost resistance in *Arabidopsis*. Science 310:1180–1183

Love AJ, Yun BW, Laval V, Loake GJ, Milner JJ (2005) Cauliflower mosaic virus, a compatible pathogen of *Arabidopsis*, engages three distinct defense-signaling pathways and activates rapid systemic generation of reactive oxygen species. Plant Physiol 139:935–948

Marla SS, Lee J, Groves JT (1997) Peroxynitrite rapidly permeates phospholipid membranes. Proc Natl Acad Sci USA 94:14243–14248

McLusky SR, Bennett MH, Beale MH, Lewis MJ, Gaskin P, Mansfield JW (1999) Cell wall alterations and localized accumulation of feruloyl-3′-methoxytyramine in onion epidermis at sites of attempted penetration by *Botrytis allii* are associated with actin polarisation, peroxidase activity and suppression of flavonoid biosynthesis. Plant J 17:523–534

Meyers BC, Kaushik S, Nandety RS (2005) Evolving disease resistance genes. Curr Opin Plant Biol 8:129–134

Mur LAJ, Kenton P, Atzorn R, Miersch O, Wasternack C (2006) The outcomes of concentration-specific interactions between salicylate and jasmonate signalling include synergy, antagonism and oxidative stress leading to cell death. Plant Physiol 140:249–262

Mur LAJ, Carver TLW, Prats E (2006) NO way to live; the various roles of nitric oxide in plant-pathogen interactions. J Exp Bot 57:489–505

Mur LAJ, Kenton P, Lloyd AJ, Ougham H, Prats E (2008) The hypersensitive response; the centenary is upon us but how much do we know? J Exp Bot 59:501–520

Navarro L, Zipfel C, Rowland O, Keller I, Robatzek S, Boller T, Jones JDG (2004) The transcriptional innate immune response to flg22. Interplay and overlap with Avr gene-dependent defense responses and bacterial pathogenesis. Plant Physiol 135:1113–1128

Neill S, Bright J, Desikan R, Hancock J, Harrison J, Wilson I (2008) Nitric oxide evolution and perception. J Exp Bot 59:25–35

Neill SJ, Desikan R, Hancock JT (2003) Nitric oxide signalling in plants. New Phytol 159:11–35

Nuhse TS, Bottrill AR, Jones AME, Peck SC (2007) Quantitative phosphoproteomic analysis of plasma membrane proteins reveals regulatory mechanisms of plant innate immune responses. Plant J 51:931–940

Nurnberger T, Brunner F, Kemmerling B, Piater L (2004) Innate immunity in plants and animals: Striking similarities and obvious differences. Immunological Rev 198:249–266

Nurnberger T, Lipka V (2005) Non-host resistance in plants: New insights into an old phenomenon. Mol Plant Pathol 6:335–345

Ogasawara Y, Kaya H, Hiraoka G, Yumoto F, Kimura S, Kadota Y, Hishinuma H, Senzaki E, Yamagoe S, Nagata K, Nara M, Suzuki K, Tanokura M, Kuchitsu K (2008) Synergistic activation of the *Arabidopsis* NADPH oxidase *AtrbohD* by Ca^{2+} and phosphorylation. J Biol Chem 283:8885–8892

Palmieri MC, Sell S, Huang X, Scherf M, Werner T, Durner J, Lindermayr C (2008) Nitric oxide-responsive genes and promoters in *Arabidopsis thaliana*: a bioinformatics approach. J Exp Bot 59:177–186

Pichorner H, Couperus A, Korori SAA, Ebermann R (1992) Plant peroxidase has a thiol oxidase function. Phytochemistry 31:3371–3376

Ramonell KM, Zhang B, Ewing RM, Chen Y, Xu D, Stacey G, Somerville S (2002) Microarray analysis of chitin elicitation in Arabidopsis thaliana. Mol Plant Pathol 3:301–311

Romero-Puertas MC, Perazzolli M, Zago ED, Delledonne M (2004) Nitric oxide signalling functions in plant–pathogen interactions. Cell Microbiol 6:795–803

Rouet MA, Mathieu Y, Lauriere C (2006) Characterization of active oxygen-producing proteins in response to hypo-osmolarity in tobacco and *Arabidopsis* cell suspensions: Identification of a cell wall peroxidase. J Exp Bot 57:1323–1332

Schmelzer E (2002) Cell polarization, a crucial process in fungal defence. Trends Plant Sci 7:411–415

Schulze-Lefert P (2004) Knocking on heaven's wall: Pathogenesis of and resistance to biotrophic fungi at the cell wall. Curr Opin Plant Biol 7:377–383

Shimada C, Lipka V, O'Connel R, Okuno T, Schulze-Lefert P, Takano Y (2006) Nonhost resistance in *Arabidopsis-colletotrichum* interactions acts at the cell periphery and requires actin filament function. Mol Plant Microbe Interact 19:270–279

Shirasu K, Nakajima H, Rajasekhar VK, Dixon RA, Lamb C (1997) Salicylic acid potentiates an agonist-dependent gain control that amplifies pathogen signals in the activation of defense mechanisms. Plant Cell 9:261–270

Simon-Plas F, Elmayan T, Blein JP (2002) The plasma membrane oxidase *NtrbohD* is responsible for AOS production in elicited tobacco cells. Plant J 31:137–147

Soylu S, Brown I, Mansfield JW (2005) Cellular reactions in *Arabidopsis* following challenge by strains of *Pseudomonas syringae*: from basal resistance to compatibility. Physiol Mol Plant Pathol 66:232–243

Sugiyama N, Nakagami H, Mochida K, Daudi A, Tomita M, Shirasu K, Ishihama Y (2008) Large-scale phosphorylation mapping reveals the extent of tyrosine phosphorylation in *Arabidopsis*. Mol Syst Biol 4:193

Svedruzic D, Jonsson S, Toyota CG, Reinhardt LA, Ricagno S, Lindqvist Y, Richards NGJ (2005) The enzymes of oxalate metabolism: Unexpected structures and mechanisms. Arch Biochem Biophys 433:176–192

Szabo C, Zingarelli B, Oconnor M, Salzman AL (1996) DNA strand breakage, activation of poly(ADP-ribose) synthetase, and cellular energy depletion are involved in the cytotoxicity in macrophages and smooth muscle cells exposed to peroxynitrite. Proc Natl Acad Sci USA 93:1753–1758

Tiwari BS, Belenghi B, Levine A (2002) Oxidative stress increased respiration and generation of reactive oxygen species, resulting in ATP depletion, opening of mitochondrial permeability transition, and programmed cell death. Plant Physiol 128:1271–1281

Torres MA, Dangl JL (2005) Functions of the respiratory burst oxidase in biotic interactions, abiotic stress and development. Curr Opin Plant Biol 8:397–403

Torres MA, Dangl JL, Jones JDG (2002) *Arabidopsis* gp91(phox) homologues *AtrbohD* and *AtrbohF* are required for accumulation of reactive oxygen intermediates in the plant defense response. Proc Natl Acad Sci USA 99:517–522

Torres MA, Jones JDG, Dangl JL (2006) Reactive oxygen species signaling in response to pathogens. Plant Physiol 141:373–378

Torres MA, Jones JDG, Dangl JL (2005) Pathogen-induced, NADPH oxidase-derived reactive oxygen intermediates suppress spread of cell death in *Arabidopsis thaliana*. Nat Genetics 37:1130–1134

Truman W, Bennett MH, Kubigsteltig I, Turnbull C, Grant M (2007) *Arabidopsis* systemic immunity uses conserved defense signalling pathways and is mediated by jasmonates. Proc Natl Acad Sci USA 104:1075–1080

Unger C, Kleta S, Jandl G, von Tiedmann A (2005) Suppression of the defence-related oxidative burst in bean leaf tissue and bean suspension cells by the necrotrophic pathogen *Botrytis cinerea*. J Phytopathol 153:15–26

Uren AG, O'Rourke K, Aravind L, Pisabarro MT, Seshagiri S, Koonin EV, Dixit VM (2000) Identification of paracaspases and metacaspases: two ancient families of caspase-like proteins, one of which plays a key role in MALT lymphoma. Mol Cell 6:961–967

van't Slot LA, Knogge W (2002) A dual role for microbial pathogen-derived effector proteins in plant disease and resistance. Crit Rev Plant Sci 21:229–271

van Loon LC, Rep M, Pieterse CMJ (2006) Significance of inducible defense-related proteins in infected plants. Annu Rev Phytopathol 44:135–162

Wojtaszek P, Trethowan J, Bolwell GP (1997) Reconstitution in vitro of the components and conditions required for the oxidative cross-linking of extracellular proteins in French bean (*Phaseolus vulgaris L*). FEBS Lett 405:95–98

Wojtaszek P, Trethowan J, Bolwell GP (1995) Specificity in the immobilization of cell-wall proteins in response to different elicitor molecules in suspension-cultured cells of french bean (*Phaseolus-vulgaris* L). Plant Mol Biol 28:1075–1087

Yoshioka H, Numata N, Nakajima K, Katou S, Kawakita K, Rowland O, Jones JDG, Doke N (2003) *Nicotiana benthamiana* gp91(phox) homologs *NbrbohA* and *NbrbohB* participate in H_2O_2 accumulation and resistance to *Phytophthora infestans*. Plant Cell 15:706–718

Zeidler D, Zahringer U, Gerber I, Dubery I, Hartung T, Bors W, Hutzler P, Durner J (2004) Innate immunity in *Arabidopsis thaliana*: lipopolysaccharides activate nitric oxide synthase (NOS) and induce defense genes. Proc Natl Acad Sci USA 101:15811–15816

Zemojtel T, Frohlich A, Palmieri MC, Kolanczyk M, Mikula I, Wyrwicz LS, Wanker EE, Mundlos S, Vingron M, Martasek P, Durner J (2006) Plant nitric oxide synthase: a never-ending story? Trends Plant Sci 11:524–525

Zhang Z, Henderson C, Gurr SJ (2004) *Blumeria graminis* secretes an extracellular catalase during infection of barley: potential role in suppression of host defence. Mol Plant Pathol 5:537–547

Zipfel C, Robatzek S, Navarro L, Oakeley EJ, Jones JDG, Felix G, Boller T (2004) Bacterial disease resistance in *Arabidopsis* through flagellin perception. Nature 428:764–767

ROS in the Legume-*Rhizobium* Symbiosis

Karine Mandon, Nicolas Pauly, Alexandre Boscari, Renaud Brouquisse, Pierre Frendo, Bruce Demple, and Alain Puppo

Abstract Plants appear to generate reactive oxygen species (ROS) as signaling molecules to control various fundamental processes. With this background, this review aims to highlight the involvement of ROS, and their possible interactions with nitric oxide (NO) and glutathione (GSH) in the symbiosis between rhizobia and leguminous plants. This compatible interaction, which is very important for sustainable agriculture, leads to the formation of a novel organ capable of fixing atmospheric nitrogen. ROS are involved in the early steps of the symbiotic interaction: their presence is essential for the development of optimal symbiosis and points to a signaling role for ROS during the symbiotic process. ROS may also regulate nodule function by interacting with NO.

1 Introduction

The symbiotic association between the roots of leguminous plants and soil rhizobia results in the development of specific organs, called nodules, whose primary function is nitrogen fixation. Nodule formation involves extensive recognition by both partners, which leads to the appearance of a newly created organ in which the bacteria differentiate into bacteroids able to reduce atmospheric nitrogen. This process involves both plant–microbe interactions in order to allow an organized journey of the bacteria though the plant and cell division and differentiation processes leading to the development of the nodule meristem. Nodules are subsequently colonized by bacteria released from infection threads formed upon infection (Long 2001; Oldroyd and Downie 2008).

K. Mandon, N. Pauly, A. Boscari, R. Brouquisse, P. Frendo, and A. Puppo (✉)
Interactions Biotiques et Santé Végétale, UMR INRA 1301/Université de Nice-Sophia Antipolis/CNRS 6243, 400, route des Chappes, BP 167, 06903 Sophia Antipolis Cedex, France
e-mail: puppo@unice.fr

B. Demple
Department of Genetics and Complex Diseases, Harvard School of Public Health, Boston, Massachusetts 02115, USA

L.A. del Río and A. Puppo (eds.), *Reactive Oxygen Species in Plant Signaling*, Signaling and Communication in Plants, DOI 10.1007/978-3-642-00390-5_8,

In plants, reactive oxygen species (ROS) are unavoidable by-products of biochemical pathways such as oxidative phosphorylation and photosynthesis. As a result, plants have evolved enzymatic and non-enzymatic antioxidant mechanisms to eliminate ROS and avoid oxidative injury (Apel and Hirt 2004). On the other hand, it is now generally accepted that ROS can play a signaling role in various cellular mechanisms (Neill et al. 2002). Indeed, accumulating experimental evidence shows that ROS are key players in fundamental processes such as cellular growth (Foreman et al. 2003), stomatal closure (Pei et al. 2000), and the regulation of gene expression (Neill et al. 2002; Vranova et al. 2002). Moreover, ROS are involved in the plant defence against pathogens (Apel and Hirt 2004).

There is increasing evidence that ROS play an important role in legume–rhizobia symbiosis (Pauly et al. 2006). Indeed, the symbiotic interaction involves infection as well as developmental processes, and ROS are involved in both. ROS are important regulators of the hypersensitive response, a well-known process in plant pathogenesis. One should note that 90% of the infection threads initially formed in *Medicago* are aborted, which is caused by a hypersensitive-like response (Vasse et al. 1993; Penmetsa and Cook 1997). On the other hand, ROS involvement in plant developmental processes has been demonstrated (Foreman et al. 2003), which suggests that nodule organogenesis may also be dependant on ROS. Within this framework, the role of ROS in the legume–*Rhizobium* symbiosis will be addressed, considering first the initial steps of the symbiotic interaction and then the situation in functioning nodules. In addition, because (1) crosstalk between ROS and NO occurs in plant-microbe interactions (Zago et al. 2006; Zaninotto et al. 2006) and (2) growing evidence suggests a model in which ROS and glutathione (γ-glutamylcysteine-glycine, GSH) interact to regulate redox homeostasis and antioxidant signaling (Foyer and Noctor 2005), the roles of this reactive nitrogen species and of the major thiol antioxidant will also be considered.

2 ROS and the Early Steps of the Interaction

The involvement of ROS in the establishment of the legume-*Rhizobium* symbiosis has been delineated. In the early stages of the symbiotic interaction, oxidation of the dye nitroblue tetrazolium (NBT) can be detected in infection threads, indicating that superoxide anion (O_2^-) is produced during the infection process (Santos et al. 2001; Ramu et al. 2002). Moreover, this production was not observed when *Medicago truncatula* plants were inoculated with a *Sinorhizobium meliloti nodD1ABC* mutant, unable to produce Nod factors (NF), suggesting an NF role in the oxidative burst (Ramu et al. 2002). In the semiaquatic legume *Sesbania rostrata*, the formation of cortical infection pockets depends on a NF and was shown to be associated with localized cell death and the production of large amounts of hydrogen peroxide (H_2O_2) (D'Haeze et al. 2003). Moreover, experiments with ROS inhibitors and a H_2O_2 scavenger showed that, in lateral root base nodulation of *S. rostrata*, ROS were required for the initiation of nodule primordia (D'Haeze et al. 2003). Recently, the generation of ROS in the cortical cells of *M. truncatula* roots after inoculation with *S. meliloti* was observed in vivo, using the 2′,7′-dichlorofluorescein probe

(Peleg-Grossman et al. 2007). Moreover, the use of diphenyleneiodonium (DPI), which inhibits flavoproteins such as the gp91phox catalytic subunit of NADPH oxidase, abolished ROS production. This result suggests the possible involvement of *M. truncatula* NADPH oxidase homologues in such ROS production (Peleg-Grossman et al. 2007). The DPI treatment also suppressed root hair curling and infection thread formation, which further underscores ROS involvement in the early steps of symbiotic interactions (Peleg-Grossman et al. 2007).

However, the situation may be different in the very earliest steps of the symbiotic interaction, where the production of H_2O_2 appears to be inhibited by the NF (Shaw and Long 2003; Lohar et al. 2007); in the same way, a *S. meliloti nodC*$^-$ mutant, defective in NF biosynthesis, showed an increase in H_2O_2 accumulation (Bueno et al. 2001). The compatible interaction between *M. sativa* and *S. meliloti* is linked, at least in part, with increased antioxidant defence (particularly catalase and lipoxygenase) during the preinfection period (Bueno et al. 2001). More recently, the role of ROS in root hair deformation in the *M. truncatula–S. meliloti* symbiosis has been shown (Lohar et al. 2007). Exogenous application of ROS prevented root hair swelling and branching induced by NF. Transient treatment of roots with DPI mimicked NF treatment and resulted in root hair branching in the absence of NF. The results suggest a role for the transient lowering of ROS accumulation in governing NF-induced root hair deformation in legumes (Lohar et al. 2007).

Rhizobia appear to have an efficient antioxidant defence, in that ROS have not been detected in the microorganisms progressing within the threads (Santos et al. 2001; Rubio et al. 2004). Indeed, *S. meliloti* possesses two superoxide dismutases that convert O_2^- to O_2 and H_2O_2 (Santos et al. 2000; Hérouart et al. 2002) and three heme *b*-containing catalases, which are able to scavenge H_2O_2 (Hérouart et al. 1996; Ardissone et al. 2004). Bacterial catalases appear to play an important role in the nodule formation process as the double *katB/katC* and *katA/katC* mutants of *S. meliloti* are strongly impaired in nodule formation (Jamet et al. 2003). These observations suggest that maintaining the H_2O_2 steady-state level below some maximum value is necessary for optimal nodulation. On the other hand, a *S. meliloti* mutant overexpressing the constitutive catalase KatB (*katB*$^{++}$) and displaying an intracellular H_2O_2 concentration below that of the wild-type strain, exhibited a delayed nodulation phenotype (Jamet et al. 2007). This effect was associated with the formation of abnormal infection threads. These last results clearly indicate that the presence of H_2O_2 is essential for optimal symbiosis development and point to a signaling role for ROS during the symbiotic process. A comparative transcriptomic study between the wild-type strain and the *katB*$^{++}$ mutant is currently in progress in the authors' laboratory.

In addition to the catalases, *S. meliloti* genome analysis suggests that this bacterium contains three thiol peroxidase genes: the alkyl hydroperoxide reductase *ahpC*-like gene and two organic hydroperoxide resistance *ohr*-like genes, named *ohr1* and *ohr2*. Both types of enzymes display biochemically equivalent functions and catalyze the reduction of organic peroxides to the corresponding less toxic organic alcohols. Moreover, AhpC might use H_2O_2 in addition to alkyl hydroperoxide as a substrate (Seaver and Imlay 2001). The expression patterns of the three genes differ, reflecting possible distinct physiological roles both in the free-living and the symbiotic states.

It must be noted that the *ohr1* gene was shown to be actively transcribed during all steps of nodule development, while *ohr2* was poorly expressed (Mandon, unpublished results). Other non-enzymatic scavengers such as glutathione (see below) or other enzymatic systems might account for the antioxidant machinery of rhizobia. In *R. etli*, for example, a microaerobiosis-induced peroxiredoxin (PrxS) that is also expressed during the symbiotic interaction is involved, in conjunction with the catalase *katG*, in the maintenance of the redox balance during the nitrogen fixation process (Dombrecht et al. 2005). Furthermore, it has been demonstrated that the levels of peroxiredoxins are decreased during nodule development and are redox-regulated during nodule senescence (Groten et al. 2006).

The question of the origin of ROS formation during rhizobial infection and nodule organogenesis is still unanswered. As mentioned above, the formation of superoxide radicals in the infection threads (Santos et al. 2001), and the correlation during the early steps of the *M. truncatula–S. meliloti* symbiosis between decreased expression of two putative NADPH oxidase genes and the decreased flux of ROS (Lohar et al. 2007), favor the hypothesis of this type of enzyme involvement. Other possible sources of H_2O_2 are cell wall peroxidases, germin-like oxalate oxidases, and diamine oxidases (Wisniewski et al. 2000). Indeed, a germin-like oxidase from *Pisum sativum* has been characterized (PsGER1). This protein has superoxide dismutase activity and is associated with nodules (Gucciardo et al. 2007). Recently, a gene encoding a functional class III plant peroxidase (Srprx1) was shown to be transiently upregulated during nodulation in the tropical legume *S. rostrata* (Den Herder et al. 2007). Srprx1 expression was not induced after wounding or pathogen attack, consistent with the peroxidase being a symbiosis-specific isoform. On the other hand, the possible involvement of nodule peroxisomes in this process should be taken into account. In any case, it would be of interest to analyze the consequences for symbiosis of modifying plant ROS-producing activities.

As mentioned above, ROS interaction with GSH is a key modulating factor of the redox homeostasis and signaling; GSH is not a passive bystander in this crosstalk, but rather functions as a key signaling compound (Foyer and Noctor 2005). Moreover, it must be noted that one of the specificities of leguminous plants is the production of a GSH homolog, homoglutathione (γ-glutamylcysteine-β-alanine, hGSH) (Frendo et al. 1999; Matamoros et al. 1999b, 2003a). This specificity is linked to the presence of a new enzyme, homoglutathione synthetase (hGSHS), which uses β-alanine as a substrate, whereas glutathione synthetase (GSHS) uses glycine exclusively. hGSHS has been suggested to derive from GSHS by gene duplication (Frendo et al. 2001); hGSH is thought to have similar roles to GSH. To examine the importance of (h)GSH during nodule formation, the concentration of these tripeptides were lowered using buthionine sulfoximine, a specific inhibitor of (h)GSH synthesis, and transgenic roots expressing antisense transcripts for GSHS and hGSHS. Both approaches resulted in significantly diminished numbers of nodules on the plants, supporting an important role of (h)GSH during nodule development (Frendo et al. 2005). Deficiency in (h)GSH did not change the number of infection events or the expression of *Rhizobium-induced Peroxidase 1*, which demonstrates that the lower number of nodules was not linked to a nonspecific general stress process as observed during ethylene treatment (Oldroyd et al. 2001). In contrast, a strong diminution of the number of nodule meristems, correlated with the lower expression of the early-nodulin genes *Mtenod12* and *Mtenod40*, was observed in (h)GSH-depleted plants. GSH therefore plays a role in the nodule meristem formation. Interestingly, this inhibition of nodule formation is correlated to a diminution in the number of lateral roots, strongly suggesting that (h)GSH deficiency generally impairs meristem initiation. Thus, (h)GSH emerges as an essential factor involved in nodule development. This result is consistent with the data obtained in *Arabidopsis*, in which GSH plays a crucial role in cell cycle regulation and root apical meristem functioning (Sanchez-Fernandez et al. 1997; Vernoux et al. 2000). In order to define the modulation of gene expression by (h)GSH deficiency, a transcriptomic analysis was performed on *M. truncatula* plants inoculated with *S. meliloti* using the cDNA-amplified fragment length polymorphism (AFLP) technology. The expression pattern of 6,149 genes was studied during the first 4 days after inoculation. Of these, 181 were identified as differentially regulated between control and (h)GSH-deficient plants. This transcriptomic analysis showed the lower expression of genes involved in meristem formation and the accumulation of transcripts corresponding to defence genes in (h)GSH-deficient plants (Pucciariello et al., submitted). These results suggest that (h)GSH and more generally the redox state may be involved in both bacterial recognition and meristem formation during the nodulation process. The roles of the genes regulated during the (h)GSH deficiency needs to be investigated to clarify their involvement in the nodulation process.

The interaction between legumes and Rhizobia involves stressful conditions for the bacteria with the production of ROS and the potential conditions of nutritional and osmotic stress present in the infection thread. A *Rhizobium tropici* mutant strain bearing a single Tn*5-luxAB* insertion in a gene with high similarity to the *Escherichia coli gshB* gene (encoding glutathione synthetase) was hypersensitive to osmotic or oxidative stress (Riccillo et al. 2000). We therefore tested whether the bacterial GSH pool plays a role in the interaction between *M. truncatula* and *S. meliloti*. In silico analysis of the bacterial genome identified the likely *gshA* and *gshB* genes responsible for GSH synthesis in Rhizobia. The deduced amino acid sequence of the ORF SMc00825 was about 75% identical to putative *gshA* genes encoding *Rhizobiales* γ-glutamyl-cysteine synthetase (γECS) and about 55% identical to plant γECS. Interestingly, the sequences of *gshA* from γ-proteobacteria (*E. coli*) and α-proteobacteria (*S. meliloti*) fall into two distinct groups (Copley and Dhillon 2002). The latter group consists primarily of α-proteobacteria such as *S. meliloti, Mesorhizobium loti*, and *Agrobacterium tumefaciens*, and flowering plants such as *A. thaliana*, *M. truncatula* and *Glycine max*. Similarly, the putative translational product of ORF SMc00419 (named *gshB1*) had 85% identity to the predicted product of the *gshB* gene encoding GSHS in *Rhizobium tropici*. In the *S. meliloti* genome, the ORF SMb21586 (named *gshB2*) is also proposed as a putative GSHS but displays only 35% identity to *R. tropici* GshB protein.

In order to validate the identity of the *S. meliloti* genes involved in GSH synthesis, and to investigate the importance of GSH in the transition from bacterial free-living growth to the interaction with the host plant, defects in γECS or GSHS

(SmgshA and SmgshB1, respectively) were engineered into the wild-type *S. meliloti* strain Rm1021. The SmgshA mutant strain lacked GSH and was unable to grow without exogenously added GSH or functional complementation of *gshsA*. The SmgshB1 mutant strain accumulated the dipeptide intermediate γEC showing that *gshB1* encodes a GSHS. The result also indicates that the *gshsB2*-encoded protein is not the major GSHS during normal growth; indeed, the SmgshB1 strain grew normally. These results are significantly different from those observed in *E. coli*. Indeed, *E. coli gshA* mutant strain devoid of glutathione not only grows normally but also shows normal resistance to H_2O_2, cumene hydroperoxide, heat, or ionizing radiation (Greenberg and Demple 1986). The *S. meliloti* results are more similar to those observed in *Saccharomyces cerevisae*, in which is a *gshA* deletion is lethal (Grant et al. 1996). The thiol redox system involves on the one hand GSH, GSH reductase and glutaredoxins, while the NADH/NADPH-coupled thioredoxin reductases and thioredoxins share a similar architecture in *E. coli* and in yeast (Toledano et al. 2007). Although these pathways are functionally redundant in *E. coli*, that does not appear to be the case in yeast (Toledano et al. 2007). The different influence of GSH deficiency on cell growth observed between *S. meliloti* and *E. coli* may arise from a divergent evolutionarily role of the two branches of the thiol redox system in these two bacteria genera as observed for yeast and *E. coli*.

The nodulation efficiency of the SmgshA and SmgshB1 strains was tested during the interaction with alfalfa. Under our experimental procedures we found no evidence that the mutant SmgshA was able to induce nodules on *M. sativa*. Since the growth of SmgshA was severely affected, it seems logical to expect that its nodulation ability will also be negatively affected. The emergence of nodules on *M. sativa* following inoculation with mutant strain SmgshB was delayed in comparison to wt Rm1021, although final nodule number was unaffected. The delay observed in nodule formation could be connected with the strong upregulation of catalase expression observed in the SmgshB1 mutant. As already mentioned, modification of catalase activity modified the nodulation efficiency.

3 ROS and the Functioning Nodule

A prolonged production of ROS occurs in the later stages of the nodulation process. H_2O_2 production was detected in ultrathin sections of mature 6-week-old nodules as an electron-dense precipitate stained with cerium chloride (Santos et al. 2001; Rubio et al. 2004). Intense cerium labeling was also observed in the cell walls of infecting cells and in some infection threads around bacteria. The highly reducing environment needed for nitrogen fixation may lead to ROS formation, in that many electron donors (e.g., ferredoxin, uricase, and hydrogenase) are susceptible to auto-oxidation resulting in superoxide formation (Dalton et al. 1991). Oxyleghemoglobin, which is present at a high concentration in nodules, can also produce superoxide, which disproportionates to hydrogen peroxide (Puppo et al. 1991). Indeed, the reduction of leghemoglobin concentration in transgenic plants leads to a strong diminution of

ROS formation in nodules (Gunther et al. 2007). The production of ROS together with the high levels of transition metals (Becana and Klucas 1992) present in the nodule may lead to the generation of highly oxidizing species such as hydroxyl radicals, which reacts with all the main cellular components (Moller et al. 2007). Under these circumstances, a strong antioxidant defence may be crucial for efficient nodule metabolism (Matamoros et al. 2003b). The significantly higher level of (h)GSH observed in root nodule compared to noninoculated roots suggests a potential role for these molecules in the nitrogen-fixing symbiotic interaction. Moreover, the concentration of (h)GSH is three- to fourfold higher in actively nitrogen-fixing nodules than in ineffective ones (Dalton et al. 1993). The correlation between nitrogen fixation efficiency and the (h)GSH pool is also true in legumes under abiotic stresses such as dark stress (Gogorcena et al. 1997; Matamoros et al. 1999a) or treatment with nitrate (Escurado et al. 1996; Matamoros et al. 1999a), during which oxidative stress seems to be involved in nodule senescence. Finally, this relationship is further observed during natural nodule senescence (Evans et al. 1999; Loscos et al. 2008). However, the connection between nodule metabolism and (h)GSH content has never been verified by testing nitrogen fixation in transgenic or mutant plants with varying (h)GSH content.

GSH also appears to play an important role at the microsymbiont level. Indeed, the nitrogen-fixing capacity of nodules containing the SmgshB1 mutant was only 75% of that for wild-type bacteria. Microscopic analysis of nodule structure correlates the decrease in nitrogen fixation efficiency with an abnormal nodule structure that suggests early senescence of the bacteroid (Harrison et al. 2005). Thus, GSH in *S. meliloti* plays a crucial role in the nitrogen-fixing activity during symbiosis. Multiple hypotheses may explain the severe phenotype of SmgshB1 bacteria. The connection observed between GSH and the enzymatic antioxidant defence may imply that a severe oxidative stress is associated with nitrogen deficiency. However, multiple thiol redox targets may also be affected by GSH deficiency via the glutaredoxin system. In this framework, multiple thioredoxin mutants have altered nitrogen fixation activity (Loferer et al. 1993; Vargas et al. 1994; Castro-Sowinski et al. 2007). An overall genetic dissection of the thiol redox system would be welcome to resolve data on the importance of this regulatory pathway in *S. meliloti*.

NO production in functioning nodules may largely interfere with the ROS signaling described above. NO also appears to control primary and adventitious root organogenesis, a developmental process that shares common features with nodule formation, including the involvement of auxins (Correa-Aragunde et al. 2004; Lanteri et al. 2006). Indeed, recent advances point out that NO and auxin may be involved in controlling indeterminate nodule formation (Pii et al. 2007). Data from Shimoda et al. (2005) suggest that rapid and transient NO production, detected with the cell-permeable NO sensing probe 4,5 diaminofluorescein diacetate (DAF-2DA), occurs in *Lotus japonicus* roots inoculated with *Mesorizobium loti* (Shimoda et al. 2005). In *M. truncatula* the presence of NO during early stages of symbiosis remains unclear; however, its production in mature nodules has been clearly demonstrated. Direct detection of NO using the DAF-2DA probe was undertaken using confocal microscopy in *M. truncatula/S. meliloti* nodules (Baudouin et al. 2006).

The NO staining was localized in the bacteroid-containing cells of the fixing zone and was impaired by the NO scavenger carboxyPTIO. These data corroborate previous studies on soybean nodules in which the presence of NO complexed with leghemoglobin was detected in soybean nodules using electron paramagnetic resonance (Mathieu et al. 1998; Herold and Puppo 2005).

In the microsymbionts, a low level of free NO compatible with nitrogenase activity could act as a signal during the low oxygen response of the nitrogen-fixing cells. NO can regulate specific bacterial transcription factors such as NnrR, which controls denitrification gene expression (Mesa et al. 2003). This activation would involve the FixLJ pathway, which is activated by low oxygen concentrations and can readily fix NO (David et al. 1988; McGongile et al. 2000). Interestingly, several NO-induced genes are controlled by FixLJ under microaerobic conditions (Bobik et al. 2006). Thus, the identity of the FixLJ-dependent gene-regulating signal in planta may be suggested: low oxygen concentration, NO or both. On the plant side, a series of genes related to the plant response to hypoxia was recently identified in a screen for *M. truncatula* genes induced upon NO treatment (Ferrarini et al. 2008). As these hypoxia-related genes are also regulated in nitrogen-fixing *L. japonicus* nodules (Colebatch et al. 2004), one may extend to both symbiotic partners the hypothesis of NO regulating the low-oxygen response. In this context, nodulation affects the transcription of genes involved in sucrose metabolism, glycolysis, dicarboxylic acid metabolism and nitrogen assimilation (Gyorgyey et al. 2000; Colebatch et al. 2004; El Yahyaoui et al. 2004; Tesfaye et al. 2006). Many of these genes [a sucrose transporter, sucrose synthase (Susy), phosphofructokinase, glyceraldehyde dehydrogenase (GaPDH), isocitrate dehydrogenase, malate dehydrogenase (MDH), glutamine synthetase (GS), asparagine synthetase, and alanine amino transferase] are regulated by NO during the *M. truncatula–S. meliloti* symbiosis (Ferrarini et al. 2008). Taken together, the data suggest that NO may act as a "metabolic switch" from aerobic conditions to microaerophilic ones, for both partners. In addition to the regulation of gene expression, NO controls the activity of numerous proteins through the S-nitrosylation of their Cys residues (Mannick and Schonhoff 2002). A recent proteomic analysis showed that, among the 100 S-nitrosylated proteins identified in *A. thaliana*, several of these are involved in carbon metabolism (Susy, GaPDH, phosphoglycerate kinase, phosphoenolpyruvate carboxylase, pyruvate kinase, MDH…) or nitrogen metabolism (GS, glutamate dehydrogenase) (Lindermayr et al. 2005). The in silico analysis of the *Arabidopsis* genome showed that hexokinase and AS are likely candidates for S-nitrosylation (Huber and Hardin 2004). At the enzymatic level, the NO-dependent modulation of the activity of most of these enzymes in extracts of *M. truncatula* roots and nodules (Brouquisse, unpublished) suggests that these enzymes might be controlled by S-nitrosylation in vivo. Thus, NO could regulate carbon and nitrogen metabolism at both the transcriptional and the post-translational level in symbiotic nodules. The identification of the NO-generating system(s) in nodules will greatly help in deciphering the role of this reactive species during the symbiotic process.

Finally, ROS involvement in nodule senescence process has been suggested (Alesandrini et al. 2003; Rubio et al. 2004), and a model in which ROS and antioxidants interact with hormones such as abscissic acid in orchestrating nodule senescence has been proposed (Groten et al. 2005; Puppo et al. 2005). Nodules have high levels of the redox buffers glutathione and ascorbate, which are important in nodule senescence. These metabolites decline with nitrogen fixation as the nodule ages, but the resultant decrease in redox buffering capacity does not necessarily lead to enhanced ROS (Puppo et al. 2005).

4 Conclusion

It appears that ROS play important roles in the establishment and functioning of the symbiotic nodule. A better definition of the spatio-temporal characteristics of ROS production is needed to define precisely their subtle signaling action in the symbiotic process. Moreover, modulating the ROS-producing system(s) will allow identification of the ROS target genes and will provide information on the nature of the processes regulated by ROS in both partners during symbiosis. ROS may also induce post-translational modifications, and the identification of the nodule "peroxidome" will be very useful to obtain valuable information on the role of ROS at the protein level. Last but not least, the identification of common H_2O_2, NO, and GSH targets (including the identification of the nodule "nitrosylome" and "glutathionylome") will provide new insight in the regulatory networks operating during the symbiotic interaction. In that symbiosis is a good model for infectious and developmental processes, this information will be of interest for the plant biologists in general.

References

Alesandrini F, Mathis R, Van de Sype G, Hérouart D, Puppo A (2003) Possible roles of a cysteine protease and hydrogen peroxide in soybean nodule development and senescence. New Phytol 158:131–138

Apel K, Hirt H (2004) Reactive oxygen species: metabolism, oxidative stress, and signal transduction. Annu Rev Plant Biol 55:373–399

Ardissone S, Frendo P, Laurenti E, Jantschko W, Obinger C, Puppo A, Ferrari RP (2004) Purification and physical-chemical characterization of the three hydroperoxidases from the symbiotic bacterium *Sinorhizobium meliloti*. Biochemistry 43:12692–12699.

Baudouin E, Pieuchot L, Engler G, Pauly N, Puppo A (2006) Nitric oxide is formed in *Medicago truncatula-Sinorhizobium meliloti* functional nodules. Mol Plant Microbe Interact 19:970–975

Becana M, Klucas RV (1992) Transition metals in legume root nodules. Iron dependant free radical production increases during nodule senescence. Proc Natl Acad Sci USA 87:7295–7299

Bobik C, Meilhoc E, Batut J (2006) FixJ: a major regulator of the oxygen limitation response and late symbiotic functions of *Sinorhizobium meliloti*. J Bacteriol 188:4890–4902

Bueno P, Soto MJ, Rodriguez-Rosales MP, Sanjuan J, Olivares J, Donaire JP (2001) Time-course of lipoxygenase, antioxidant enzyme activities and H_2O_2 accumulation during early stages of *Rhizobium* legume symbiosis. New Phytol 152:91–96

Castro-Sowinski S, Matan O, Bonafede P, Okon Y (2007) A thioredoxin of *Sinorhizobium meliloti* CE52G is required for melanin production and symbiotic nitrogen fixation. Mol Plant Microbe Interact 20:986–993

Colebatch G, Desbrosses G, Ott T, Krusell L, Montanari O, Kloska S, Kopka J, Udvardi MK (2004) Global changes in transcription orchestrate metabolic differentiation during symbiotic nitrogen fixation in *Lotus japonicus*. Plant J 39:487–512

Copley SD, Dhillon JK (2002) Lateral gene transfer and parallel evolution in the history of glutathione biosynthesis genes. Genome Biol 3:research0025

Correa-Aragunde N, Graziano M, Lamattina L (2004) Nitric oxide plays a central role in determining lateral root development in tomato. Planta 218:900–905

D'Haeze W, De Rycke R, Mathis R, Goormachtig S, Pagnotta S, Verplancke C, Capoen W, Holsters M (2003) Reactive oxygen species and ethylene play a positive role in lateral root base nodulation of a semiaquatic legume. Proc Natl Acad Sci USA 100:11789–11794

Dalton DA, Langeberg L, Treneman NC (1993) Correlations between the ascorbate-glutathione pathway and effectiveness in legume root nodules. Physiol Plant 84:365–370

Dalton DA, Post CJ, Langeberg L (1991) Effects of ambient oxygen and of fixed nitrogen on concentrations of glutathione, ascrobate, and associated enzymes in soybean root nodules. Plant Physiol 96:812–818

David M, Daveran ML, Batut J, Dedieu A, Domergue O, Ghai J, Hertig C, Boistard P, Kahn D (1988) Cascade regulation of nif gene expression in *Rhizobium meliloti*. Cell 54:671–683

Den Herder J, Lievens S, Rombauts S, Holsters M, Goormachtig S (2007) A symbiotic plant peroxidase involved in bacterial invasion of the tropical legume *Sesbania rostrata*. Plant Physiol 144:717–727

Dombrecht B, Heusdens C, Beullens S, Verreth C, Mulkers E, Proost P, Vanderleyden J, Michiels J (2005) Defence of *Rhizobium etli* bacteroids against oxidative stress involves a complexly regulated atypical 2-Cys peroxiredoxin. Molecular microbiology 55:1207–1221

El Yahyaoui F, Kuster H, Ben Amor B, Hohnjec N, Puhler A, Becker A, Gouzy J, Vernie T, Gough C, Niebel A, Godiard L, Gamas P (2004) Expression profiling in Medicago truncatula identifies more than 750 genes differentially expressed during nodulation, including many potential regulators of the symbiotic program(1[w]). Plant Physiol 136:3159–3176

Escurado PR, Minchin FR, Gogoncena Y, Iturbe-Ormaetxe I, Klucas RV, Becana M (1996) Involved of activated oxygen in nitrate induced senescence of pea root nodules. Plant Physiol 110:1187–1195

Evans PJ, Gallesi D, Mathieu C, Hernandez MJ, de Felipe M, Halliwell B, Puppo A (1999) Oxidative stress occurs during soybean nodule senescence. Planta 208:73–79

Ferrarini A, De Stefano M, Baudouin E, Pucciariello C, Polverari A, Puppo A, Delledonne M (2008) Expression of *Medicago truncatula* genes responsive to nitric oxide in pathogenic and symbiotic conditions. Mol Plant Microbe Interact 21:781–790

Foreman J, Demidchik V, Bothwell JH, Mylona P, Miedema H, Torres MA, Linstead P, Costa S, Brownlee C, Jones JD, Davies JM, Dolan L (2003) Reactive oxygen species produced by NADPH oxidase regulate plant cell growth. Nature 422:442–446

Foyer CH, Noctor G (2005) Redox homeostasis and antioxidant signaling: a metabolic interface between stress perception and physiological responses. Plant Cell 17:1866–1875

Frendo P, Gallesi D, Turnbull R, Van de Sype G, Hérouart D, Puppo A (1999) Localisation of glutathione and homoglutathione in Medicago truncatula is correlated to a differential expression of genes involved in their synthesis. Plant J 17:215–219

Frendo P, Harrison J, Norman C, Hernandez Jimenez MJ, Van de Sype G, Gilabert A, Puppo A (2005) Glutathione and homoglutathione play a critical role in the nodulation process of *Medicago truncatula*. Mol Plant Microbe Interact 18:254–259

Frendo P, Jimenez MJ, Mathieu C, Duret L, Gallesi D, Van de Sype G, Hérouart D, Puppo A (2001) A *Medicago truncatula* homoglutathione synthetase is derived from glutathione synthetase by gene duplication. Plant Physiol 126:1706–1715

Gogorcena Y, Gordon AJ, Escuredo PR, Minchin FR, Witty JF, Moran JF, Becana M (1997) N2 fixation, carbon metabolism and oxidative damage in nodules of dark stressed Common Bean plants. Plant Physiol 113:1193–1201

Grant CM, MacIver FH, Dawes IW (1996) Glutathione is an essential metabolite required for resistance to oxidative stress in the yeast *Saccharomyces cerevisiae*. Curr Genet 29:511–515

Greenberg JT, Demple B (1986) Glutathione in Escherichia coli is dispensable for resistance to H_2O_2 and gamma radiation. J Bacteriol 168:1026–1029

Groten K, Dutilleul C, van Heerden PD, Vanacker H, Bernard S, Finkemeier I, Dietz KJ, Foyer CH (2006) Redox regulation of peroxiredoxin and proteinases by ascorbate and thiols during pea root nodule senescence. FEBS Lett 580:1269–1276

Groten K, Vanacker H, Dutilleul C, Bastian F, Bernard S, Carzaniga R, Foyer CH (2005) The roles of redox processes in pea nodule development and senescence. Plant Cell Environ 28:1293–1304

Gucciardo S, Wisniewski JP, Brewin NJ, Bornemann S (2007) A germin-like protein with superoxide dismutase activity in pea nodules with high protein sequence identity to a putative rhicadhesin receptor. J Exp Bot 58:1161–1171

Gunther C, Schlereth A, Udvardi M, Ott T (2007) Metabolism of reactive oxygen species is attenuated in leghemoglobin-deficient nodules of Lotus japonicus. Mol Plant Microbe Interact 20:1596–1603

Gyorgyey J, Vaubert D, Jimenez-Zurdo JI, Charon C, Troussard L, Kondorosi A, Kondorosi E (2000) Analysis of *Medicago truncatula* nodule expressed sequence tags. Mol Plant Microbe Interact 13:62–71

Harrison J, Jamet A, Muglia CI, Van de Sype G, Aguilar OM, Puppo A, Frendo P (2005) Glutathione plays a fundamental role in growth and symbiotic capacity of *Sinorhizobium meliloti*. J Bacteriol 187:168–174

Herold S, Puppo A (2005) Oxyleghemoglobin scavenges nitrogen monoxide and peroxynitrite: a possible role in functioning nodules? J Biol Inorg Chem 10:935–945

Hérouart D, Baudouin E, Frendo P, Harrison J, Santos R, Jamet A, Van de Sype G, Touati D, Puppo A (2002) Reactive oxygen species, nitric oxide and glutathione: a key role in the establishment of the legume – *Rhizobium* symbiosis. Plant Physiol Biochem 40:619–624

Hérouart D, Sigaud S, Moreau S, Frendo P, Touati D, Puppo A (1996) Cloning and characterization of the *katA* gene of *Rhizobium meliloti* encoding a hydrogen peroxide-inducible catalase. J Bacteriol 178:6802–6809.

Huber SC, Hardin SC (2004) Numerous posttranslational modifications provide opportunities for the intricate regulation of metabolic enzymes at multiple levels. Curr Opin Plant Biol 7:318–322

Jamet A, Mandon K, Puppo A, Herouart D (2007) H_2O_2 is required for optimal establishment of the *Medicago sativa/Sinorhizobium meliloti* symbiosis. J Bacteriol 189:8741–8745

Jamet A, Sigaud S, Van de Sype G, Puppo A, Hérouart D (2003) Expression of the bacterial catalase genes during *Sinorhizobium meliloti-Medicago sativa* symbiosis and their crucial role during the infection process. Mol Plant Microbe Interact 16:217–225

Lanteri ML, Pagnussat GC, Lamattina L (2006) Calcium and calcium-dependent protein kinases are involved in nitric oxide- and auxin induced adventitious root formation in cucumber. J Exp Bot 57:1341–1351

Lindermayr C, Saalbach G, Durner J (2005) Proteomic identification of S-nitrosylated proteins in *Arabidopsis*. Plant Physiol 137:921–930

Loferer H, Bott M, Hennecke H (1993) Bradyrhizobium japonicum TlpA, a novel membrane-anchored thioredoxin-like protein involved in the biogenesis of cytochrome aa3 and development of symbiosis. EMBO J 12:3373–3383

Lohar DP, Haridas S, Gantt JS, VandenBosch KA (2007) A transient decrease in reactive oxygen species in roots leads to root hair deformation in the legume-rhizobia symbiosis. New Phytol 173:39–49

Long SR (2001) Genes and signals in the *Rhizobium*-legume symbiosis. Plant Physiol 125:69–72

Loscos J, Matamoros MA, Becana M (2008) Ascorbate and homoglutathione metabolism in common bean nodules under stress conditions and during natural senescence. Plant Physiol 146:1282–1292

Mannick JB, Schonhoff CM (2002) Nitrosylation: the next phosphorylation? Arch Biochem Biophys 408:1–6

Matamoros MA, Baird LM, Escuredo PR, Dalton DA, Minchin FR, Iturbe-Ormaetxe I, Rubio MC, Moran JF, Gordon AJ, Becana M (1999a) Stress induced legume root nodule senescence. Physiological, biochemical and structural alterations. Plant Physiol 121:97–112

Matamoros MA, Clemente MR, Sato S, Asamizu E, Tabata S, Ramos J, Moran JF, Stiller J, Gresshoff PM, Becana M (2003a) Molecular analysis of the pathway for the synthesis of thiol tripeptides in the model legume *Lotus japonicus*. Mol Plant Microbe Interact 16:1039–1046

Matamoros MA, Dalton DA, Ramos J, Clemente MR, Rubio MC, Becana M (2003b) Biochemistry and molecular biology of antioxidants in the rhizobia–legume symbiosis. Plant Physiol 133:499–509

Matamoros MA, Moran JF, Iturbe-Ormaetxe I, Rubio MC, Becana M (1999b) Glutathione and homoglutathione synthesis in legume root nodules. Plant Physiol 121:879–888

Mathieu C, Moreau S, Frendo P, Puppo A, Davies MJ (1998) Direct detection of radicals in intact soybean nodules: presence of nitric oxide leghemoglobin complexes. Free Rad Bio Med 24:1242–1249

McGongile B, Keeler SJ, Lau SMC, Koeppe MJ, O'Keefe DP (2000) A genomics approach to the comprehensive analysis of the glutathione S-transferase gene family in soybean and maize. Plant Physiol 124:1105–1120

Mesa S, Bedmar EJ, Chanfon A, Hennecke H, Fischer HM (2003) *Bradyrhizobium japonicum* NnrR, a denitrification regulator, expands the FixLJ-FixK2 regulatory cascade. J Bacteriol 185:3978–3982

Moller IM, Jensen PE, Hansson A (2007) Oxidative modifications to cellular components in plants. Annu Rev Plant Biol 58:459–481

Neill SJ, Desikan R, Clarke A, Hurst RD, Hancock JT (2002) Hydrogen peroxide and nitric oxide as signalling molecules in plants. J Exp Bot 53:1237–1247

Oldroyd GE, Downie JA (2008) Coordinating nodule morphogenesis with rhizobial infection in legumes. Annu Rev Plant Biol 59:519–546

Oldroyd GED, Engstrom EM, Long SR (2001) Ethylene inhibits the nod factor signal transduction pathway of *Medicago truncatula*. Plant Cell 13:1835–1849

Pauly N, Pucciariello C, Mandon K, Innocenti G, Jamet A, Baudouin E, Herouart D, Frendo P, Puppo A (2006) Reactive oxygen and nitrogen species and glutathione: key players in the legume–*Rhizobium* symbiosis. J Exp Bot 57:1769–1776

Pei ZM, Murata Y, Benning G, Thomine S, Klusener B, Allen GJ, Grill E, Schroeder JI (2000) Calcium channels activated by hydrogen peroxide mediate abscisic acid signalling in guard cells. Nature 406:731–734

Peleg-Grossman S, Volpin H, Levine A (2007) Root hair curling and Rhizobium infection in *Medicago truncatula* are mediated by phosphatidylinositide-regulated endocytosis and reactive oxygen species. J Exp Bot 58:1637–1649

Penmetsa RV, Cook DR (1997) A legume ethylene-insensitive mutant hyperinfected by its rhizobial symbiot. Science 275:527–530

Pii Y, Crimi M, Cremonese G, Spena A, Pandolfini T (2007) Auxin and nitric oxide control indeterminate nodule formation. BMC Plant Biol 7:21

Puppo A, Groten K, Bastian F, Carzaniga R, Soussi M, Lucas MM, de Felipe MR, Harrison J, Vanacker H, Foyer CH (2005) Legume nodule senescence: roles for redox and hormone signalling in the orchestration of the natural aging process. New Phytol 165:683–701

Puppo A, Herrada G, Rigaud J (1991) Lipid peroxidation in peribacteroid membranes from french-bean nodules. Plant Physiol 96:826–830

Ramu SK, Peng HM, Cook DR (2002) Nod factor induction of reactive oxygen species production is correlated with expression of the early nodulin gene *rip1* in *Medicago truncatula*. Mol Plant Microbe Interact 15:522–528

Riccillo PM, Muglia CI, de Bruijn FJ, Roe AJ, Booth IR, Aguilar OM (2000) Glutathione is involved in environmental stress responses in *Rhizobium tropici*, including acid tolerance. J Bacteriol 182:1748–1753

Rubio MC, James EK, Clemente MR, Bucciarelli B, Fedorova M, Vance CP, Becana M (2004) Localization of superoxide dismutases and hydrogen peroxide in legume root nodules. Mol Plant Microbe Interact 17:1294–1305

Sanchez-Fernandez R, Fricker M, Corben LB, White NS, Sheard N, Leaver CJ, Van Montagu M, Inzé D, May MJ (1997) Cell proliferation and hair tip growth in the *Arabidopsis* root are under mechanistically different forms of redox control. Proc Natl Acad Sci USA 94:2745–2750

Santos R, Hérouart D, Puppo A, Touati D (2000) Critical protective role of bacterial superoxide dismutase in *Rhizobium*-legume symbiosis. Mol Microbiol 38:750–759

Santos R, Hérouart D, Sigaud S, Touati D, Puppo A (2001) Oxidative burst in alfalfa-*Sinorhizobium meliloti* symbiotic interaction. Mol Plant Microbe Interact 14:86–89

Seaver LC, Imlay JA (2001) Hydrogen peroxide fluxes and compartmentalization inside growing *Escherichia coli*. J Bacteriol 183:7182–7189

Shaw SL, Long SR (2003) Nod factor inhibition of reactive oxygen efflux in a host legume. Plant Physiol 132:2196–2204

Shimoda Y, Nagata M, Suzuki A, Abe M, Sato S, Kato T, Tabata S, Higashi S, Uchiumi T (2005) Symbiotic *Rhizobium* and nitric oxide induce gene expression of non-symbiotic hemoglobin in *Lotus japonicus*. Plant Cell Physiol 46:99–107

Tesfaye M, Samac DA, Vance CP (2006) Insights into symbiotic nitrogen fixation in *Medicago truncatula*. Mol Plant Microbe Interact 19:330–341

Toledano MB, Kumar C, Le Moan N, Spector D, Tacnet F (2007) The system biology of thiol redox system in *Escherichia coli* and yeast: differential functions in oxidative stress, iron metabolism and DNA synthesis. FEBS Lett 581:3598–3607

Vargas C, Wu G, Davies AE, Downie JA (1994) Identification of a gene encoding a thioredoxin-like product necessary for cytochrome c biosynthesis and symbiotic nitrogen fixation in *Rhizobium leguminosarum*. J Bacteriol 176:4117–4123

Vasse J, De Billy F, Truchet G (1993) Abortion of infection during the *Rhizobium meliloti*-alfalfa symbiotic interaction is accompanied by a hypersensitive reaction. Plant J 4:555–566

Vernoux T, Wilson RC, Seeley KA, Reichheld JP, Muroy S, Brown S, Maughan SC, Cobbett CS, Van Montagu M, Inze D, May MJ, Sung ZR (2000) The root meristemless1/cadmium sensitive2 gene defines a glutathione-dependent pathway involved in initiation and maintenance of cell division during postembryonic root development. Plant Cell 12:97–110

Vranova E, Inze D, Van Breusegem F (2002) Signal transduction during oxidative stress. J Exp Bot 53:1227–1236

Wisniewski JP, Rathbun EA, Knox JP, Brewin NJ (2000) Involvement of diamine oxidase and peroxidase in insolubilization of the extracellular matrix: implications for pea nodule initiation by *Rhizobium leguminosarum*. Mol Plant Microbe Interact 13:413–420

Zago E, Morsa S, Dat JF, Alard P, Ferrarini A, Inze D, Delledonne M, Van Breusegem F (2006) Nitric oxide- and hydrogen peroxide-responsive gene regulation during cell death induction in tobacco. Plant Physiol 141:404–411

Zaninotto F, La Camera S, Polverari A, Delledonne M (2006) Cross talk between reactive nitrogen and oxygen species during the hypersensitive disease resistance response. Plant Physiol 141:379–383

Hydrogen Peroxide-Responsive Genes in Stress Acclimation and Cell Death

Sandy Vanderauwera, Frank A. Hoeberichts, and Frank Van Breusegem

Abstract Reactive oxygen species (ROS) are key signalling molecules that regulate growth and development and coordinate responses to biotic and abiotic stresses. ROS homeostasis is controlled through a complex network of ROS production and scavenging enzymes. Recently, the first genes involved in ROS perception and signal transduction have been identified and, currently, we are facing the challenge to uncover the other players within the ROS regulatory gene network. The specificity of ensuing cellular responses depends on the type of ROS and their subcellular production sites. Various experimental systems, including catalase-deficient plants, in combination with genome-wide expression studies demonstrated that increased hydrogen peroxide (H_2O_2) levels significantly affect the transcriptome of plants and are capable of launching both defence responses and cell death events. A comparative analysis between H_2O_2-induced transcriptional changes and those provoked by different environmental stresses, not only consolidated a prominent role for H_2O_2 signalling in stress acclimation, but also allowed the identification of new candidate regulatory genes within the plant's abiotic stress response.

1 Dual Face of Reactive Oxygen Species

Reactive oxygen species (ROS) are partially reduced or activated derivatives of oxygen that are formed during normal cell metabolism in different subcellular compartments. Adverse environmental conditions lead to an increased production of ROS, resulting in oxidative stress. At higher concentrations, ROS cause oxidative injuries by reacting with different cellular components, including DNA, proteins and lipids, and become toxic to the cell. In order to avoid ROS-dependent cellular damage, a fine-tuned balance between ROS production and scavenging is essential. In *Arabidopsis thaliana*, this balance is governed by a complex network that comprises

S. Vanderauwera, F.A. Hoeberichts, and F. Van Breusegem (✉)
Department of Plant Systems Biology, VIB, and Department of Plant Biotechnology and Genetics, Ghent University, 9052 Gent, Belgium
e-mail: frank.vanbreusegem@psb.vib-ugent.be

L.A. del Río and A. Puppo (eds.), *Reactive Oxygen Species in Plant Signaling*, Signaling and Communication in Plants, DOI 10.1007/978-3-642-00390-5_9,

at least 152 genes encoding both ROS-producing and ROS-scavenging enzymes (Mittler et al. 2004). The different components of the antioxidant system are diversified in their scavenging capacity not only in terms of specificity and affinity towards the different ROS species, but also by their distinct subcellular location in the vicinity of the ROS production sites. Peroxisomes, chloroplasts and mitochondria are organelles with highly oxidising metabolic activities or with intense rates of electron flow and are, hence, major sources of ROS production in plant cells. Therefore, it is not surprising that these subcellular compartments contain various ROS-scavenging enzymes (Mittler et al. 2004). The tight regulation of ROS homeostasis also creates a baseline on which ROS spikes can signal in different cellular processes. Although the underlying signalling function of ROS has long been masked by the inherent toxic nature of these oxidants, their triggering ability during various developmental processes and environmental stress responses is now widely recognised (Foyer and Noctor 2005; Torres and Dangl 2005; Gapper and Dolan 2006; Mullineaux et al. 2006; Fig. 1). ROS with documented signalling functions include hydrogen peroxide (H_2O_2), singlet oxygen ($^1O_2$), hydroxyl radical (OH•), and superoxide anion radical ($O_2^{\bullet -}$). How ROS and ROS-derived signals are perceived, integrated and transduced to eventually control a broad range of biological processes remains largely elusive. ROS might interact selectively with specific target molecules that thereby sense and translate increased ROS levels into a transcriptional response (Apel and Hirt 2004). Intensity, duration and subcellular localisation of ROS production are likely crucial in the cellular and intracellular transduction of the ROS signals (Bailey-Serres and Mittler 2006).

2 ROS Perception and Signal Transduction

Increased ROS levels in different subcellular compartments can result in changes in nuclear gene expression, indicating a ROS-dependent intraorganellar signalling (Vanderauwera et al. 2005; Rhoads et al. 2006; Koussevitzky et al. 2007). Recently,

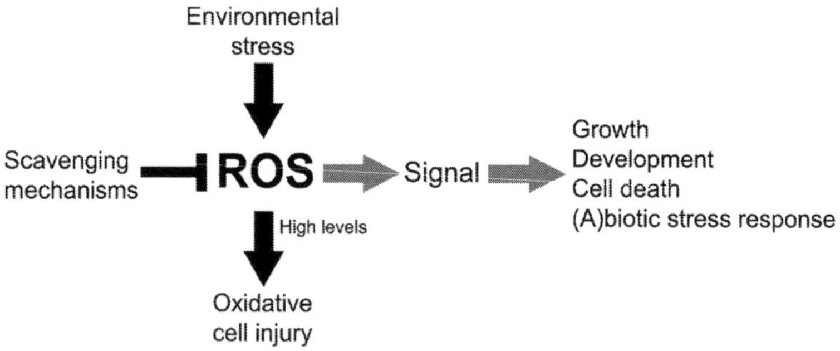

Fig. 1 The dual face of ROS. Unfavourable environmental conditions result in excessive ROS production that leads to oxidative cell injuries at high concentrations. To prevent ROS-dependent cellular damage, plants have a broad range of antioxidant mechanisms at their disposal to use ROS simultaneously as a signal within various biological processes, such as growth and development, defence responses to biotic and abiotic stimuli and programmed cell death

the first genes involved in ROS perception and signal transduction have been identified (Rentel et al. 2004; Kaminaka et al. 2006; Lee et al. 2007), but it remains a challenge to identify other players in the gene regulatory networks and to unravel their mode of action in ROS signal perception and transduction.

One likely mechanism for cells to perceive the presence of ROS is through ROS-dependent protein modifications. The oxidising nature of H_2O_2 and $O_2{\bullet}^-$ allows the direct modification of thiol groups in proteins, although other amino acids, such as tyrosine, tryptophan and histidine, may be oxidised as well. Oxidation of cysteine (Cys) residues might lead to disulphide bridging and subsequent conformational changes or, alternatively, to a variable degree of oxidation that depends on the residue's redox potential and the intracellular redox state (Hancock et al. 2006). The alteration of key Cys residues of transcription factors provides a direct and powerful means to influence nuclear gene expression. For example, in *Escherichia coli*, the transcription factor OxyR is activated by H_2O_2 (Zheng et al. 1998), whereas in *Saccharomyces cerevisiae*, oxidation of YAP1 following oxidative stress results in its nuclear accumulation and subsequent activation of antioxidant gene expression (Delaunay et al. 2000). In plants, redox regulation has been reported for the nonexpressor of pathogenesis-related 1 (NPR1) and the leucine zipper transcription factor TGA1 that are both important mediators of systemic acquired resistance during plant-pathogen interactions. Reduction of key Cys residues in these proteins relocates NPR1 to the nucleus and modulates the DNA-binding activity of the NPR1/TGA1 protein complex, thereby decisively affecting downstream gene expression (Després et al. 2003; Mou et al. 2003). The activity of the transcription factor Rap2.4a that controls plastid antioxidant gene expression is also redox regulated. Rap2.4a is activated by a dithiol/disulphide transition of regulatory Cys residues and subsequent changes in the quaternary structure (Shaikhali et al. 2008).

Transcriptional modifications may also require more upstream sensing and transduction of ROS and ROS-derived signals. In yeast, histidine kinases (HKs) of two-component signalling systems have been reported to function as sensors of oxidative stress (Singh 2000). In *Arabidopsis*, the HK of the ethylene receptor ETR1 appears to be essential for H_2O_2 perception leading to stomatal closure. Mutation of a Cys residue in the N-terminal region of ETR1 disrupted the H_2O_2 perception and diminished the stomatal closure in response to H_2O_2, indicating that this thiol group is important for H_2O_2 signalling. As the kinase domain of ETR1 was not required for H_2O_2 signalling, the H_2O_2-provoked signalling through ETR1 was seemingly unrelated to its well-described role as an ethylene receptor (Desikan et al. 2005). In addition, the *Arabidopsis* HK5 plays a crucial role in mediating H_2O_2-dependent processes in stomatal guard cells that are induced by both environmental and hormonal signals (Desikan et al. 2008).

A mitogen-activated protein kinase (MAPK) cascade and several protein phosphatases are involved in relaying H_2O_2 signals in *Arabidopsis*. Both ANP1, a MAPK kinase kinase, and OXI1, a serine/threonine protein kinase, are important for H_2O_2 sensing and the activation of a MAPK cascade (Kovtun et al. 2000; Rentel et al. 2004). H_2O_2 also strongly activates the nucleotide diphosphate kinase 2 (NDPK2) that specifically interacts with MPK3 and MPK6, thereby regulating the cellular redox state (Moon et al. 2003). MPK6 can also be activated via H_2O_2-mediated inhibition of the protein tyrosine phosphatase PTP1 (Gupta and Luan 2003), indicating that protein phosphatases might act as negative regulators of ROS-induced MAPK pathways (Schweighofer et al. 2004).

Taken together, these studies give a glimpse of the complex signalling network that underlies the cellular response to elevated levels of ROS and, more particularly, H_2O_2. However, specific signalling effects have also been reported for singlet oxygen and superoxide anion, involving the two closely related chloroplastic proteins EXECUTER1 and EXECUTER2, and LESION SIMULATING DISEASE 1 (LSD1), respectively (Mateo et al. 2004; Wagner et al. 2004; Kaminaka et al. 2006; Lee et al. 2007).

3 ROS-Dependent Gene Expression

The integrated cellular response to oxidative stress is particularly suited for a genomic approach: ROS levels can be modulated relatively easily in an organism by exposing it to increased dosages of oxidants or by depleting or perturbing its antioxidant capacities (Table 1). Genome-wide expression tools provided the means to assess the overall effect of ROS signalling on gene expression in different organisms. Transcriptomic and proteomic surveys in bacterial, yeast and mammalian cell lines have provided significant insights into the cellular responses towards oxidative stress (Causton et al. 2001; Mostertz et al. 2004; Murray et al. 2004; Vandenbroucke et al. 2008). In plants, the first report on a genome-wide expression analysis was provided by Desikan et al. (2001). Using a cDNA microarray, at least 113 and 62 transcripts were shown to be induced and repressed, respectively, in *Arabidopsis* cell suspensions treated with a specific dose of H_2O_2. A substantial number of these genes were predicted to function within the defence response, cell signalling, transcription and cell death. In tobacco (*Nicotiana tabacum*), a combination of differential display and cDNA macroarray analysis identified 53 transcripts involved in stress acclimation responses (Vranová et al. 2002). The two experimental set-ups relied on the exogenous addition of H_2O_2 and a superoxide generator, respectively. An alternative approach to modulate ROS levels in planta is through the use of loss-of-function mutants in specific antioxidant genes. The cDNA-amplified fragment length polymorphism (AFLP) technology revealed that more than 700 transcripts were differentially expressed upon increase of photorespiratory H_2O_2 (Vandenabeele et al. 2003). Later, the advent of different microarray platforms that covered the majority of the annotated *Arabidopsis* genes allowed the assessment of the transcriptome of available mutants and transgenic plants perturbed in the antioxidant gene network (op den Camp et al. 2003; Rizhsky et al. 2003; Davletova et al. 2005a; Umbach et al. 2005; Vanderauwera et al. 2005) (Table 1). These surveys provided not only detailed inventories of ROS-dependent gene expression in plants, but also evidence for the specific signalling capacities of different ROS that allow them to act efficiently in various developmental processes and environmental stress responses. A meta-analysis of oxidative stress-responsive transcripts in *Arabidopsis* revealed both common and specific responses towards the different ROS signals

Table 1 Overview of studies describing the impact of different ROS on gene expression

	Major ROS	Localisation	Plant system	Experimental details	Expression profiling platform	Reference
Exogenous ROS						
H_2O_2	H_2O_2	Unknown	Cell suspension	1.5, 3 h 20 mM	cDNA microarray (8K)	Desikan et al. (2001)
H_2O_2	H_2O_2	Unknown	Cell suspension	5 h 10 mM	cDNA microarray (7K)	Takahashi et al. (2004)
H_2O_2	H_2O_2	Unknown	2-week-old plants	1, 3, 6, 12 h 5 mM (liquid medium)	cDNA microarray (26K)	Kim et al. (2005)
H_2O_2	H_2O_2	Unknown	5-day-old plants	1 h 20 mM (liquid medium)	Affymetrix ATH1 microarray (22K)	Davletova et al. (2005b)
Aminotriazol	H_2O_2	Peroxisomes	4-week-old plants	7 h 20 mM (spray)	Agilent Arab2 microarray (22K)	Gechev et al. (2005)
AAL toxin	H_2O_2	Unknown	4-week-old plants	7, 24, 48, 72 h 200 nM (infiltration)	Agilent Arab2 microarray (22K)	Gechev et al. (2004)
Ozone	O_3, $O_2{\bullet}^-$, H_2O_2	Apoplast	4-week-old plants	1, 4, 8, 12, 24 h 300 nL L^{-1} (6 h exposure)	Operon microarray (26K)	Mahalingam et al. (2006)
			4-week-old plants	3, 6, 12, 24 h 300 ppb (6 h exposure)	Affymetrix ATH1 microarray (22K)	Tosti et al. (2006)
			4-week-old plants	3, 6 h 350 ppb (6 h exposure)	Affymetrix ATH1 microarray (22K)	Ludwikow et al. (2004)
			Rosette stage	8-12 days free air concentration enrichment	Operon microarray (26K)	Li et al. (2006)
Endogenous ROS						
cat2 RNAi	H_2O_2	Peroxisomes	6-week-old plants	3, 8 h from 100 to 1,800 $\mu mol\ m^{-2}\ s^{-1}$	cDNA microarray (6K)	Vandenabeele et al. (2004)
				3, 8 h from 100 to 1,800 $\mu mol\ m^{-2}\ s^{-1}$	Affymetrix ATH1 microarray (22K)	Vanderauwera et al. (2005)

(continued)

Table 1 (continued)

	Major ROS	Localisation	Plant system	Experimental details	Expression profiling platform	Reference
cat2 KO	H_2O_2	Peroxisomes	5-week-old plants	1 week from 4,500 ppm CO_2 to ambient air	cDNA-AFLP (2K)	Queval et al. (2007)
apx1 KO	H_2O_2	Cytosol	2-week-old plants	0, 1, 48 h from 100 to 425 µmol m^{-2} s^{-1}	Affymetrix microarray (8K)	Pnueli et al. (2003)
			3-week-old plants	0, 0.25, 0.5, 1.5, 3, 6, 24 h from 25 to 250 µmol m^{-2} s^{-1}	Affymetrix ATH1 microarray (22K)	Davletova et al. (2005a)
flu mutant	$^1O_2$	Chloroplasts	3-week-old plants	from dark to 100 µmol m^{-2} s^{-1}	Affymetrix ATH1 microarray (22K)	op den Camp et al. (2003)
AOX1a AS	$O_2{\bullet}^-$, H_2O_2	Mitochondria	3-week-old plants	Standard growth conditions	Affymetrix ATH1 microarray (22K)	Umbach et al. (2005)
SOD KD	$O_2{\bullet}^-$, H_2O_2	Chloroplasts	3-week-old plants	Standard growth conditions	Affymetrix ATH1 microarray (22K)	Rizhsky et al. (2003)

Plants were treated with exogenously applied ROS and ROS-generating agents, whereas endogenous ROS accumulation was the result of perturbed levels of individual ROS-scavenging enzymes. *AS* antisense, *KD* knockdown, *KO* knockout

(Gadjev et al. 2006). Besides several hallmark transcripts of the general oxidative stress response, new candidate regulatory transcripts that could orchestrate specific transcriptomic signatures were detected as well as antagonistic effects triggered by the different ROS. Anthocyanin biosynthesis transcripts were induced in Cu/Zn superoxide dismutase (SOD) knockdown plants, while their induction during high-light (HL) exposure was compromised in catalase-deficient plants (Gadjev et al. 2006). This inverse correlation between the effect of $O_2\bullet^-$ and H_2O_2 on anthocyanin accumulation clearly demonstrates specific signalling capacities of particular ROS.

4 Photorespiratory H_2O_2-Dependent Gene Expression

Photorespiration is the process by which ribulose-1,5-bisphosphate carboxylase uses oxygen to oxygenate ribulose-1,5-biphosphate instead of carboxylating carbon dioxide (CO_2). This oxygenation results in two glycolates that are subsequently transported from the chloroplast to the peroxisomes and catalysed by glycolate oxidase, yielding H_2O_2. Photorespiration occurs at high rates in the leaves of C_3 plants (Noctor et al. 2002), most probably acts as an alternative electron sink, and increases under adverse environmental conditions that, either limit the CO_2 availability (for instance, through sustained stomatal closure during drought stress) or increase the light energy influx. The considerable energy used in photorespiratory carbon– and nitrogen-recycling lowers the quantum yield of photosynthesis, thereby utilising light less efficiently in CO_2 fixation. This effect could be physiologically advantageous because it reduces the probability of photoinhibition during abiotic stresses and, thereby, attenuates ROS production in the chloroplasts. A complementary function of increased photorespiratory flux during abiotic stress follows from the increased H_2O_2 production in the peroxisomes: photorespiratory H_2O_2 production was estimated to account for approximately 70% of the total H_2O_2 formed at any given irradiance intensity (Noctor et al. 2002). Increased levels of photorespiratory H_2O_2 are counteracted by peroxisomal catalases that are tetrameric heme-containing enzymes converting H_2O_2 into oxygen and water. Catalases are the principal peroxisomal H_2O_2-scavenging enzymes and are an important sink for photorespiratory H_2O_2 (Dat et al. 2001; Noctor et al. 2002; Mateo et al. 2004). Perturbation of catalase activities, by mutation or gene silencing, results in decreased H_2O_2 scavenging during HL irradiation or low CO_2 availability in different C_3 plants. These catalase-deficient plants were more sensitive to a variety of environmental stresses, such as chilling, ozone and salt stress (Kendall et al. 1983; Willekens et al. 1997; Vandenabeele et al. 2004), and revealed that photorespiratory H_2O_2 is able to trigger both local and systemic defence responses and activate a cell death programme that shares several features with programmed cell death. Cell death occurred both in the mesophyll and palisade parenchyma and spatially correlated with increased levels of H_2O_2 (Chamnongpol et al. 1998; Dat et al. 2003; Vandenabeele et al. 2004).

The importance of peroxisomal H_2O_2 scavenging by catalase was also corroborated in catalase-deficient *Arabidopsis* plants. Transgenic lines, containing antisense or

RNAi constructs that target the catalase 2 gene (*CAT2*), were more sensitive to HL irradiation and ozone, and subsequently provoked cell death (Fig. 2a). Both the timing and extent of cell death were inversely correlated to the residual catalase

Fig. 2 Photorespiratory H_2O_2-induced transcriptional changes and cell death in catalase-deficient plants upon HL irradiation. (**a**) Control and catalase-deficient plants exposed to HL for 23 h. Leaves of catalase-deficient plants developed cell death after 8 h of HL, while in the control plants no cell death was visible even after 23 h of HL. (**b**) Scatter plots representing temporal expression changes in catalase-deficient versus control plants during a HL time course (0, 3, 8 and 23 h). Genes that fall above or below the diagonal are at least twofold upregulated (*red*) or downregulated (*blue*), respectively. Samples were analysed with Affymetrix MAS5.0 software and the Silicon Genetics GeneSpring version 5.1

activity (Vandenabeele et al. 2004). To validate the photorespiratory nature of the elevated H_2O_2 levels in HL-treated catalase-deficient plants, plants were exposed to HL under ambient air conditions (400 ppm, 21% O_2) and under high CO_2 levels (1,500 ppm, 21% O_2). As expected, leaf cell death could be prevented in a CO_2-enriched environment (Vandenabeele et al. 2004).

Recently, *Arabidopsis cat2* knockout mutants were used to define the photoperiod as an additional regulatory factor of H_2O_2-induced cell death and defence gene expression. By growing *cat2* at different daylengths in ambient air, H_2O_2-induced lesion development was found to depend on long days. Unexpectedly, perturbed intracellular redox state and oxidative signalling pathway induction were more prominent in short days than in long days, as evidenced by the glutathione status and the induction of defence genes and oxidative stress-responsive transcripts. These data indicated that photoperiod is a critical determinant in the acclimation versus cell death decision during stress (Queval et al. 2007). In a second work, the *cat2* mutant uncovered a new crosstalk between oxidative stress, cation homeostasis and ethylene signalling. While investigating the effects of catalase deficiency on cation homeostasis, *cat2* plants appeared to be more tolerant to lithium, as a result of lithium-induced insensitivity to the inhibitory ethylene and reduced capability for ethylene production. The accumulation of H_2O_2 during stress responses was postulated to alleviate the negative effects of excess ethylene generated (Bueso et al. 2007).

Catalase loss-of-function mutants are an ideal in planta system to examine the consequences of increased availability of endogenous H_2O_2, because perturbation in H_2O_2 homeostasis can be sustained over time, no invasive techniques are needed, and physiologically relevant levels of H_2O_2 are obtained. Transcriptome analysis of catalase-deficient plants demonstrated that perturbation of the H_2O_2-scavenging capacity significantly affects nuclear gene expression after HL irradiation, altered photoperiod or low CO_2 levels (Vandenabeele et al. 2003, 2004; Vanderauwera et al. 2005; Queval et al. 2007) (Fig. 2b). By monitoring different time points after HL exposure, the dynamics of the transcriptional response could be visualised, and early expressed candidate receptor and signal transducing genes could be distinguished, from later expressed effector genes, such as those involved in defence responses and cell death (Vandenabeele et al. 2003, 2004; Vanderauwera et al. 2005).

Under non-stressed conditions, several stress-related genes, including glutathione S-transferases, heat shock proteins and UDP-glycosyltransferases, were upregulated in the catalase-deficient plants. Almost all the 51 induced transcripts, including a WRKY (At5g24110) and a bHLH (At1g10585) transcription factor, also responded highly to HL and to various other environmental stresses, such as wounding, heat, salt and osmotic and genotoxic stress (Vanderauwera et al. 2005). Both bHLH and WRKY transcription factors have been shown to be involved in abiotic stress responses in plants (Eulgem et al. 2000; Wang et al. 2003; Lorenzo et al. 2004). Therefore, these early responsive genes might be implicated in the primary cellular stress responses mediated by increased levels of H_2O_2.

Exposure of both control and catalase-deficient plants to HL allowed HL- and H_2O_2-responsive transcripts to be distinguished. Of the 906 transcripts that were exclusively regulated by HL, 379 were upregulated and 527 were downregulated.

In response to H_2O_2, 349 transcripts were induced and 88 were repressed in the catalase-deficient plants (Vanderauwera et al. 2005).

The Hsp17.6 class of small heat shock proteins and the heat shock transcription factors (HSFs) HsfA2, HsfB2b and HsfA7a were strongly responsive to photorespiratory H_2O_2. A rapid induction of two other HSFs, HsfA4a and HsfA8, in response to light stress has been reported in knockout ascorbate peroxidase 1 plants (Pnueli et al. 2003). HsfA4a and HsfA8 were, however, not induced during HL stress in the catalase-deficient plants, indicating that the induction of these HSFs is not part of a general response to oxidative damage but rather the outcome of a response to increased levels of specific ROS. HSFs also respond to many different biotic and abiotic stresses, and HsfA2 has indeed been reported as a key regulator in response to heat, HL and osmotic stress (Nishizawa et al. 2006; Ogawa et al. 2007). Recent studies in yeast and mammalian cells provided considerable evidence for redox-dependent regulation of HSFs (Kotak et al. 2007), further feeding the hypothesis that oxidative stress-responsive HSFs might also function in plants as H_2O_2 sensors involved in the regulation of oxidative stress-responsive genes (Miller and Mittler 2006).

As mentioned above, oxidative stimuli provoke large-scale changes in the transcriptome of plants. One approach to analyse this transcriptional response is to identify co-regulated genes by making intersections with other related datasets or by experimentally merging several perturbations and, thereby, trying to dissect the transcriptional response into smaller functional entities (Ma and Bohnert 2008). By combining a genetic (catalase deficiency) with an environmental (HL) perturbation, a transcriptional regulon was identified that was strongly and rapidly induced by HL in wild-type plants, but whose induction was impaired in catalase-deficient plants. This cluster comprised the complete anthocyanin biosynthetic and regulatory pathway, together with several genes of hitherto unknown function (Vanderauwera et al. 2005). This molecular phenotype also correlated with the observed phenology of the catalase-deficient plants under HL stress: wild-type plants accumulated significantly more anthocyanins than catalase-deficient plants after exposure to HL. Because anthocyanins have the ability to reduce the potential to oxidative damage via light attenuation (Steyn et al. 2002), failure of anthocyanin accumulation in the catalase-deficient plants might contribute to their increased sensitivity to HL irradiation (Hoch et al. 2003). How H_2O_2 impinges negatively on this HL-induced anthocyanin regulon remains currently elusive and is the subject of further investigations.

By assessing the expression of H_2O_2-induced transcripts within three important environmental stresses (cold, heat and drought), an important intersection was revealed that not only confirmed a prominent role for photorespiratory H_2O_2-dependent signalling but also allowed the identification of new candidate regulatory genes within the plant's abiotic stress response (Vanderauwera et al. 2005). Among the 20 genes that were commonly induced by H_2O_2 and at least two of the selected stresses, two transcription factors, DREB2A and ZAT12, were identified. DREB2A is known to be an important regulator of water- and heat-responsive gene expression (Shinozaki and Yamaguchi-Shinozaki 2000; Sakuma et al. 2006a,b, 2006), whereas ZAT12 is thought to be involved in cold and oxidative stress signalling (Rizhsky et al. 2004; Davletova et al. 2005b; Vogel et al. 2005). These examples illustrate the importance of the systematic perturbation of H_2O_2-regulated transcription factors and subsequent functional analysis to unravel the hierarchical structure of the network that governs the H_2O_2 response in plants. Such an integrative survey of transcriptional footprints and stress tolerance characteristics of new transgenic plants and mutants will provide a view on how these different regulatory factors interconnect with each other and what their role is in the mechanisms by which ROS signals are integrated to eventually regulate specific biological processes.

5 Perspectives

Molecular phenotyping of different ROS-responsive mutants and transgenic plants have resulted in comprehensive inventories of transcripts involved in production, scavenging, perception and signal transduction of ROS. Not only was the signalling role of ROS during various environmental cues validated, but a better view was also gained on the shape of the complex regulatory network implicated in ROS signal transduction. Next, a more detailed analysis is required to clarify the function of each of these genes within the ROS signalling network. The most informative results are expected from in-depth functional analyses of transgenic plants perturbed in signal transducers and transcription factors, which will identify downstream targets and organise the oxidative stress response into various transcriptional regulons.

Currently, microarrays are the most popular technology for large-scale expression profiling because they allow the simultaneous detection of tens of thousands of transcripts at a reasonable cost. Most microarrays used in today's research are based on spotted oligonucleotides that are designed to question the levels of individual transcripts and standard sets covering various parts of the genome are available (CATMA, ATH1). The major limitation of such microarrays is that they rely on current genome annotations, which precludes the identification of novel or very small transcription units. Tiling arrays, on which the entire genome (including intergenic regions) is represented by evenly spaced probes, provide a novel and unbiased means of transcript identification (Rensink and Buell 2005). Tiling arrays will make it possible to identify novel transcription units involved in H_2O_2-induced responses, including antisense transcripts, small non-coding RNAs and previously unidentified genes (Busch and Lohmann 2007; Gregory et al. 2008). Recent breakthroughs in sequencing technologies have led to innovations, such as pyrosequencing (Margulies et al. 2005), sequencing by synthesis (Steemers and Gunderson 2005), SOLiD sequencing by oligonucleotide ligation (Cloonan et al. 2008) and true single molecule sequencing (Harris et al. 2008). The unprecedented number of base pairs that can be read in one run by high-throughput sequencers based on these techniques allows hypothesis-free, quantitative analysis of the entire transcriptome. As a result, so-called digital gene expression analysis will soon become an attractive alternative for microarrays.

Both tiling arrays and deep-sequencing approaches can be used to detect small RNAs, a class of non-coding transcripts that have been shown to control different aspects of plant development as well as plant stress responses (Sunkar et al. 2007).

An interesting example is the upregulation of two Cu/ZnSOD genes that depended on changes in the levels of miR398 (Sunkar et al. 2006). Hence, small RNAs may prove to be yet another level in the regulatory networks that orchestrate ROS-directed gene expression.

Integration of transcriptional networks with other "omics" disciplines, such as proteomics and metabolomics will become increasingly important in the near future. Such "poly-omics" approaches are well suited to analyse the consequences of ROS-provoked changes at both the cellular tissue and whole plant levels, and, for example, allow the comparison of the relative importance of transcriptional and post-transcriptional regulation of the plant's response to ROS. Eventually, this will provide us with a holistic view on the networks involved in various ROS-mediated plant processes.

Acknowledgments The authors thank Dr Martine De Cock for help in preparing the manuscript. This work was supported by a grant from the Research Fund of the Ghent University (Geconcerteerde Onderzoeksacties grant no. 12051403). S.V. is a Postdoctoral Fellow of the Research Foundation-Flanders.

References

Apel K, Hirt H (2004) Reactive oxygen species: metabolism, oxidative stress, and signal transduction. Annu Rev Plant Biol 55:373–399

Bailey-Serres J, Mittler R (2006) The roles of reactive oxygen species in plant cells. Plant Physiol 141:311–311

Bueso E, Alejandro S, Carbonell P, Perez-Amador MA, Fayos J, Bellés JM, Rodriguez PL, Serrano R (2007) The lithium tolerance of the Arabidopsis *cat2* mutant reveals a cross-talk between oxidative stress and ethylene. Plant J 52:1052–1065

Busch W, Lohmann JU (2007) Profiling a plant: expression analysis in *Arabidopsis*. Curr Opin Plant Biol 10:136–141

Causton HC, Ren B, Koh SS, Harbison CT, Kanin E, Jennings EG, Lee TI, True HL, Lander ES, Young RA (2001) Remodeling of yeast genome expression in response to environmental changes. Mol Biol Cell 12:323–337

Chamnongpol S, Willekens H, Moeder W, Langebartels C, Sandermann Jr H, Van Montagu M, Inzé D, Van Camp W (1998) Defense activation and enhanced pathogen tolerance induced by H_2O_2 in transgenic plants. Proc Natl Acad Sci USA 95:5818–5823

Cloonan N, Forrest ARR, Kolle G, Gardiner BBA, Faulkner GJ, Brown MK, Taylor DF, Steptoe AL, Wani S, Bethel G, Robertson AJ, Perkins AC, Bruce SJ, Lee CC, Ranade SS, Peckham HE, Manning JM, McKernan KJ, Grimmond SM (2008) Stem cell transcriptome profiling via massive-scale mRNA sequencing. Nat Methods 5:613–619

Dat JF, Inzé D, Van Breusegem F (2001) Catalase-deficient tobacco plants: tools for *in planta* studies on the role of hydrogen peroxide. Redox Rep 6:37–42

Dat JF, Pellinen R, Beeckman T, Van De Cotte B, Langebartels C, Kangasjärvi J, Inzé D, Van Breusegem F (2003) Changes in hydrogen peroxide homeostasis trigger an active cell death process in tobacco. Plant J 33:621–632

Davletova S, Rizhsky L, Liang H, Zhong S, Oliver DJ, Coutu J, Shulaev V, Schlauch K, Mittler R (2005a) Cytosolic ascorbate peroxidase 1 is a central component of the reactive oxygen gene network of Arabidopsis. Plant Cell 17:268–281

Davletova S, Schlauch K, Coutu J, Mittler R (2005b) The zinc-finger protein Zat12 plays a central role in reactive oxygen and abiotic stress signaling in Arabidopsis. Plant Physiol 139:847–856

Delaunay A, Isnard A-D, Toledano MB (2000) H_2O_2 sensing through oxidation of the Yap1 transcription factor. EMBO J 19:5157–5166

Desikan R, Hancock JT, Bright J, Harrison J, Weir I, Hooley R, Neill SJ (2005) A role for ETR1 in hydrogen peroxide signaling in stomatal guard cells. Plant Physiol 137:831–834

Desikan R, Horák J, Chaban C, Mira-Rodado V, Witthöft J, Elgass K, Grefen C, Cheung M-K, Meixner AJ, Hooley R, Neill SJ, Hancock JT, Harter K (2008) The histidine kinase AHK5 integrates endogenous and environmental signals in *Arabidopsis* guard cells. PLoS ONE 3:e2491.1–e2491.15

Desikan R, Mackerness SA-H, Hancock JT, Neill SJ (2001) Regulation of the Arabidopsis transcriptome by oxidative stress. Plant Physiol 127:159–172

Després C, Chubak C, Rochon A, Clark R, Bethune T, Desveaux D, Fobert PR (2003) The Arabidopsis NPR1 disease resistance protein is a novel cofactor that confers redox regulation of DNA binding activity to the basic domain/leucine zipper transcription factor TGA1. Plant Cell 15:2181–2191

Eulgem T, Rushton PJ, Robatzek S, Somssich IE (2000) The WRKY superfamily of plant transcription factors. Trends Plant Sci 5:199–206

Foyer CH, Noctor G (2005) Redox homeostasis and antioxidant signaling: a metabolic interface between stress perception and physiological responses. Plant Cell 17:1866–1875

Gadjev I, Vanderauwera S, Gechev TS, Laloi C, Minkov IN, Shulaev V, Apel K, Inzé D, Mittler R, Van Breusegem F (2006) Transcriptomic footprints disclose specificity of reactive oxygen species signaling in Arabidopsis. Plant Physiol 141:436–445

Gapper C, Dolan L (2006) Control of plant development by reactive oxygen species. Plant Physiol 141:341–345

Gechev TS, Gadjev IZ, Hille J (2004) An extensive microarray analysis of AAL-toxin-induced cell death in *Arabidopsis thaliana* brings new insights into the complexity of programmed cell death in plants. Cell Mol Life Sci 61:1185–1197

Gechev TS, Minkov IN, Hille J (2005) Hydrogen peroxide-induced cell death in Arabidopsis: transcriptional and mutant analysis reveals a role of an oxoglutarate-dependent dioxygenase gene in the cell death process. IUBMB Life 57:181–188

Gregory BD, Yazaki J, Ecker JR (2008) Utilizing tiling microarrays for whole-genome analysis in plants. Plant J 53:636–644

Gupta R, Luan S (2003) Redox control of protein tyrosine phosphatases and mitogen-activated protein kinases in plants. Plant Physiol 132:1149–1152

Hancock J, Desikan R, Harrison J, Bright J, Hooley R, Neill S (2006) Doing the unexpected: proteins involved in hydrogen peroxide perception. J Exp Bot 57:1711–1718

Harris TD, Buzby PR, Babcock H, Beer E, Bowers J, Braslavsky I, Causey M, Colonell J, DiMeo J, Efcavitch JW, Giladi E, Gill J, Healy J, Jarosz M, Lapen D, Moulton K, Quake SR, Steinmann K, Thayer E, Tyurina A, Ward R, Weiss H, Xie Z (2008) Single-molecule DNA sequencing of a viral genome. Science 320:106–109

Hoch WA, Singsaas EL, McCown BH (2003) Resorption protection. Anthocyanins facilitate nutrient recovery in autumn by shielding leaves from potentially damaging light levels. Plant Physiol 133:1296–1305

Kaminaka H, Näke C, Epple E, Dittgen J, Schütze K, Chaban C, Holt BF III, Merkle T, Schäfer E, Harter K, Dangl JL (2006) bZIP10-LSD1 antagonism modulates basal defense and cell death in *Arabidopsis* following infection. EMBO J 25:4400–4411

Kendall AC, Keys AJ, Turner JC, Lea PJ, Miflin BJ (1983) The isolation and characterization of a catalase-deficient mutant of barley (*Hordeum vulgare*). Planta 159:505–511

Kim HS, Yu Y, Snesrud EC, Moy LP, Linford LD, Haas BJ, Nierman WC, Quackenbush J (2005) Transcriptional divergence of the duplicated oxidative stress-responsive genes in the Arabidopsis genome. Plant J 41:212–220

Kotak S, Larkindale J, Lee U, von Koskull-Döring P, Vierling E, Scharf K-D (2007) Complexity of the heat stress response in plants. Curr Opin Plant Biol 10:310–316

Koussevitzky S, Nott A, Mockler TC, Hong F, Sachetto-Martins G, Surpin M, Lim J, Mittler R, Chory J (2007) Multiple signals from damaged chloroplasts converge on a common pathway to regulate nuclear gene expression. Science 316:715–719 [Err. postdate 22 June 2007]

Kovtun Y, Chiu W-L, Tena G, Sheen J (2000) Functional analysis of oxidative stress-activated mitogen-activated protein kinase cascade in plants. Proc Natl Acad Sci USA 97:2940–2945

Lee KP, Kim C, Landgraf K, Apel K (2007) EXECUTER1- and EXECUTER2-dependent transfer of stress-related signals from the plastid to the nucleus of *Arabidopsis thaliana*. Proc Natl Acad Sci USA 104:10270–10275

Li P, Mane SP, Sioson AA, Vasquez Robinet C, Heath LS, Bohnert HJ, Grene R (2006) Effects of chronic ozone exposure on gene expression in *Arabidopsis thaliana* ecotypes and in *Thellungiella halophila*. Plant Cell Environ 29:854–868

Lorenzo O, Chico JM, Sánchez-Serrano JJ, Solano R (2004) *JASMONATE-INSENSITIVE1* encodes a MYC transcription factor essential to discriminate between different jasmonate-regulated defense responses in Arabidopsis. Plant Cell 16:1938–1950

Ludwikow A, Gallois P, Sadowski J (2004) Ozone-induced oxidative stress response in *Arabidopsis*: transcription profiling by microarray approach. Cell Mol Biol Lett 9:829–842

Ma S, Bohnert HJ (2008) Gene networks in *Arabidopsis thaliana* for metabolic and environmental functions. Mol BioSyst 4:199–204

Mahalingam R, Jambunathan N, Gunjan SK, Faustin E, Weng H, Ayoubi P (2006) Analysis of oxidative signalling induced by ozone in *Arabidopsis thaliana*. Plant Cell Environ 29:1357–1371

Margulies M, Egholm M, Altman WE, Attiya S, Bader JS, Bemben LA, Berka J, Braverman MS, Chen Y-J, Chen Z, Dewell SB, Du L, Fierro JM, Gomes XV, Godwin BC, He W, Helgesen S, Ho CH, Hutchison SK, Irzyk GP, Jando SC, Alenquer MLI, Jarvie TP, Jirage KB, Kim J-B, Knight JR, Lanza JR, Leamon JH, Lefkowitz SM, Lei M, Li J, Lohman KL, Lu H, Makhijani VB, McDade KE, McKenna MP, Myers EW, Nickerson E, Nobile JR, Plant R, Puc BP, Ronan MT, Roth GT, Sarkis GJ, Simons JF, Simpson JW, Srinivasan M, Tartaro KR, Tomasz A, Vogt KA, Volkmer GA, Wang SH, Wang Y, Weiner MP, Yu P, Begley RF, Rothberg JM (2005) Genome sequencing in microfabricated high-density picolitre reactors. Nature 437:376–380 [Err. 439, 502; Err. 441, 120]

Mateo A, Mühlenbock P, Rustérucci C, Chi-Chen Chang C, Miszalski Z, Karpinska B, Parker JE, Mullineaux PM, Karpinski S (2004) *LESION SIMULATING DISEASE 1* is required for acclimation to conditions that promote excess excitation energy. Plant Physiol 136:2818–2830

Miller G, Mittler R (2006) Could heat shock transcription factors function as hydrogen peroxide sensors in plants? Ann Bot 98:279–288

Mittler R, Vanderauwera S, Gollery M, Van Breusegem F (2004) The reactive oxygen gene network in plants. Trends Plant Sci 9:490–498

Moon H, Lee B, Choi G, Shin D, Prasad DT, Lee O, Kwak S-S, Kim DH, Nam J, Bahk J, Hong JC, Lee SY, Cho MJ, Lim CO, Yun D-H (2003) NDP kinase 2 interacts with two oxidative stress-activated MAPKs to regulate cellular redox state and enhances multiple stress tolerance in transgenic plants. Proc Natl Acad Sci USA 100:358–363

Mostertz J, Scharf C, Hecker M, Homuth G (2004) Transcriptome and proteome analysis of *Bacillus subtilis* gene expression in response to superoxide and peroxide stress. Microbiology 150:497–512

Mou Z, Fan W, Dong X (2003) Inducers of plant systemic acquired resistance regulate NPR1 function through redox changes. Cell 113:935–944

Mullineaux PM, Karpinski S, Baker NR (2006) Spatial dependence for hydrogen peroxide-directed signaling in light-stressed plants. Plant Physiol 141:346–350

Murray JI, Whitfield ML, Trinklein ND, Myers RM, Brown PO, Botstein D (2004) Diverse and specific gene expression responses to stresses in cultured human cells. Mol Biol Cell 15:2361–2374

Nishizawa A, Yabuta Y, Yoshida E, Maruta T, Yoshimura K, Shigeoka S (2006) Arabidopsis heat shock transcription factor A2 as a key regulator in response to several types of environmental stress. Plant J 48:535–547

Noctor G, Veljovic-Jovanovic S, Driscoll S, Novitskaya L, Foyer CH (2002) Drought and oxidative load in the leaves of C_3 plants: a predominant role for photorespiration? Ann Bot 89:841–850

Ogawa D, Yamaguchi K, Nishiuchi T (2007) High-level overexpression of the *Arabidopsis HsfA2* gene confers not only increased thermotolerance but also salt/osmotic stress tolerance and enhanced callus growth. J Exp Bot 58:3373–3383 op den Camp RGL, Przybyla D, Ochsenbein C, Laloi C, Kim C, Danon A, Wagner D, Hideg E, Göbel C, Feussner I, Nater M, Apel K (2003) Rapid induction of distinct stress responses after the release of singlet oxygen in Arabidopsis. Plant Cell 15:2320–2332

Pnueli L, Liang H, Rozenberg M, Mittler R (2003) Growth suppression, altered stomatal responses, and augmented induction of heat shock proteins in cytosolic ascorbate peroxidase (*Apx1*)-deficient *Arabidopsis* plants. Plant J 34:187–203

Queval G, Issakidis-Bourguet E, Hoeberichts FA, Vandorpe M, Gakière B, Vanacker H, Miginiac-Maslow M, Van Breusegem F, Noctor G (2007) Conditional oxidative stress responses in the Arabidopsis photorespiratory mutant *cat2* demonstrate that redox state is a key modulator of daylength-dependent gene expression, and define photoperiod as a crucial factor in the regulation of H_2O_2-induced cell death. Plant J 52:640–657

Rensink AW, Buell CR (2005) Microarray expression profiling resources for plant genomics. Trends Plant Sci 10:603–609

Rentel MC, Lecourieux D, Ouaked F, Usher SL, Petersen L, Okamoto H, Knight H, Peck SC, Grierson CS, Hirt H, Knight MR (2004) OXI1 kinase is necessary for oxidative burst-mediated signalling in *Arabidopsis*. Nature 427:858–861

Rhoads DM, Umbach AL, Subbaiah CC, Siedow JN (2006) Mitochondrial reactive oxygen species. Contribution to oxidative stress and interorganellar signaling. Plant Physiol 141:357–366

Rizhsky L, Liang H, Mittler R (2003) The water-water cycle is essential for chloroplast protection in the absence of stress. J Biol Chem 278:38921–38925

Rizhsky L, Liang H, Shuman J, Shulaev V, Davletova S, Mittler R (2004) When defense pathways collide. The response of Arabidopsis to a combination of drought and heat stress. Plant Physiol 134:1683–1696

Sakuma Y, Maruyama K, Qin F, Osakabe Y, Shinozaki K, Yamaguchi-Shinozaki K (2006a) Dual function of an *Arabidopsis* transcription factor DREB2A in water-stress-responsive and heat-stress-responsive gene expression. Proc Natl Acad Sci USA 103:18822–18827

Sakuma Y, Maruyama K, Osakabe Y, Qin F, Seki M, Shinozaki K, Yamaguchi-Shinozaki K (2006b) Functional analysis of an *Arabidopsis* transcription factor, DREB2A, involved in drought-responsive gene expression. Plant Cell 18:1292–1309

Schweighofer A, Hirt H, Meskiene I (2004) Plant PP2C phosphatases: emerging functions in stress signaling. Trends Plant Sci 9:236–243

Shaikhali J, Heiber I, Seidel T, Ströher E, Hiltscher H, Birkmann S, Dietz K-J, Baier M (2008) The redox-sensitive transcription factor Rap24a controls nuclear expression of 2-Cys peroxiredoxin A and other chloroplast antioxidant enzymes. BMC Plant Biol 8:48.1–48.4

Shinozaki K, Yamaguchi-Shinozaki K (2000) Molecular responses to dehydration and low temperature: differences and cross-talk between two stress signaling pathways. Curr Opin Plant Biol 3:217–223

Singh KK (2000) The *Saccharomyces cerevisiae* Sln1p-Ssk1p two-component system mediates response to oxidative stress and in an oxidant-specific fashion. Free Rad Biol Med 29:1043–1050

Steemers FJ, Gunderson KL (2005) Illumina, Inc. Pharmacogenomics 6:777–782

Steyn WJ, Wand SJE, Holcroft DM, Jacobs G (2002) Anthocyanins in vegetative tissues: a proposed unified function in photoprotection. New Phytol 155:349–361

Sunkar R, Chinnusamy V, Zhu J, Zhu J-K (2007) Small RNAs as big players in plant abiotic stress responses and nutrient deprivation. Trends Plant Sci 12:301–309

Sunkar R, Kapoor A, Zhu J-K (2006) Posttranscriptional induction of two Cu/Zn superoxide dismutase genes in *Arabidopsis* is mediated by downregulation of miR398 and important for oxidative stress tolerance. Plant Cell 18:2051–2065

Takahashi S, Seki M, Ishida J, Satou M, Sakurai T, Narusaka M, Kamiya A, Nakajima M, Enju A, Akiyama K, Yamaguchi-Shinozaki K, Shinozaki K (2004) Monitoring the expression profiles of genes induced by hyperosmotic, high salinity, and oxidative stress and abscisic acid treatment in Arabidopsis cell culture using a full-length cDNA microarray. Plant Mol Biol 56:29–55

Torres MA, Dangl JL (2005) Functions of the respiratory burst oxidase in biotic interactions, abiotic stress and development. Curr Opin Plant Biol 8:397–403

Tosti N, Pasqualini S, Borgogni A, Ederli L, Falistocco E, Crispi S, Paolocci F (2006) Gene expression profiles of O_3-treated *Arabidopsis* plants. Plant Cell Environ 29:1686–1702
Umbach AL, Fiorani F, Siedow JN (2005) Characterization of transformed Arabidopsis with altered alternative oxidase levels and analysis of effects on reactive oxygen species in tissue. Plant Physiol 139:1806–1820
Vandenabeele S, Van Der Kelen K, Dat J, Gadjev I, Boonefaes T, Morsa S, Rottiers P, Slooten L, Van Montagu M, Zabeau M, Inzé D, Van Breusegem F (2003) A comprehensive analysis of hydrogen peroxide-induced gene expression in tobacco. Proc Natl Acad Sci USA 100:16113–16118
Vandenabeele S, Vanderauwera S, Vuylsteke M, Rombauts S, Langebartels C, Seidlitz HK, Zabeau M, Van Montagu M, Inzé D, Van Breusegem F (2004) Catalase deficiency drastically affects gene expression induced by high light in *Arabidopsis thaliana*. Plant J 39:45–58
Vandenbroucke K, Robbens S, Vandepoele K, Inzé D, Van de Peer Y, Van Breusegem F (2008) H_2O_2-induced gene expression across kingdoms: a comparative analysis. Mol Biol Evol 25:507–516
Vanderauwera S, Zimmermann P, Rombauts S, Vandenabeele S, Langebartels C, Gruissem W, Inzé D, Van Breusegem F (2005) Genome-wide analysis of hydrogen peroxide-regulated gene expression in Arabidopsis reveals a high light-induced transcriptional cluster involved in anthocyanin biosynthesis. Plant Physiol 139:806–821
Vogel JT, Zarka DG, Van Buskirk HA, Fowler SG, Thomashow MF (2005) Roles of the CBF2 and ZAT12 transcription factors in configuring the low temperature transcriptome of Arabidopsis. Plant J 41:195–211
Vranová E, Inzé D, Van Breusegem F (2002) Signal transduction during oxidative stress. J Exp Bot 53:1227–1236
Wagner D, Przybyla D, op den Camp R, Kim C, Landgraf F, Lee KP, Würsch M, Laloi C, Nater M, Hideg E, Apel K (2004) The genetic basis of singlet oxygen-induced stress responses of *Arabidopsis thaliana*. Science 306:1183–1185
Wang Y-J, Zhang Z-G, He X-J, Zhou H-L, Wen Y-X, Dai J-X, Zhang J-S, Chen S-Y (2003) A rice transcription factor OsbHLH1 is involved in cold stress response. Theor Appl Genet 107:1402–1409
Willekens H, Chamnongpol S, Davey M, Schraudner M, Langebartels C, Van Montagu M, Inzé D, Van Camp W (1997) Catalase is a sink for H_2O_2 and is indispensable for stress defence in C_3 plants. EMBO J 16:4806–4816
Zheng M, Åslund F, Storz G (1998) Activation of the OxyR transcription factor by reversible disulfide bond formation. Science 279:1718–1721

The ROS Signaling Network of Cells

Yael Harir and Ron Mittler

Abstract Reactive oxygen species (ROS) are toxic derivatives of atmospheric oxygen used by plant cells to control many different biological processes, including growth, development, and response to biotic and abiotic stimuli. Because of their toxicity, as well as their important signaling role, the steady-state level of ROS in cells is tightly regulated by a network of genes termed the "ROS gene network". In the flowering plant *Arabidopsis thaliana*, the ROS gene network includes more than 150 genes that manage the level of ROS in cells. The ROS network is highly dynamic and redundant, and encodes for ROS-scavenging as well as ROS-producing proteins. Recent studies have unraveled some of the key players of the network and shed light on some of the questions related to its mode of regulation, its protective roles, and its modulation of signaling networks that control growth, development, and stress response. In this chapter we will describe some of these findings.

1 Introduction

Ever since the introduction of molecular oxygen (O_2) into the earth's atmosphere by O_2-evolving photosynthetic organisms about 2.7 billion years ago, reactive oxygen species (ROS) have been the unwelcome companions of aerobic life (Halliwell 2006; Halliwell and Gutteridge 1999). In contrast to O_2, these partially reduced or activated derivatives of oxygen ($^1O_2$, O_2^-, H_2O_2, and HO•) are highly reactive and toxic, and can lead to the oxidative destruction of cells (Asada 2006; Asada and Takahashi 1987). Consequently, the successful evolution of plants, as well as all aerobic organisms, on the earth has been dependent upon the development of efficient ROS-scavenging mechanism. These allow organisms to overcome ROS toxicity and to use some of these toxic molecules as signal transduction molecules (Van Breusegem et al. 2008; Mittler et al. 2004; Bailey-Serres and Mittler 2006). ROS, therefore,

Y. Harir and R. Mittler (✉)
Department of Biochemistry and Molecular Biology, University of Nevada,
Mail Stop 200, Reno, NV 89557
e-mail: ronm@unr.edu

L.A. del Río and A. Puppo (eds.), *Reactive Oxygen Species in Plant Signaling*,
Signaling and Communication in Plants, DOI 10.1007/978-3-642-00390-5_10,

play a dual role in plants as both toxic compounds and key regulators of many biological processes (Foyer and Noctor 2005; Apel and Hirt 2004; Mittler et al. 2004).

The use of ROS as signaling molecules by plant cells suggests that, during the course of evolution, plants were able to achieve a high degree of control over ROS toxicity and are now using ROS as signaling molecules. The delicate balance between ROS production and scavenging that allows this duality in function to exist in plants is thought to be orchestrated by a large network of genes termed the "ROS gene network", which includes more than 150 genes in the model plant *Arabidopsis thaliana*, tightly regulating ROS production and scavenging (Mittler et al. 2004).

2 Production of ROS in Plants

Organelles with a highly oxidizing metabolic activity or with an intense rate of electron flow, such as chloroplasts, mitochondria, and peroxisomes, are a major source of ROS production in plant cells. Localized ROS production in these organelles and in specific regions, e.g., apoplastic space, or apex of polarized cells, may trigger different signaling cascades. Production of ROS by the Mehler reaction and the antenna pigments in the chloroplasts is enhanced in plants by conditions limiting CO_2 fixation, such as drought, salt, and temperature stress, as well as by the combination of these conditions with high-light stress (Miller et al. 2008). During pathogenesis or wounding, ROS production is mainly apoplastic, whereas during salt stress, ROS can also be produced from internalized membranes (endosomes) by NADPH oxidase in a phosphatidyl inositol 3 kinase (PI3K)-dependent manner (Leshem et al. 2006). In C3 plants, limiting CO_2 conditions can also activate the photorespiratory pathway (del Río et al. 2006). As part of this pathway, H_2O_2 is generated in peroxisomes by the enzymatic activity of glycolate oxidase. In mitochondria, over-reduction of the electron transport chain is the main source of O_2^- production under specific stress conditions (Møller 2001). Additional sources of ROS in plant cells are the detoxifying reactions catalyzed by cytochromes in both the cytoplasm and the endoplasmic reticulum, as well as superoxide production by xanthine oxidase in peroxisomes (Corpas et al. 2008; Mittler et al. 2004; Van Breusegem et al. 2008).

Plasma membrane NADPH-dependent oxidases have been the subject of intense investigation (Torres and Dangl 2005; Torres et al. 2006). They are thought to play a key role in ROS signaling, and contain a multimeric flavocytochrome that forms an electron transport chain capable of reducing O_2 to superoxide. In addition to NADPH oxidases, pH-dependent cell wall peroxidases, germin-like oxalate oxidases and amine oxidases have been proposed to generate ROS at the apoplast (Mittler et al. 2004; Van Breusegem et al. 2008). Although much attention has been given to NADPH oxidases, other ROS-producing mechanisms in the mitochondria, apoplast, and peroxisomes are likely to play a role in ROS signaling in response to different stimuli or developmental signals. For example, a newly identified extracellular peroxidase and two type III peroxidases play an active role in H_2O_2 production and subsequent cell death in the local and systemic responses to pathogen attack (Bindschedler et al. 2006; Choi et al. 2007). Intriguingly, a previously unidentified nuclear source of ROS production has been observed during elicitor treatment (Ashtamker et al. 2007). These reports add additional subcellular sources of ROS production in plant cells, thereby upgrading the complexity of the ROS signaling network.

3 Scavenging of ROS in Plants

ROS-scavenging enzymes such as superoxide dismutase (SOD), ascorbate peroxidase (APX), catalase (CAT), glutathione peroxidase (GPX), and peroxiredoxin (PrxR) together with the antioxidants ascorbic acid and glutathione, provide cells with highly efficient machinery for detoxifying O_2^- and H_2O_2. (Foyer and Noctor 2005; Apel and Hirt 2004; Mittler et al. 2004). These mechanisms, together with the sequestering of metal ions such as free iron by ferritin and other metal-binding proteins, prevents the formation of the highly toxic HO• via the metal-dependent Haber–Weiss or the Fenton reactions (Asada 2006; Asada and Takahashi 1987). The cellular pools of the antioxidants ascorbic acid and glutathione are maintained in their reduced state by a set of enzymes capable of using NAD(P)H to regenerate oxidized glutathione or ascorbic acid (e.g., monodehydroascorbate reductase, dehydroascorbate reductase, and glutathione reductase). In addition, monodehydroascorbate radicals can be reduced back into ascorbic acid via ferredoxin using electrons diverted from the photosynthetic apparatus in the water–water cycle in chloroplasts (Asada 2006). Scavenging of H_2O_2 can also be mediated in plants by "classical" plant peroxidases (class III) using a variety of reductants. Membranes are highly susceptible to oxidative stress. In plant cells, they are protected by the activity of specific phospholipid glutathione peroxidases and by α-tocopherol (vitamin E), which is kept at its reduced state by the pool of reduced ascorbic acid. Protection of cells against singlet oxygen is generally believed to be mediated by carotenoids (Asada and Takahashi 1987).

The importance of peroxiredoxins, glutaredoxins, and thioredoxins as scavengers of ROS has gained significant support in recent years (Cheng et al. 2006; Dos Santos and Rey 2006). A novel function was recently assigned to the peroxiredoxin PrxII E in detoxifying ONOO–, a potent oxidizing and nitrating species formed in a diffusion-limited reaction between NO and O_2^-, suggesting a key role for peroxiredoxins in mediating the crosstalk between NO and ROS (Romero-Puertas et al. 2007; Hong et al. 2008; Wilson et al. 2008). Moreover, peroxiredoxins were also reported to function as redox sensors, linking the redox signaling and ROS networks of cells (Dietz 2008). These studies suggest integration of ROS, NO and redox signaling in cells and provide an excellent platform for future studies that will unravel the complexity of these networks. Insight into the crosstalk between different ROS and ROS scavenging mechanisms was also gained from studies of double or triple mutants that lack key ROS-scavenging enzymes in different subcellular locations (Giacomelli et al. 2007; Miller et al. 2007) and exploration of crosstalk between distinct ROS such as singlet oxygen and H_2O_2 (Laloi et al. 2007). These studies not only expose the redundancy of the ROS-scavenging network, but also suggest that different antioxidant enzymes and different ROS in the same, or different compartments, mediate signature signals that control chloroplast function and plant response to various environmental stimuli.

4 ROS Signaling and its Modulation by the ROS Gene Network

Plants constantly sense and assess the level of ROS in cells and reprogram their enzymatic activities and gene expression to optimally respond and acclimate to the changing conditions in their environment. ROS signaling is thought to be controlled by two opposing processes of production and scavenging. At least three different mechanisms were proposed for sensing of ROS in plant cells: (1) receptor proteins, (2) redox-sensitive transcription factors, and (3) direct inhibition of phosphatases by ROS (Mittler 2002; Mittler et al. 2004; Apel and Hirt 2004; Miller et al. 2008). In addition, ROS can be sensed due to their effect on different metabolic pathways and the state of oxidation or redox potential of specific cellular proteins. These could be membrane-associated or soluble, and may be present in different cellular compartments. Different developmental or environmental signals feed into the ROS signaling network and perturb ROS homeostasis in a compartment-specific or even a cell-specific manner. As described above, perturbed ROS levels are perceived by different proteins, enzymes, or receptors, and modulate different developmental, metabolic, and defense pathways. ROS production is mediated by different cellular pathways, including respiration and photosynthesis, as well as by different proteins and enzymes, e.g., NADPH oxidases, amine oxidases, and xanthine oxidase (Mittler et al. 2004). In contrast, ROS scavenging is mediated by different ROS-scavenging enzymes and antioxidants that include ascorbate peroxidases, catalases, peroxiredoxins, and superoxide dismutases (Mittler 2002; Mittler et al. 2004; Apel and Hirt 2004). The intensity, duration, and localization of the different ROS signals in cells are therefore determined by interplay between these two opposing forces, i.e., ROS scavenging and ROS production, and the decoding of this signal will determine the cellular response to the original cue, modulating different developmental, metabolic, and/or defense pathways. The process described above requires a tight mode of regulation and might involve amplification and/or feedback inhibition loops. In addition to regulating the intensity and duration of the different ROS signals, the ROS-scavenging pathways are also responsible for maintaining a low steady-state "base line" of ROS on which the different signals can be registered.

It is possible that the use of ROS as versatile signaling molecules originated from their proposed use to sense biotic or abiotc stress. Most forms of stress disrupt the metabolic balance of cells, resulting in enhanced production of ROS. Simple organisms, such as bacteria or yeast, sense the enhanced production of ROS by redox-sensitive transcription factors and other molecular sensors, activate different ROS-defense pathways, and regulate their metabolic pathways to lower the production rate of ROS (Mittler et al. 2004). This "basic cycle" of ROS metabolism maintains a low steady-state level of ROS in cells. Variations on this pathway could have originated during evolution and contributed to the use of ROS as signaling molecules to control more specialized processes, such as plant growth and defense, hormonal signaling, and development. For example, pathogen infection could alter plant metabolism causing the accumulation of ROS due to suppression or activation of different pathways (Mittler 2002). A specialized pathway used for pathogen sensing via ROS could have evolved and resulted in the pathway we now know in which the identification of a pathogen via a plant receptor will trigger an *R* gene-dependent pathway that will result in the enhanced production of ROS by plasma membrane-localized NADPH oxidases (Torres et al. 2006). This will result in a ROS signal that will activate several different defense pathways. The activity of ROS-scavenging enzymes is important in this case because these enzymes can modulate the ROS signal and determine the intensity, duration or even the type of response (Mittler et al. 1999). Similarly, an abiotic stress such as osmotic stress will result in enhanced ROS production due to altered metabolic balance (Miller et al. 2008; Mittler 2006). A signaling pathway that senses and responds to osmotic stress was recently shown to involve the activation of ROS production via NADPH oxidases and other ROS-producing signals (Mittler 2002, 2006).

5 A Model for ROS Signaling Developed from the Study of Plants Lacking Apx1

Mutants lacking cytosolic ascorbate peroxidase 1 (*apx1*) have been used in the past several years to study ROS signaling in *Arabidopsis*. Using a time-course microarray analysis comparing these plants to wild-types under conditions of moderate light stress, we have identified several key regulators involved in ROS signaling in plants, and using different mutants we studied several of these proteins, independently assessing their role in ROS signaling and responses to abiotic stress conditions (Rizhsky et al. 2004; Pnueli et al. 2003; Davletova et al. 2005a, b; Suzuki et al. 2005, 2008; Mittler et al. 2006; Ciftci-Yilmaz et al. 2007). Our analysis suggests that cytosolic APX1 functions as a buffer to control the levels of ROS that reach the nuclei and activate gene expression (Fig. 1). Thus, in the absence of APX1, the expression of different light stress response transcripts such as that of the key regulator Zat12 was facilitated (Davletova et al. 2005a).

Our model suggests that, in the absence of APX1, ROS that accumulates in the cytosol is sensed by different redox-response transcription factors, such as heat shock transcription factors (HsfA4a), and triggers a cascade of different transcription factors that include members of the zinc finger protein Zat family (Zat12, 10, and 7) and members of the WRKY transcription factor family. Additional players

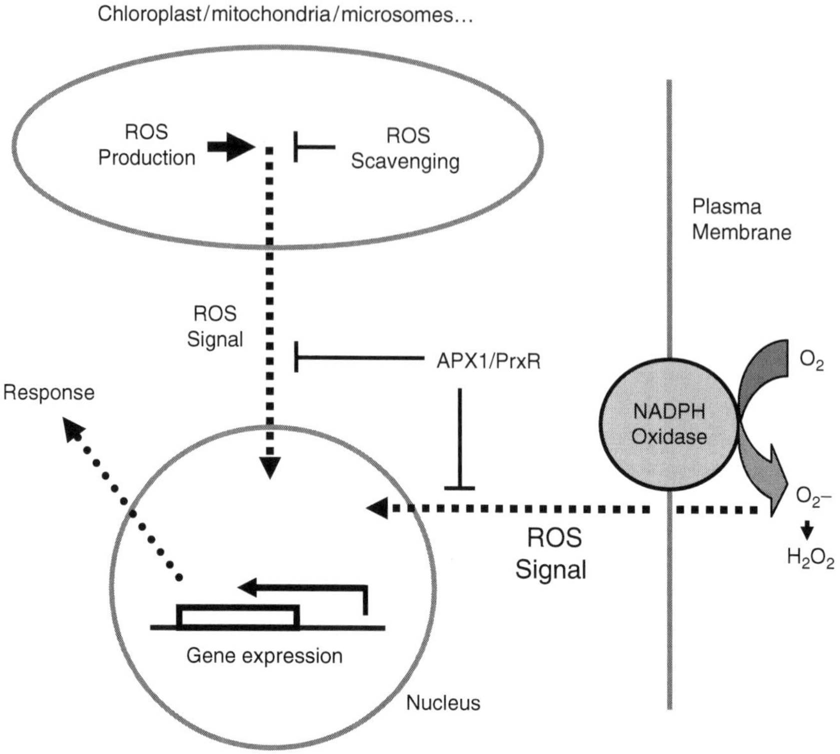

Fig. 1 A hypothetical model showing the role of cytosolic ROS-scavenging systems in the regulation of ROS signaling in plant cells. Different ROS signals originating from different organelles, or at the plasma membrane, e.g., by NADPH oxidases, reach the nucleus and regulate gene expression. Cytosolic ROS-scavenging systems such as cytosolic ascorbate peroxidase 1 (APX1) or cytosolic peroxiredoxins (PrxR) act as buffers to attenuate these signals and control the amount of ROS that reach the nucleus

in this pathway could include MBF1c that is a transcriptional coactivator and NADPH oxidase (RbohD) that is likely to be involved in amplifying the ROS signal (Rizhsky et al. 2004; Pnueli et al. 2003; Davletova et al. 2005a, b; Suzuki et al. 2005, 2008; Mittler et al. 2006; Ciftci-Yilmaz et al. 2007).

The model presented in Fig. 1 could be extended to include additional sources of ROS such as mitochondria or apoplast that generate ROS in response to specific stimuli, and include additional cytosolic ROS-scavenging mechanisms such as preoxiredoxins, glutathione peroxidase, and thioredoxins. The ROS signal generated in the different cellular compartments is hypothesized to reach the nuclei and activate gene expression in a process that will induce a response to the perceived stimuli. In the presence of APX1, the levels of ROS that reach the nuclei will be attenuated to generate the correct response, preventing, for example, misactivation of cell death or any additional unwanted responses. Additional studies are underway in our laboratory to study this model in more detail.

6 Coordination of the ROS Network

The different scavenging and producing enzymes encoded by the ROS gene network can be found in many different subcellular compartments. In addition, more than one enzymatic activity per a specific ROS can usually be found in each of the different compartments (Mittler et al. 2004). Because ROS such as H_2O_2 can diffuse between different cellular compartments (Bienert et al. 2007), ROS metabolism in a particular compartment can effect or alter the ROS homeostasis/signaling of a neighboring compartment or reach the nuclei and activate gene expression (Fig. 1). Transporters for the antioxidants ascorbic acid and glutathione are likely to be central in determining the specific concentrations of these compounds and the redox potential in the different cellular compartments. An anonymous player in the ROS signaling network is the vacuole. Its ROS-scavenging and ROS-producing potentials are unknown at present. It is possible that this organelle, because of its relatively large cellular volume, plays an essential role in the control of ROS metabolism in plants. Recent studies in *Arabidopsis* have suggested that the mode of coordination between different components of the ROS removal network of plants is complex. For example, the application of light stress to *Arabidopsis* resulted in the induction of cytosolic and not chloroplastic ROS-defense enzymes (Mittler et al. 2004), even though most ROS produced during light stress are thought to be generated in chloroplasts and/or peroxisomes. The cytosolic ROS-scavenging pathways were further shown to be required for the protection of chloroplasts during light stress (Davletova et al. 2005a). In a different study, a double mutant deficient in cytosolic ascorbate peroxidase 1 and peroxisomal catalase 1 was found to be more tolerant to light stress compared to wild-type or single mutants deficient in ascorbate peroxidase 1 or catalase 1 (Rizhsky et al. 2002). This finding was very surprising because it suggested that different cellular pathways are activated in cells in response to enhanced ROS production in the cytosol or peroxisomes. Activation of both the cytosolic and peroxisomal pathways further results in the generation of a new signal that is different from that activated by the two individual cytosolic or peroxisomal signals. How ROS metabolism and signaling are coordinated between different organelles in cells is largely unknown at present.

7 NADPH Oxidases, a Possible Link Between Calcium and ROS Signaling

NADPH oxidases play a key role in ROS signaling and plant defense responses to pathogen infection, abiotic stress and injury. They generate $O_2^{\bullet-}$ by oxidizing NADPH and transferring the electron to O_2. They represent the plant homologs to the mammalian phagocyte NADPH oxidase subunit $gp91^{phox}$ (Torres and Dangl 2005). The *Arabidopsis* genome contains ten classical NADPH oxidase genes, all of which contain a presumably cytosolic 300 amino-acid amino-terminal extension with two EF-hands that binds Ca^{2+} and at least one key phosphorylation site.

It was recently shown that activation of the plasma membrane localized NADPH oxidases involves phosphorylation of two N-terminal serines by a calcium-dependent protein kinase (CDPK) as well as interaction with Rho-like GTPase (ROP). NADPH oxidase phosphorylation as well as binding to calcium synergizes its activation, raising the possibility that it may function as a calcium sensor (Ogasawara et al. 2008; Takeda et al. 2008). The NADPH oxidase/ROP interaction is regulated by the binding of calcium to two EF-hand motifs at the amino terminus of the oxidase (Wong et al. 2007). The consequence of NADPH oxidase activation is localized production of $O_2{\bullet}^-$, which is rapidly converted to H_2O_2, presumably in the apoplastic space. The ROS signal produced could reach the nuclei (Fig. 1), activating gene expression in response to a given stimuli and generating a plant response to this stimuli. In root tips, a complex interaction between calcium, pH, and ROS oscillation was recently reported to control elongation (Monshausen et al. 2007; Van Breusegem et al. 2008).

8 Concluding Remarks

Although ROS were initially considered to be toxic byproducts of aerobic metabolism, in recent years it became obvious that plants can cope with ROS toxicity to the degree of using ROS as signal transduction molecules. ROS signaling was shown to be involved in the regulation of basic biological processes and responses to biotic and abiotic stimuli. ROS signaling and ROS toxicity are kept in check by the ROS gene network of plants. This network includes ROS-scavenging and ROS-producing enzymes that modulate the level of ROS in cells. Thus, the overall level of ROS is always kept under control and ROS are allowed to accumulate and/or oscillate for the purpose of signaling in a highly controlled manner (Mittler et al. 2004). This process is achieved by a tightly controlled balance between ROS production and ROS scavenging in the different cellular compartments. The interplay between ROS scavenging and production in the different cellular compartments, therefore, determines the intensity, duration, and localization of ROS signals, and the decoding of these signals determine the plant's response, or developmental and/or growth adaptations. A complex relationship between divergent ROS signals from different cell components can also result in cell reprogramming depending on the nature of the signals and their origin.

References

Apel K, Hirt H (2004) Reactive oxygen species: metabolism, oxidative stress, and signal transduction. Annu Rev Plant Biol 55, 373–399.

Asada K (2006) Production and scavenging of reactive oxygen species in chloroplasts and their functions. Plant Physiol 141, 391–396.

Asada K, Takahashi M (1987) Production and scavenging of active oxygen in photosynthesis. In: Kyle DJ, Osmond CB, Arentzen CJ (eds), Photoinhibition, Elsevier, Amsterdam, pp 227–287.

Ashtamker C, Kiss V, Sagi M, Davydov O, Fluhr R (2007) Diverse subcellular locations of cryptogein-induced reactive oxygen species production in tobacco Bright Yellow-2 cells. Plant Physiol 143, 1817–1826.

Bailey-Serres J, Mittler R (2006) The roles of reactive oxygen species in plant cells. Plant Physiol. 141, 311.

Bienert GP, Møller ALB, Kristiansen KA, Schulz A, Møller IM, Schjoerring JK, Jahn TP (2007) Specific aquaporins facilitate the diffusion of hydrogen peroxide across membranes. J Biol Chem 282, 1183–1192.

Bindschedler LV, Dewdney J, Blee KA, Stone JM, Asai T, Plotnikov J, Denoux C, Hayes T, Gerrish C, Davies DR, Ausubel FM, Bolwell GP (2006) Peroxidase-dependent apoplastic oxidative burst in Arabidopsis required for pathogen resistance. Plant J 47, 851–863.

Cheng NH, Liu JZ, Brock A, Nelson RS, Hirschi KD (2006) AtGRXcp, an Arabidopsis chloroplastic glutaredoxin, is critical for protection against protein oxidative damage. J Biol Chem 281, 26280–26288.

Choi HW, Kim YJ, Lee SC, Hong JK, Hwang BK (2007) Hydrogen peroxide generation by the pepper extracellular peroxidase CaPO2 activates local and systemic cell death and defense response to bacterial pathogens. Plant Physiol 145, 890–904.

Ciftci-Yilmaz S, Morsy MR, Song L, Coutu A, Krizek BA, Lewis MW, Warren D, Cushman J, Connolly EL, Mittler R (2007) The EAR-motif of the Cys2/His2-type zinc finger protein Zat7 plays a key role in the defense response of *Arabidopsis* to salinity stress. J Biol Chem 282, 9260–9268.

Corpas FJ, Palma JM, Sandalio LM, Valderrama R, Barroso JB, del Río LA (2008) Peroxisomal xanthine oxidoreductase: Characterization of the enzyme from pea (*Pisum sativum* L.) leaves. J Plant Physiol 165, 1319–1330.

Davletova S, Rizhsky L, Liang H, Shuman J, Shulaev V, Oliver D, Mittler, R (2005a) Cytosolic ascorbate peroxidase 1 is a central component of the reactive oxygen gene network of arabidopsis. Plant Cell 17, 268–281.

Davletova S, Schlauch K, Coutu J, Mittler R (2005b) The zinc-finger protein Zat12 plays a central role in reactive oxygen and abiotic stress signaling in Arabidopsis. Plant Physiol 139, 847–856.

del Río LA, Sandalio LM, Corpas FJ, Palma JM, Barroso JB (2006) Reactive oxygen species and reactive nitrogen species in peroxisomes. Production, scavenging, and role in cell signaling. Plant Physiol 141, 330–335.

Dietz K-J (2008) Redox signal integration: From stimulus to networks and genes. Physiol Plant 133, 459–468.

Dos Santos C, Rey P (2006) Plant thioredoxins are key actors in the oxidative stress response. Trends Plant Sci 11, 329–334.

Foyer CH, Noctor G (2005) Redox homeostasis and antioxidant signaling: A metabolic interface between stress perception and physiological responses. Plant Cell 17, 1866–1875.

Giacomelli L, Masi A, Ripoll DR, Lee MJ, van Wijk KJ (2007) Arabidopsis thaliana deficient in two chloroplast ascorbate peroxidases shows accelerated light-induced necrosis when levels of cellular ascorbate are low. Plant Mol Biol 65, 627–644

Halliwell B (2006) Reactive species and antioxidants. Redox biology is a fundamental theme of aerobic life. Plant Physiol 141, 312–322.

Halliwell B, Gutteridge JMC (1999). Free radicals in biology and medicine, 3rd Edition. Clarendon, Oxford.

Hong JK, Yun BW, Kang JG, Raja MU, Kwon E, Sorhagen K, Chu C, Wang Y, Loake GJ (2008) Nitric oxide function and signalling in plant disease resistance. J Exp Bot 59, 147–154.

Laloi C, Stachowiak M, Pers-Kamczyc E, Warzych E, Murgia I, Apel K (2007) Cross-talk between singlet oxygen- and hydrogen peroxide-dependent signaling of stress responses in Arabidopsis thaliana. Proc Natl Acad Sci USA 104, 672–677.

Leshem Y, Melamed-Book N, Cagnac O, Ronen G, Nishri Y, Solomon M, Cohen G, Levine A (2006) Suppression of *Arabidopsis* vesicle-SNARE expression inhibited fusion of H_2O_2-containing vesicles with tonoplast and increased salt tolerance. Proc Natl Acad Sci USA 103, 18008–18013.

Miller G, Shulaev V, Mittler R (2008) Reactive oxygen signaling and abiotic stress. Physiol Plant 133, 481–489.

Miller G, Suzuki N, Rizhsky L, Hegie A, Koussevitzky S, Mittler R (2007) Double mutants deficient in cytosolic and thylakoid ascorbate peroxidase reveal a complex mode of interaction between reactive oxygen species, plant development, and response to abiotic stresses. Plant Physiol 144, 1777–1785

Mittler R, Herr EH, Orvar BL, van Camp W, Willekens H, Inze, D, Ellis BE (1999) Transgenic tobacco plants with reduced capability to detoxify reactive oxygen intermediates are hyperresponsive to pathogen infection. Proc Natl Acad Sci USA 96, 14165–14170.

Mittler R (2002) Oxidative stress, antioxidants and stress tolerance. Trends Plant Sci 7, 405–410.

Mittler R, Vanderauwera S, Gollery M, Van Breusegem F (2004) Reactive oxygen gene network of plants. Trends Plant Sci 9, 490–498.

Mittler R (2006) Abiotic stress, the field environment and stress combination. Trends Plant Sci 11, 15–19.

Mittler R, Song L, Coutu J, Coutu A, Ciftci S, Kim YS, Lee H, Stevenson B, Zhu, J-K (2006) Gain- and loss-of-function mutations in Zat10 enhance the tolerance of plants to abiotic stress. FEBS Lett 580, 6537–6542.

Møller IM (2001) Plant mitochondria and oxidative stress: Electron transport, NADPH turnover, and metabolism of reactive oxygen species. Annu Rev Plant Physiol Plant Mol Biol 52, 561–591.

Monshausen GB, Bibikova TN, Messerli MA, Shi C, Gilroy S (2007) Oscillations in extracellular pH and reactive oxygen species modulate tip growth of Arabidopsis root hairs. Proc Natl Acad Sci USA 104, 20996–201001.

Ogasawara Y, Kaya H, Hiraoka G, Yumoto F, Kimura S, Kadota Y, Hishinuma H, Senzaki E, Yamagoe S, Nagata K, Nara M, Suzuki K, Tanokura M, Kuchitsu K (2008) Synergistic activation of the arabidopsis NADPH oxidase AtrbohD by Ca^{2+} and phosphorylation. J Biol Chem 283, 8885–8892.

Pnueli L, Hongjian L, Mittler R (2003) Growth suppression, abnormal guard cell response, and augmented induction of heat shock proteins in cytosolic ascorbate peroxidase (Apx1) – deficient *Arabidopsis* plants. Plant J 34, 187–203.

Rizhsky L, Davletova S, Liang H, Mittler R (2004) The zinc-finger protein Zat12 is required for cytosolic ascorbate peroxidase 1 expression during oxidative stress in *Arabidopsis*. J Biol Chem 279, 11736–11743.

Rizhsky L, Hallak-Herr E, Van Breusegem F, Rachmilevitch S, Rodermel S, Inzé D, Mittler R (2002) Double antisense plants with suppressed expression of ascorbate peroxidase and catalase are less sensitive to oxidative stress than single antisense plants with suppressed expression of ascorbate peroxidase or catalase. Plant J 32, 329–342.

Romero-Puertas MC, Laxa M, Mattè A, Zaninotto F, Finkemeier I, Jones AME, Perazzolli M, Vandelle E, Dietz K-J, Delledonne M (2007) S-nitrosylation of peroxiredoxin II E promotes peroxynitrite-mediated tyrosine nitration. Plant Cell 19, 4120–4130.

Suzuki N, Rizhsky L, Liang H, Shuman J, Shulaev V, Mittler R (2005) Enhanced tolerance to environmental stresses in transgenic plants expressing the transcriptional co-activator MBF1. Plant Physiol 139, 1313–1322.

Suzuki N, Bajad S, Shuman J, Shulaev V, Mittler R (2008) The transcriptional co-activator MBF1c is a key regulator of thermotolerance in *Arabidopsis thaliana*. J Biol Chem 283, 9269–9275.

Takeda S, Gapper C, Kaya H, Bell E, Kuchitsu K, Dolan L (2008) Local positive feedback regulation determines cell shape in root hair cells. Science 319, 1241–1244.

Torres MA, Dangl JL (2005) Functions of the respiratory burst oxidase in biotic interactions, abiotic stress and development. Curr Opin Plant Biol 8, 397–403.

Torres MA, Jones JD, Dangl JL (2006) Reactive oxygen species signaling in response to pathogens. Plant Physiol 141, 373–378.

Van Breusegem F, Bailey-Serres J, Mittler R (2008) Unraveling the tapestry of networks involving reactive oxygen species in plants. Plant Physiol 147, 978–984.

Wilson ID, Neill SJ, Hancock JT (2008) Nitric oxide synthesis and signalling in plants. Plant Cell Environ 31, 622–631.

Wong HL, Pinontoan R, Hayashi K, Tabata R, Yaeno T, Hasegawa K, Kojima C, Yoshioka H, Iba K, Kawasaki T, Shimamoto K (2007) Regulation of rice NADPH oxidase by binding of Rac GTPase to its N-terminal extension. Plant Cell 19, 4022–4034.

Reactive Oxygen Species and Signaling in Cadmium Toxicity

Luisa M. Sandalio, María Rodríguez-Serrano, Luis A. del Río, and María C. Romero-Puertas

Abstract The toxicity of heavy metals in living organisms has become a major focus of research in recent decades as a result of the increased environmental pollution in industrial areas. Cadmium is one of the most dangerous heavy metals due to its high mobility in plants. This metal produces malfunctions in membranes, photosynthesis rate, and water-nutrient balance, and also causes oxidative damages. By contrast with the enormous number of publications on the tolerance and accumulation of cadmium in plants, there is a remarkable lack of knowledge on the molecular mechanisms and signaling events underlying plant responses to Cd toxicity, especially those involving reactive oxygen species (ROS). The dual role of ROS in heavy metal toxicity as both oxidative damage inducers and signaling molecules has been demonstrated in recent years and will be discussed in this chapter. The contribution of oxidative damage to Cd toxicity and the mechanisms involved in the cellular response to this metal, such as antioxidant regulation, protein defenses, and the role of NO and hormones, will also be analyzed.

1 Introduction

The toxicity of heavy metals in plants, animals, and humans has become a major area of toxicological research in recent decades. Heavy metals, such as Cd, Hg, Pb, and Al, are important environmental pollutants, particularly in industrial areas, as a result of anthropogenic activities, such as metal-working industries, cement factories, smelting plants and refineries, waste incineration, traffic, and heating systems, while Cd can also enter the soil through phosphate fertilizers (Sanitá di Toppi and Gabbrielli 1999). In plants, heavy metals affect essential metabolic pathways by reducing growth rates and even causing death. The toxicity of these metals can be explained by their

L.M. Sandalio (✉), M. Rodríguez-Serrano, L.A. del Río, and M.C. Romero-Puertas
Departamento de Bioquímica, Biología Celular y Molecular de Plantas,
Estación Experimental del Zaidín (CSIC), Apartado de correos 419,
E-18080 Granada, Spain
e-mail: luisamaria.sandalio@eez.csic.es

L.A. del Río and A. Puppo (eds.), *Reactive Oxygen Species in Plant Signaling*,
Signaling and Communication in Plants, DOI 10.1007/978-3-642-00390-5_11,

chemical characteristics. Thus, heavy metals such as Cd, Hg, and Pb can affect the activity of several enzymes by binding to SH groups or displacing metals from the enzyme (Van Assche and Clijsters 1990).

Toxic metal ions uptaken by plant roots are potentially dangerous for human health as their accumulation in crop plants is the principal route of toxic metal entry into the food-chain. Cadmium is one of the most dangerous heavy metals due to its highly mobile nature, the ease with which it is absorbed by plant roots, and the low concentrations required to adversely affect the plant. This chapter will focus on Cd because it has been one of the most widely studied metals in animals and plants during the last two decades. Cd is not commonly found in earth and is mainly associated with Zn (Sanitá di Toppi and Gabbrielli 1999). Unpolluted soil solutions contain Cd concentrations ranging from 0.04 to 0.32 μM, and soils containing 1 μM can be regarded as moderately polluted. In soils containing higher Cd concentrations, such as 35 μM, the only plant species that can grow are the "hyperaccumulators", mainly belonging to the *Thlaspi* genera (Sanitá di Toppi and Gabbrielli 1999). This chapter provides an update of the different mechanisms involved in Cd toxicity, with special emphasis on those associated with oxidative stress and the regulation of plant responses to Cd.

2 Effect of Cadmium on Plant Metabolism

The toxic effects of Cd on several plant species have been reported by different authors (Sanitá di Toppi and Gabbrielli 1999; Sandalio et al. 2001; Schutzendübel et al. 2001; Benavides et al. 2005), although the mechanisms involved in Cd toxicity are not yet fully understood. Cd produces alterations in membrane functionality by inducing changes in lipid composition (Ouariti et al. 1997; Hernández and Cooke 1997) and by affecting the enzymatic activities associated with membranes such as the H^+-ATPase (Fodor et al. 1995). Photosynthesis is also sensitive to Cd, which targets chlorophyll and enzymes involved in CO_2 fixation (Sandalio et al. 2001; Herbette et al. 2006), and inhibits PSII photoactivation by binding to essential Ca^{+2} sites (Faller et al. 2005). The plant–water balance is also disturbed by Cd, and the stomatal opening is inhibited (Poschenrieder et al. 1989; Sandalio et al. 2001; Perfus-Barbeoch et al. 2002). Cd toxicity is associated with disturbances in the uptake and distribution of macro- and micronutrients (Hernández et al. 1998; Rogers et al. 2000; Sandalio et al. 2001; Tsyganov et al. 2007) and, therefore, can compete with other cations for protein and transporter binding sites (Clemens 2006). Cd uptake takes place through the same plasma membrane transporters as those used for other cations, such as K^+, Ca^{2+}, Mg^{2+}, Fe^{2+}, Mn^{2+}, or Cu^2 (Clemens 2006). Members of the ZIP family of Fe and Zn transporters contribute to Mn and Cd uptake, and the Ca channels are permeable to Cd ions (Clemens 2006). The Nramp family of metal transporters, located in the vacuole membrane, is also potentially involved in the Cd entry into plant cells (Clemens 2006). Cd leads to a reduction in Ca^{2+} content (Rivetta et al. 1997; Sandalio et al. 2001), which can then affect the activity of calmodulin-dependent proteins (Rivetta et al. 1997; Rodriguez-Serrano et al. 2009).

A relationship between Cd tolerance and Ca^{2+} homeostasis in a Cd-resistant pea mutant (SGECd^{t}) has been observed (Tsyganov et al. 2007), and in radish and *Arabidopsis* seedlings calcium has been reported to alleviate Cd toxicity by reducing Cd uptake (Rivetta et al. 1997; Suzuki 2005).

Plants have developed defense strategies to avoid metal toxicity through the exudation by roots of metal-complexing agents (citrates and phytosiderophores) to prevent entry of metals and also through the metal immobilization in pectic sites and hystidyl groups in the cell wall (Sanitá di Toppi and Gabbrielli 1999; Clemens 2006). Once Cd has entered the cytosol, induction of specific peptides, called phytochelatins (PCs), in order to complexate the metal, takes place. PC-Cd complexes are transported into the vacuole, preventing the free circulation of Cd ions in the cytosol (Cobett 2000). The isolation of an *Arabidopsis cad1* mutant, which is defective in PC activity and hypersensitive to Cd, has demonstrated the significance of this mechanism in the defense against Cd (Howden et al. 1995). Cd and other metals can be also complexed by metallotioneins and nicotianamine (Sharma and Dietz 2006).

3 Plant Transcriptional Responses to Cadmium

By contrast with the large number of publications in the field of Cd toxicity, there is a glaring lack of knowledge about the molecular mechanisms and signaling events underlying plant transcriptional responses to Cd. In order to fill this gap, a number of studies, involving both small-scale experiments and whole-genome approaches, have been carried out in recent years. Some of these data suggest that the regulation of gene expression in response to Cd is time-regulated rather than dose-regulated (Herbette et al. 2006), and the genes regulated by Cd can be categorized into different groups of proteins in terms of: signal transduction and transcriptional regulation; metal transport; hydric balance; photosynthetic processes; cell wall metabolism; cellular defenses, reactive oxygen species (ROS) detoxification and repair; sulfate and GSH metabolism.

Numerous transcription factors from different families, such us WRKY, bZip, and MYB (Herbette et al. 2006; Weber et al. 2006; Van de Mortel et al. 2008) and genes involved in hormone signaling (Herbette et al. 2006; Minglin et al. 2005) have also been shown to be regulated in response to Cd. In addition, several genes associated with cellular detoxification and repair have been shown to be induced by Cd treatment, especially chitinases and heat shock proteins (HSPs), regarded as a second line of defense against metal stress (Metwally et al. 2003; Békésiová et al. 2008; Rodriguez-Serrano et al. 2009). Transgenic plants expressing fungal chitinases actually show enhanced tolerance to metals (Dana et al. 2006), and chitinase isoforms are differentially modified by different metals (Békésiová et al. 2008). HSPs are upregulated by heat stress and act as molecular chaperones to enable protein to be transported to organelles and to prevent protein aggregation (Ma et al. 2006). Induction of HSPs by Cd toxicity has been observed in different plant species (Sanitá di Toppi and Gabbrielli 1999; Rodriguez-Serrano et al. 2009) and may be regulated by H_2O_2 overproduction since the transcription factors involved can act as H_2O_2 sensors (Mittler and Mittler 2006; Rodriguez-Serrano et al. 2009).

Some of the genes regulated by Cd are involved in Cd transport like AtPcr1, (Song et al. 2004) and those belonging to the ABC family. AtPDR8 is a Cd extrusion pump, and AtOSA1 could be involved in the signal transduction pathway in response to oxidative stress (Kim et al. 2007; Jasinski et al. 2008), while Cd binding proteins like the Cdl19 could be involved in the maintenance of heavy metal homeostasis and/or detoxification (Suzuki et al. 2002).

Although Cd is a bivalent cation unable to participate in redox reactions in the cell, most transcriptome studies show upregulation of genes encoding proteins involved in the defense against oxidative stress and ROS production. These results, which are supported by a number of studies on the effect of Cd on different plant species (Gallego et al. 1996; Sandalio et al. 2001; Romero-Puertas et al. 2002, 2004; Schützendübel and Polle 2002), suggest that oxidative stress is one of the primary effects of Cd exposure. However, ROS, and particularly H_2O_2, is a two-faced molecule, as a large quantity of H_2O_2 could lead to extensive oxidative damage, while low levels of H_2O_2 could be involved in signaling processes (del Río et al. 2006; Mittler et al. 2004; Van Breusegem et al. 2008). The control of ROS production, enabling these species to act as signaling molecules, seems to require a large gene network, which in *Arabidopsis* plants is composed of at least 152 genes (Mittler et al. 2004).

4 Oxidative Stress Induced by Cadmium

ROS occur as by-products of the normal aerobic metabolism, such as respiration and photosynthesis, and their steady-state levels are determined by the interplay between different ROS-producing and ROS-scavenging mechanisms. An excess of ROS is dangerous mainly due to reactions with lipids, proteins, and nucleic acids giving rise to lipid peroxidation, membrane leakage, enzyme inactivation, and DNA breaks or mutations, which can produce severe damage to cell viability (Halliwell and Gutteridge 2007). Cd has been found to increase the lipid peroxidation rate in different plant species (Dixit et al. 2001; Sandalio et al. 2001; Schutzendübel et al. 2001; Djebali et al. 2008) due to an increase in ROS production or the induction of lipoxygenase activity (Skórzynska-Polit et al. 2006; Smeets et al. 2008). Oxidative damage to proteins has also been observed in different plant species exposed to Cd ions (Sandalio et al. 2001; Djebali et al. 2008; Paradiso et al. 2008), and some of the target proteins of oxidative stress in pea leaves have been identified, and these include catalase (CAT), glutathione reductase (GR), Rubisco and Mn-SOD (Romero-Puertas et al. 2004). Besides, an increase in proteolytic activity of leaves following Cd treatment has been observed, as the oxidized proteins are more efficiently degraded (McCarthy et al. 2001; Romero-Puertas et al. 2004). Similar results were obtained with *Helianthus annuss* (Pena et al. 2006) and *Solanum lycopersicum* (Djebali et al. 2008), although some contradictory results were reported concerning the activity of proteasome 20S (Pena et al. 2006; Djebali et al. 2008). A proteomic study carried out on *Arabidopsis thaliana* cells has also reported an increase in several proteases after Cd treatment (Sarry et al. 2006). The remobilization of oxidized proteins may be a protective mechanism under stress conditions in order to prevent further damage to other macromolecules and to facilitate the recycling of amino acids for protein biosynthesis.

4.1 Cadmium Effect on Non-Enzymatic Antioxidants

Depletion of glutathione appears to be the main mechanism in short-term heavy metal toxicity because of its consumption in phytochelatins (PCs) production (Schützendübel and Polle 2002). The ability to maintain the GSH pool appears to be very important in protecting against Cd. In this respect, a close correlation between glutathione content and the tolerance index has been observed in ten pea genotypes with different Cd sensitivity rates (Metwally et al. 2005). In Ni-tolerant *T. goesingese*, a strong correlation between GSH, Cys and *O*-acetyl-L-serine content and the resistance to both growth-inhibitory and oxidative stress induced by the metal has also been reported (Freeman et al. 2004). However, the role of GSH in Cd stress seems to be plant-specific since protection by high GSH content has not been observed in all cases (Arisi et al. 2000). The induction of the sulfur metabolism by Cd has been described and involves a coordinated transcriptional regulation of genes for sulfate uptake and assimilation, as well as GSH and PCs biosynthesis (Howarth et al. 2003; Nocito et al. 2006). Mutants deficient in *O*-acetylserine (thyol) lyase (OASTL), an enzyme involved in Cys biosynthesis, actually showed greater sensitivity to Cd, although this appears to be due to disturbances in H_2O_2 homeostasis rather than to a different chelation capability (López-Martín et al. 2008).

Apart from GSH, ASC also plays an important role in protecting against oxidative stress. Cd toxicity reduces ASC content in different plant species (Rodríguez-Serrano et al. 2006; Romero-Puertas et al. 2007a; Paradiso et al. 2008), and the supply of ASC precursors prevented both the accumulation of H_2O_2 and lipid peroxides and the changes observed in antioxidative enzymes in wheat plants treated with Cd (Zhao et al. 2005; Paradiso et al. 2008). Vitamin E also contributes to the defense against heavy metals. Enzymes involved in vitamin E biosynthesis are upregulated in response to Cu and Cd in *Arabidopsis* plants, and vitamin E-deficient mutants (*vte1*) showed enhanced oxidative stress and sensitivity to both metals (Collin et al. 2008).

4.2 Cadmium Effect on Enzymatic Antioxidants

There is an enormous amount of data on changes in the antioxidative enzyme activity due to Cd stress, although results differ depending on the plant species, the age of the plant, the period of treatment, metal concentration, and tissue or organ studied (for review, see Sanitá di Toppi and Gabbrielli 1999; Schützendübel and Polle 2002; Benavides et al. 2005). Despite these different responses, some common response patterns can be observed. Thus, low Cd concentrations and short treatment periods generally produce an increase in antioxidative enzymes (Smeets et al. 2005), while longer periods of treatment induce a reduction in antioxidants (Sandalio et al. 2001, Rodríguez-Serrano et al. 2006). SOD has been observed to respond differently in the roots and leaves of pea plants following exposure to different Cd concentrations, and it also varied depending on length of exposure (Dixit et al. 2001; Sandalio et al. 2001; Zhang et al. 2005; Rodríguez-Serrano et al. 2006). Pepper cultivars with different levels of sensitivity to Cd have also been observed to respond differently to this heavy metal (León et al. 2002). Analysis of SOD isoforms in pea leaves has shown a dose-dependent reduction in SOD, with the CuZn-SOD being the most sensitive isoform (Sandalio et al. 2001). Analysis of transcript expression showed a reduction in steady-state levels mainly of CuZn-SOD and, to a lesser extent, Mn-SOD, along with a reduction in both activity and protein content (Romero-Puertas et al. 2007a). Transcript levels of chloroplastic CuZn-SOD (*CSD2*) were also lower in *Arabidopsis* plants after Cd exposure, and no significant differences were observed between Fe-SOD and Mn-SOD in leaves, while an induction of both Fe-SOD and Mn-SOD isoforms was observed in roots (Smeets et al. 2008). In *Arabidopsis* plants, CuZn-SOD regulation under oxidative and heavy metal stress is mediated by microRNAs (miR398) (Sunkar et al. 2006). CAT activity is reduced by Cd^{2+} ions in pea (Dixit et al. 2001; Rodríguez-Serrano et al. 2006) and pine (Schützendübel et al. 2001) roots, and also in pea leaves (Sandalio et al. 2001), sunflower (Laspina et al. 2005), and *Arabidopsis* leaves (Cho and Seo 2004), while the opposite effect was observed in radish roots (Vitória et al. 2001). Cd regulates CAT at both transcriptional and posttranscriptional level, being mediated by light and H_2O_2 (Rodríguez-Serrano et al. 2006; Azpilicueta et al. 2007; Romero-Puertas et al. 2004, 2007a).

The ascorbate (ASC)–glutathione (GSH) cycle is another important antioxidant mechanism involved in H_2O_2 detoxification. Fluctuations in ascorbate peroxidase (APX) activity were observed in *Pinus sylvestris* roots depending on the period of treatment (Schützendübel et al. 2001). Most studies have shown an increase in GR activity under Cd exposure depending on dosage and exposure time (Dixit et al. 2001; Smeets et al. 2005), while GR does not change significantly in pea leaves either in terms of activity or mRNA expression (Romero-Puertas et al. 2007a). A reduction in monodehydroascorbate reductase (MDAR) activity by Cd has been observed in *Pinus* plants (Schützendübel et al. 2001), although an increase in both activity and transcript content was observed in pea leaves (Romero-Puertas et al. 2007a). In *Ceratophyllum demersum* L. exposed to Cd, a reduction in MDAR and DAR activities was observed (Aravind et al. 2005). Transcriptome analysis of most of the antioxidative enzymes in the leaves of pea plants grown over a long period of time with Cd suggested that Ca^{2+} channels, phosphorylation/dephosphorylation processes, nitric oxide, cGMP, salicylic acid (SA), and H_2O_2 were involved in some steps between the Cd signal and transcript expression of CuZn-SOD, CAT, and MDAR (Romero-Puertas et al. 2007a). On the other hand, the contribution of thioredoxins and peroxiredoxins to the defense against heavy metals has been reported in *Arabidopsis* and *Chlamydomonas* (Lemaire et al. 1999; Finkemeier et al. 2005).

In pepper plants with different levels of sensitivity to Cd, tolerance to this heavy metal was more dependent on the availability of NADPH than on its antioxidant capacity (León et al. 2002). The activity of glucose 6-P dehydrogenase (G6PDH), malic enzyme (ME), and NADP isocitrate dehydrogenase (NADP-ICDH) is stimulated by Ni, Zn, and Cd (Van Assche and Clijsters 1990; León et al. 2002).

Guayacol peroxidases are also affected by exposure to Cd, and the response varies considerably depending on the plant species (Schützendübel et al. 2001; Sandalio et al. 2001; León et al. 2002). This activity is associated with cell wall lignification, which is a well-documented response to metal stress (Sanitá di Toppi and Gabbrielli 1999).

4.3 Sources of ROS Production Under Cadmium Toxicity

The histochemical and subcellular location of ROS production during Cd stress (Fig 1a–d) and the mechanisms involved in this process have been studied in different plant species. In peroxisomes purified from pea leaves, a Cd-dependent increase in the H_2O_2 concentration was observed which was due to the activation of glycolate oxidase (Romero-Puertas et al. 1999). In pea leaves, using a cytochemical approach, it was demonstrated that Cd-dependent ROS production occurs in peroxisomes, in the outer mitochondrial membrane, and mainly in the plasma membrane, where the NADPH oxidase is the main source of ROS (Romero-Puertas et al. 2004) (Fig. 1c, d). H_2O_2 was present in peroxisomes in locations in close contact with other organelles, which suggests possible crosstalk with other cell compartments (Romero-Puertas et al. 2004). In mitochondria, the Cd-dependent H_2O_2 produced could be due to an increased $O_2^{\bullet-}$ production at complex III of the electron transport

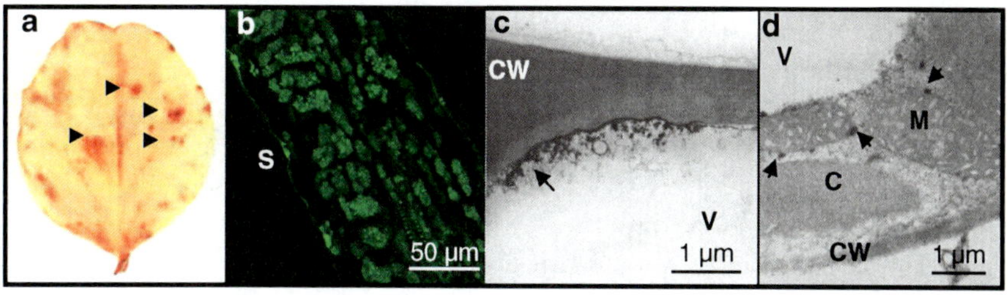

Fig. 1 Imaging of H_2O_2 production in leaves of pea plants exposed to 50 μm $CdCl_2$ using different histochemical approaches. (**a**) H_2O_2 was detected by staining whole leaves with 3,3′-diaminobenzidine (DAB) (Romero-Puertas et al. 2004). (**b**) Analysis of H_2O_2 accumulation in leaf cross-sections by confocal laser microscopy using 2′,7′-dichlorofluorescin diacetate (DCF-DA) staining. The green fluorescence is due to H_2O_2 (Rodríguez-Serrano et al. 2006). (**c**) and (**d**) Cytochemical detecction of H_2O_2 in epidermal and mesophyll cells, respectively, using $CeCl_3$ (Romero-Puertas et al. 2004). *Arrows* indicate H_2O_2 accumulation, *S* stomata, *CW* cell wall, *V* vacuole, *M* mitochondrion, *C* chloroplast

chain, as has been reported in animals treated with Cd (Wang et al. 2004). The role of NADPH oxidase as the main source of ROS under Cd stress has also been demonstrated in tobacco cell cultures (Olmos et al. 2003; Garnier et al. 2006; Horemans et al. 2007) and alfalfa roots (Ortega-Villasante et al. 2005). In *Arabidopsis* plants, the analysis of transcript levels of different NADPH oxidases showed a transient increase in the expression of *rboh*F in response to Cd, while the expression of *rboh*C and *rboh*D did not change (Horemans et al. 2007). In tobacco cell cultures, Cd induced cell death, which was preceded by three successive waves of ROS. The first wave was due to an NADPH oxidase followed by an accumulation of $O_2^{\bullet -}$ in mitochondria and fatty acid hydroperoxides accumulation (Garnier et al. 2006). In addition, the Cd-induced oxidative burst was regulated by Ca^{+2}, phosphorylation/de-phosphorylation processes, and calmodulin or calmodulin-related proteins (Olmos et al. 2003; Romero-Puertas et al. 2004; Garnier et al. 2006).

4.4 Cadmium Effect on Nitric Oxide Production

The gaseous nitric oxide (NO) free radical is a common intracellular and intercellular messenger with a broad spectrum of regulatory functions in many physiological processes and responses to biotic and abiotic stresses by acting as a signaling molecule (del Río et al. 2006; Delledonne 2005; Wilson et al. 2007). NO can also modulate the activity of certain proteins through the *S*-nitrosylation of cystein residues (Lyndermayr et al. 2006; Romero-Puertas et al. 2007b, 2008).

Some studies of NO production during plant exposure to heavy metals have reached opposed conclusions. In *Hibiscus* and *Arabidopsis* roots, Al was found to reduce NOS activity and NO production (Illéš et al. 2006; Tian et al. 2007), while an increase in NO production was observed in *Arabidopsis* cell suspension cultures (Arnaud et al. 2006) and *Chlamydomonas reinhardtii* (Zhang et al. 2008) treated with iron and copper, respectively. However, in plants exposed to Cd, an increase in NO production was observed in soybean cell cultures (Kopyra et al. 2006) and pea roots (Bartha et al. 2005), while in pea leaves and roots, long exposure to Cd produced a reduction in NO accumulation (Rodríguez-Serrano et al. 2006, 2009). The reduction in the number of lateral roots of pea plants by Cd has been correlated with the Cd-dependent reduction in NO accumulation. The discrepancies observed in the results mentioned could be due to differences in Cd exposure times, with a short Cd treatment period promoting an NO burst and a long treatment period directly or indirectly inhibiting NO production.

Exogenously supplied NO has been demonstrated to attenuate heavy metal toxicity (Kopyra and Gwozdz 2003; Hsu and Kao 2004; Yu et al. 2005; Wang and Yang 2005; Laspina et al. 2005), probably due to its ability to act as antioxidant scavenging ROS (Wang and Yang 2005; Rodríguez-Serrano et al. 2006; Singh et al. 2008) or by promoting increases in cytosolic Ca^{2+} concentrations by regulating Ca^{2+} channels and transporters which might be involved in the signaling cascade that regulates gene expression under stress conditions (Besson-Bard et al. 2008).

5 Cadmium, ROS, and Hormones Balance

Salicylic acid (SA) acts as an important signaling element in plants with a wide and divergent range of effects on plant response to pathogen attack and plant stress acclimation (Loake and Grant 2007). In barley, soybean, and rice plants (Metwally et al. 2003; Drazic and Mihailovic 2005; Guo et al. 2007), SA alleviates Cd-induced growth inhibition and oxidative damage by mitigating the decline in antioxidant defenses, such as CAT, POX, and by enhancing GSH and non-protein thiols (Guo et al. 2007). Although the mechanism is not well understood, it has been suggested that SA might induce H_2O_2 signals involved in Cd tolerance, such as repair processes or Cd binding and compartmentation (Metwally et al. 2003). An increase in SA-dependent GSH was reported in the *Thlaspi goesingense* hyperaccumulator through the activation of serine acetyltransferase (SAT) (Freeman et al. 2005). High concentrations of SA are associated with tolerance to Ni in different species of *Thlaspi* (Freeman et al. 2005), and in cucumber SA also alleviates the oxidative stress induced by excess Mn by modifications of antioxidants and the transport of Mn (Shi and Zhu 2008).

JA is an oxylipin which acts as a signaling compound in different defensive situations such as responses to pathogen and herbivore attack and abiotic stresses (Devoto and Turner 2005). JA is produced from linolenic acid, and its generation is associated with lipid peroxidation and oxidative stress (Mithöfer et al. 2004; Montillet et al 2004). JA content increases in response to heavy metals in different plant species (Wang and Wu 2005; Rodríguez-Serrano et al. 2006, 2009), and in *Arabidopsis* plants under Cd treatment, JA regulates genes involved in GSH and phytochelatins synthesis (Xiang and Olivier 1998).

Ethylene (ET) plays a pleiotropic role in plant growth and development, and is involved in a number of processes, including germination, senescence, and defense responses (Guo and Ecker 2004). The Cd stimulation of ethylene biosynthesis has been reported in different plant species (Sanita di Toppi and Gabbrielli 1999; Rodríguez-Serrano et al. 2006). However, the molecular relationships between ethylene biosynthesis and Cd stress have not yet been well established, although transcriptomic studies in *Arabidopsis* plants have demonstrated Cd-dependent upregulation of ACC oxidase and ACC synthase (Herbette et al. 2006). The ET signaling pathways seem also be involved in the response to Cd because genes encoding the ethylene responsive factors ERF2 and ERF5 were upregulated by Cd (Herbette et al. 2006).

6 Conclusion

On the basis of the results mentioned in this chapter, the different plant responses to Cd and their regulation is summarized in Fig. 2. Cd promotes an increase in ROS production in different cell compartments, which are possibly involved in cell wall lignification in order to prevent Cd entry, and induce oxidative damages to lipids and proteins. However, ROS can also act as signaling molecules by directly regulating the

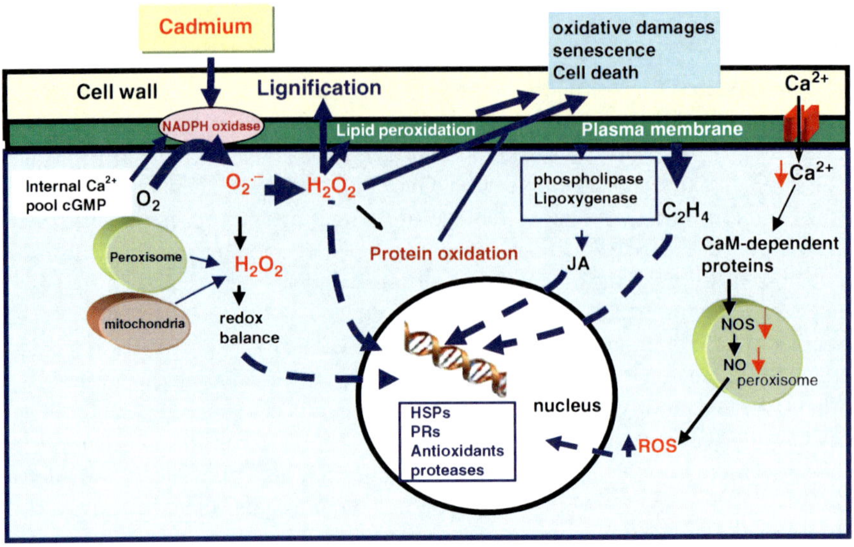

Fig. 2 Hypothetical model of ROS production in leaves and its role under long-term exposure of plants to cadmium. Cadmium induces an increase in ROS production in different cell compartments which can participate in cell wall lignification and oxidative processes with damage to lipids and proteins. ROS can also act as signaling molecules directly regulating the defense response to the metal or in conjunction with ET and JA. *Red arrows* mean reduction and *blue arrows* activation, *CaM* calmodulin, *cGMP* guanosine -5′mono-phosphate, *HSPs* heat-shock proteins, *JA* jasmonic acid, C_2H_4, ethylene, *NOS* nitric oxide synthase, *PRs* pathogenesis related proteins

defensive response to the metal or through changes in the cell's redox balance. In this scenario, calcium could play an important role in both Cd toxicity and the regulation of cell responses to the metal by activating/deactivating CaM-dependent proteins (Rodriguez-Serrano et al. 2009). The Cd-dependent oxidative damage to membranes induces the production of JA and ET which, in turn, could modulate the defensive response in conjunction with ROS (Rodriguez-Serrano et al. 2009). The increased levels of JA and ET, the reduction of NO accumulation under heavy metal stress, as well as the oxidative processes and enhanced proteolytic activity associated with exposure to Cd, suggest the induction of an accelerated senescence process (McCarthy et al. 2001; Rodríguez-Serrano et al. 2006). Several genes responsive to heavy metal stress are also induced during leaf senescence and pathogen attack (Obregón et al. 2001; Ouelhadj et al. 2006), which indicates an overlap in the regulatory mechanisms underlying these processes, with ROS production being a common event in these situations. Further studies will be necessary to understand the network involved in plant defense against heavy metal stress and the role of ROS and NO in regulating both ion homeostasis and cellular responses to heavy metals.

Acknowledgments This work was supported by the Ministry of Education and Science, Spain (Grant BIO2005–03305) and *Junta de Andalucía* (project P06-CVI-01820).

References

Aravind P, Narasimha M, Prasad V (2005) Modulation of cadmium-induced oxidative stress in *Ceratophyllum demersum* by zinc involves ascorbate-glutathione cycle and glutathione metabolism. Plant Physiol Biochem 43:107–116

Arisi AC, Mocquo B, Lagriffiould A, Mench M, Foyer CH, Jouanin L (2000). Responses to cadmium in leaves of transformed poplars overexpressing glutamylcysteine synthase. Physiol Plant 109:143–149

Arnaud N, Murgia I, Boucherez J, Briat JF, Cellier F, Gaymard F (2006) An iron-induced nitric oxide burst preceedes ubiquitin-dependent protein degradation for *Arabidopsis* AtFer1 ferritin gene expression. J Biol Chem 281:23579–23588

Azpilicueta CE, Benavides MP, Tomaro ML, Gallego S (2007) Mechanism of CATA3 induction by cadmium in sinflower leaves. Plant Physiol Biochem 45:589–595

Bartha B, Kolbert Z, Erdei L (2005) Nitric oxide production induced by heavy metals in *Brassica juncea* L. Czern. and *Pisum sativum* L. Acta Biol Szeg 49:9–12

Békésiová B, Hraška S, Libantová J, Moravcikova J, Matusšiková I (2008) Heavy-metal stress induced accumulation of chitinase isoforms in plants. Mol Biol Rep 35:579–588

Benavides MP, Gallego SM, Tomaro M (2005) Cadmium toxicity in plants. Brazil J Plant Physiol 17:21–34

Besson-Bard A, Pugin A, Wendehenne D (2008) New insights into nitric oxide signaling in plants. Ann Rev Plant Biol 59:21–39

Cho U, Seo N (2004) Oxidative stress in *Arabidopsis thaliana* exposed to cadmium is due to hydrogen peroxide accumulation. Plant Sci 168:113–120

Clemens S (2006) Toxic metal accumulation, responses to exposure and mechanisms of tolerance in plants. Biochimie 88:1707–1719

Cobett CS (2000) Phytochelatins and their roles in heavy metal detoxification. Plant Physiol 123:825–832

Collin V, Eymery F, Genty B, Rey P, Havaux M (2008) Vitamin E is essential for the tolerance of *Arabidopsis thaliana* to metal-induced oxidative stress. Plant Cell Environ 31:244–257

Dana MM, Pintor-Toro JA, Cubero B (2006) Transgenic tobacco plants overexpressing chitinases of fungal origin show enhanced resistance to biotic and abiotic stress agents. Plant Physiol 142:722–730

Delledonne M (2005) NO news is good news for plants. Curr Opin Plant Biol 8:390–396

Devoto A, Turner JG (2005) Jasmonate-regulated *Arabidopsis* stress signaling network. Physiol Plant 123:161–172

Dixit V, Pandey V, Shymar R (2001). Differential antioxidative responses to cadmium in roots and leaves of pea (*Pisum sativum* L. cv. Azad). J Exp Bot 52:1101–1109

Djebali W, Gallusci P, Polge C, Boulila L, Galtier N, Raymond P, Chaibi W, Brouquisse R (2008) Modifications in endopeptidase and 20S proteasome expression and activities in cadmium treated tomato (*Solanum lycopersicum* L.) plants. Planta 227:625–639

Drazic G, Mihailovic N (2005) Modification of cadmium toxicity in soybean seedlings by salicylic acid. Plant Sci 168:511–517

Faller P, Kienzler K, Krieger-Liszkay A (2005) Mechanism of Cd^{2+} inhibits photoactivation of photosyntem II by competitive binding to the essential Ca^{2+} site. Biochim Biophys Acta 1706:158–164

Finkemeier I, Goodman M, Lamkemeyer P, Kandlbinder A, Sweetlove LJ, Dietz KJ (2005) The mitochondrial type II peroxiredoxin F is essential for redox homeostasis and root growth of *Arabidopsis thaliana* under stress. J Biol Chem 280:12168–12180

Fodor A, Szabó-Nagy A, Erdei l (1995) The effects of cadmium on the fluidity and H^+-ATPase activity of plasma membrane from sunflower and wheat roots. J Plant Physiol 147:87–92

Freeman JL, Persana MW, Nieman K, Albretch C, Peer W, Pickering IJ, Salt DE (2004) Increased glutathione biosynthesis plays a role on nickel tolerance in *Thlaspi* nickel hyperaccumulators. Plant Cell 16:2176–2191

Freeman JL, Garcia D, Kim D, Hopf A, Salt DE (2005) Constitutively elevated salicylic acid signals glutathione-mediated nickel tolerance in *Thlaspi* nickel hyperaccumulators. Plant Physiol 137:1082–1091

Gallego SM, Benavides MO, Tomaro ML (1996) Effects of heavy-metal ion excess in sunflower leaves: evidences for involvement of oxidative stress. Plant Sci 121:151–159

Garnier L, Simon-Plas F, Thuleau P, Agnel JP, Blein JP, Ranjeva R, Montillet JL (2006) Cadmium affects tobacco cells by a series of three waves of reactive oxygen species that contribute to cytotoxicity. Plant Cell Environ 29:1956–1969

Guo H, Ecker JR (2004) The ethylene signaling pathway: new insights. Curr Opin Plant Biol 7:40–49

Guo B, Liang YC, Zhu YG, Zhao FJ (2007) Role of salicylic acid in alleviating oxidative damage in rice roots (*Oryza sativa*) subjected to cadmium stress. Environ Pollut 147:743–749

Halliwell B, Gutteridge JMC (2007) Free Radicals in Biology and Medicine, 4th edition. Oxford University Press, London

Herbette S, Taconnat L, Hugouvieux V, Piette L, Magniette ML, Cuine S, Auroy P, Richaud P, Forestier C, Bourguignon J, Renou JP, Vavasseur A, Leonhardt N (2006) Genome-wide transcriptome profiling of the early cadmium response of *Arabidopsis* roots and shoots. Biochimie 88:1751–1765

Hernández LE, Cooke DT (1997) Modification of the root plasma membrane lipid composition of cadmium-treated *Pisum sativum*. J Exp Bot 48:1375–1381

Hernández LE, Lozano-Todrógue z E, Gárate A, Carpena-Ruíz R (1998) Influence of cadmium on the uptake, tissue accumulation and subcellular distribution of manganese in pea seedlings. Plant Sci 132:139–151

Horemans N, Raeymaekers T, Van Beek K, Nowocin A, Blust R, Broos K, Cuypers A, Vangronsveld J, Guisez Y (2007). Dehydroascorbate uptake is impaired in the early response of *Arabidopsis* plant cell cultures to cadmium. J Exp Bot 16:4307–4317

Howarth JR, Domínguez-Solís JR, Gutiérrez-Alcalá G, Wray JL, Romero LC, Gotor C (2003) The serine acetyltransferase gene family in *Arabidopsis thaliana* and the regulation of its expression by cadmium. Plant Mol Biol 51:589–598

Howden R, Goldsbrough PB, Andersen CS, Cobbett CS (1995) Cadmium-sensitive, cad1 mutants of *Arabidopsis thaliana* are phytochelatin deficient. Plant Physiol 107:1059–1066

Hsu YT, Kao CH (2004) Cadmium toxicity is reduced by nitric oxide in rice leaves. Plant Growth Regul 42:227–238

Illéš P, Schlicht M, Pavlovkin J, Lichtscheidl I, Baluška F, Ovecka M (2006) Aluminium toxicity in plants: internalization of aluminium into cells of the transition zone in *Arabidopsis* roots apices related to changes in plasma membrane potential, endosomal behavior, and nitric oxide production. J Exp Bot 57:4201–4213

Jasinski M, Sudre D, Schansker G, Schellenberg M, Constant S, Martinoia E, Bovet L (2008) AtOSA1, a member of the Abc1-like family, as a new factor in cadmium and oxidative stress response. Plant Physiol 147:719–731

Kim DY, Bovet L, Maeshima M, Martinoia E, Lee Y (2007) The ABC transporter AtPDR8 is a cadmium extrusion pump conferring heavy metal resistance. Plant J 50:207–218

Kopyra M, Gwóždž EA (2003) Nitric oxide stimulates seed germination and counteracts the inhibitory effect of heavy metals and salinity on root growth of *Lupinus luteus*. Plant Physiol Biochem 41:1011–1017

Kopyra M, Stachín-Wilk M, Gwóždž EA (2006) Effect of exogenous nitric oxide on the antioxidant capacity of cadmium-treated soybean cell suspension. Acta Physiol Plant 28:525–536

Laspina NV, Groppa MD, Tomaro ML, Benavides MP (2005) Nitric oxide protects sunflower leaves against Cd-induced oxidative stress. Plant Sci 169:323–330

Lemaire S, Keryer E, Stein M, Schepens I, Issakidis-Bourguet E, Gérard-Hirme C, Miginiac-Maslow M, Jacquot JP (1999). Heavy-metal regulation of thioredoxin gene expression in *Chlamydomonas reinhardti*. Plant Physiol 120:773–778

Lindermayr C, Saalbach G, Bahnweg G, Durner J (2006) Differential inhibition of *Arabidopsis* methionine adenosyltransferases by protein *S*-nitrosylation. J Biol Chem 281: 4285–4291

León AM, Palma JM, Corpas FJ, Gomez M, Romero-Puertas MC, Chatterjee D, Mateos RM, del Río LA, Sandalio LM (2002) Antioxidative enzymes in cultivars of pepper plants with different sensitivity to cadmium. Plant Physiol Biochem 40:813–820

Loake G, Grant M (2007) Salicylic acid in plant defense - the players and protagonists. Curr Opin Plant Biol 10:466–472

López-Martín MC, Becana M, Romero LC, Gotor C (2008) Knocking out cytosolic cysteine synthesis compromises the antioxidant capacity of the cytosol to maintain discrete concentrations of hydrogen peroxide in *Arabidopsis*. Plant Physiol 147:562–572

Ma CH, Haslbeck M, Babujee L, Jahn O, Reumann S (2006) Identification and characterization of a stress-inducible and a constitutive small heat-shock protein targeted to the matrix of plant peroxisomes. Plant Physiol 141:47–60

McCarthy I, Romero-Puertas MC, Palma JM, Sandalio LM, Corpas FJ, Gómez M, del Río LA (2001) Cadmium induces senescence symptoms in leaf peroxisomes of pea plants. Plant Cell Environ 24:1065–1073

Metwally A, Finkemeier I, Georgi M, Dietz KJ (2003). Salicylic acid alleviates the cadmium toxicity in barley seedling. Plant Physiol 132:272–281

Metwally A, Safronova VI, Belimov AA, Dietz KJ (2005) Genotypic variation of the response to cadmium toxicity in *Pisum sativum*. J Exp Bot 56:167–178

Minglin L, Yuxiu Z, Tuanyao C (2005) Identification of genes up-regulated in response to Cd exposure in *Brassica juncea* L. Gene 363:151–158

Mithöfer A, Schulze B, Boland W (2004) Biotic and heavy metal stress response in plants: evidence for common signals. FEBS Lett 566:1–5

Mittler R, Vanderauwera S, Gollery M, Van Breusegem F (2004) Reactive oxygen gene network of plants. Trends Plant Sci 9:490–498

Mittler G, Mittler R (2006) Could heat shock transcription factor function as hydrogen peroxide sensor in plants? Ann Bot 98:279–288

Montillet JL, Cacas JL, Garnier L, Montané MH, Douki T, Bessoule JJ, Polkowska-Kowalczyk L, Maciejewska U, Agnel JP, Vial A, Triantaphylidès V (2004) The usptream oxylipin profile of *Arabidopsis thaliana*: a tool to scan for oxidative stresses. Plant J 40:439–451

Nocito FF, Lancilli C, Crema B, Fourcroy P, Davidian JC, Sacchi GA (2006) Heavy metal stress and sulfate uptake in maize roots. Plant Physiol 141:1138–1148

Obregón P, Marín R, Sanz A, Castresana C (2001) Activation of defense-related genes during senescence: a correlation between gen expression and cellular damage. Plant Mol Biol 46:67–77

Olmos E, Martinez-Solano JR, Piqueras A, Hellin E (2003) Early steps in the oxidative burst induced by cadmium in cultured tobacco cells (BY-2 line). J Exp Bot 54:291–301

Ortega-Villasante C, Rellán-Álvarez ZZ, Del Campo FF, Carpena-Ruiz RO, Hernández LE (2005) Cellular damage induced by cadmium and mercury in *Medicago sativa*. J Exp Bot 56:2239–2251

Ouariti O, Boussama N, Zarrouk M, Cherif A, Ghorbal MH (1997) Cadmium and copper-induced changes in tomato membrane lipids. Phytochem 45:1343–1350

Ouelhadj A, Kuschk P, Humbeck K (2006) Heavy metal stress and leaf senescence induce the barley gene HvC2d1 encoding a calcium-dependent novel C2 domain-like protein. New Phytol 170:261–273

Paradiso A, Berardino R, de Pinto MC, Sanitá di Toppi L, Srotelli MM, Tommasi F, de Gara L (2008) Increase in the ascorbate-glutathione metabolism as local and precocious systemic responses induced by cadmium in durum wheat plants. Plant Cell Physiol 49:362–374

Pena LB, Pasquini LA, Tomaro ML, Gallego SM (2006) Proteolytic system in sunflower (*Helianthus annuus* L.) leaves under cadmium stress. Plant Sci 171:531–537

Perfus-Barbeoch L, Leonhardt N, Vavasseur A, Forestier C (2002) Heavy metal toxicity: cadmium permeates through calcium channels and disturbs the plant water status. Plant J 32:539–548

Poschenrieder C, Gunsé B, Barceló J (1989) Influence of cadmium on water relations, stomatal resistance, and abscisic acid content in expanding bean leaves. Plant Physiol 90:1365–1371

del Río LA, Sandalio LM, Corpas FJ, Palma JM, Barroso JB (2006). Reactive oxygen species and reactive nitrogen species in peroxisomes. Production, scavenging and role in cell signaling. Plant Physiol 141:330–335

Rivetta A, Negrini N, Cocucci M (1997) Involvement of Ca^{2+}-calmodulin in Cd^{2+} toxicity during the early phases of radish (*Raphanus sativus* L.) seed germination. Plant Cell Environ 20:600–608

Rodríguez-Serrano M, Romero-Puertas MC, Zabalza A, Corpas FJ, Gómez M, del Río LA, Sandalio, LM (2006) Cadmium effect on oxidative metabolism of pea (*Pisum sativum* L.) roots. Imaging of reactive oxygen species and nitric oxide accumulation *in vivo*. Plant Cell Environ 29:1532–1544

Rodríguez-Serrano M, Romero-Puertas MC, Pazmiño DM, Testillano PS, Risueño MC, del Río LA, Sandalio LM (2009) Cellular Response of Pea Plants to Cadmium Toxicity: Cross Talk between Reactive Oxygen Species, Nitric Oxide, and Calcium. Plant Physiol 150:229–243

Rogers EE, Eide DJ, Guerinot ML (2000) Altered selectivity in an *Arabidopsis* metal transporter. Proc Natl Acad Sci USA 97:12356–12360

Romero-Puertas MC, McCarthy I, Sandalio LM, Palma JM, Corpas FJ, Gómez M, del Río LA. (1999) Cadmium toxicity and oxidative metabolism of pea leaf peroxisomes. Free Rad Res 31:S25–S32

Romero-Puertas MC, Palma JM, Gómez M, del Río LA, Sandalio LM (2002) Cadmium causes the modification of proteins in pea plants. Plant Cell Environ 25:677–686

Romero-Puertas MC, Rodríguez-Serrano M, Corpas FJ, Gómez M, del Río LA, Sandalio LM (2004) Cadmium-induced subcellular accumulation of $O_2\cdot^-$ and H_2O_2 in pea leaves. Plant Cell Environ 27:1122–1134

Romero-Puertas MC, Corpas FJ, Rodriguez-Serrano M, Gomez M, del Río LA, Sandalio LM (2007a) Differential expression and regulation of antioxidative enzymes by cadmium in pea plants. J Plant Physiol 164:1346–1357

Romero-Puertas MC, Laxa M, Mattè A, Zanninotto F, Finkemeier I, Jones AME, Perazzolli M, Vandelle E, Dietz KJ, Delledonne M (2007b) *S*-nitrosylation of peroxiredoxin II E promotes peroxynitrite-mediated tyrosine nitration. Plant Cell 19:4120–4130

Romero-Puertas MC, Campostrini N, Matte A, Righetti PG, Perazzolli M, Zolla L, Roepstorff P, Delledonne M (2008) Proteomic analysis of *S*-nitrosylated proteins in *Arabidopsis thaliana* undergoing hypersensitive response. Proteomics 8:1459–1469

Sandalio LM, Dalurzo HC, Gomez M, Romero-Puertas MC, del Río LA (2001) Cadmium-induced changes in the growth and oxidative metabolism of pea plants. J Exp Bot 52:2115–2126

Sanitá di Toppi L, Gabbrielli R (1999) Response to cadmium in higher plants. Environ Exp Bot 41:105–130

Sarry JE, Kuhn L, Ducruix C, Lafaye A, Junot C, Hugouvieux V, Jourdain A, Bastien O,Fievet J, Vailhen D, Amekraz B, Moulin C, Ezan C, Garin J, Bourguignon J (2006) The early response of *Arabidopsis thaliana* cells to cadmium exposure explored by protein and metabolite profiling analysis. Proteomics 6:2180–2198

Schützendübel A, Polle A (2002) Plant responses to abiotic stresses: heavy metals-induced oxidative stress and protection by mycorrhization. J Exp Bot 53:1351–1365

Schützendübel A, Schwanz P, Terchmann T, Gross K, Langenfeld-Heyger R, Godbold DL, Polle A (2001) Cadmium-induced changes in antioxidative systems, hydrogen peroxide content, and differentiation in scots pine roots. Plant Physiol 127:887–898

Sharma SS, Dietz KJ (2006) The significance of amino acids and amino acid-derived molecules in plant responses and adaptation to heavy metal stress. J Exp Bot 57:711–726

Shi QH, Zhu ZJ (2008) Effects of exogenous salicylic acid on manganese toxicity, element contents and antioxidative system in cucumber. Environ Exp Bot 63:1–3

Singh HP, Batish DR, Kaur G, Arora K, Kohli RK (2008) Nitric oxide (as sodium nitroprusside) supplementation ameliorates Cd toxicity in hydroponically grown wheat roots. Environ Exp Bot 63:158–167

Skorzynska-Polit E, Krupa Z (2006) Lipid peroxidation in cadmium-treated *Phaseolus coccineus* plants. Arch Environ Contam Toxicol 50:482–487

Smeets K, Cuypers A, Lambrechts A, Semane B, Hoet P, Van Laere A, Vangronsveld J (2005). Induction of oxidative stress and antioxidative mechanisms in *Phaseolus vulgaris* after Cd application. Plant Physiol Biochem 43:437–444

Smeets K, Ruytinx J, Semane B, Van Belleghem F, Remans T, Van Saden S, Vanginsveld J, Cuypers A (2008) Cadmium-induced transcriptional and enzymatic alterations related to oxidative stress. Environ Exp Bot 63:1–8

Song WY, Martinoia E, Lee J, Kim D, Kim DY, Vogt E, Shim D, Choi KS, Hwang I, Lee Y (2004) A novel family of cys-rich membrane proteins mediates cadmium resistance in *Arabidopsis*. Plant Physiol 135:1027–1039

Sunkar R, Kapoot A, Zhu J-K (2006). Posttranscripctional induction of two Cu/Zn superoxide dismutase genes in *Arabidopsis* is mediated by down-regulation of miT398 and important for oxidative stress tolerance. Plant Cell 18:2051–2065

Suzuki N, Yamaguchi Y, Koizumi N, Sano H (2002) Functional characterization of a heavy metal binding protein CdI19 from *Arabidopsis*. Plant J 32:165–173

Suzuki N (2005) Alleviation by calcium of cadmium-induced root growth inhibition in *Arabidopsis* seedlings. Plant Biotechnol 22:19–25

Tian QY, Sun DH, Zhao MG, Zhang WH (2007) Inhibition of nitric oxide synthase (NOS) underlies aluminium-induced inhibition of root elongation in *Hibiscus moscheutos*. New Phytol 174:322–331

Tsyganov VE, Belimov AA, Borisov AY, Safronova VI, Georgi M, Dietz KJ, Tikhonovich IA (2007) A chemically induced new pea (*Pisum sativum*) mutant SGECdt with increased tolerance to, and accumulation of, cadmium. Annal Bot 99:227–237

Van Assche F, Clijsters H (1990) Effects of metals on enzyme activity in plants. Plant Cell Environ 13:195–206

Van Breusegem F, Bailey-Serres J, Mittler R (2008) Unraveling the tapestry of networks involving reactive oxygen species in plants. Plant Physiol 147:978–984

Van de Mortel JE, Schat H, Moerland PD, Ver Loren van Themaat E, van der Ent S, Blankestijn H, Ghandilyan A, Tsiatsiani S, Aarts MG (2008) Expression differences for genes involved in lignin, glutathione and sulphate metabolism in response to cadmium in *Arabidopsis thaliana* and the related Zn/Cd-hyperaccumulator *Thlaspi caerulescens*. Plant Cell Environ 31:301–324

Vitória AP, Lea PJ, Azevedo RA (2001) Antioxidant enzymes responses to cadmium in radish tissues. Phytochem 57:701–710

Wang Y, Fang J, Leonard SS, Rao KM (2004) Cadmium inhibits the electron transfer chain and induces reactive oxygen species. Free Radic Biol Med 36:1434–1443

Wang JW, Wu JY (2005) Nitric oxide is involved in methyl jasmonate-induced defense responses and secondary metabolism activities of *Taxus xells*. Plant Cell Physiol 46:923–930

Wang YS, Yang ZM (2005) Nitric oxide reduces aluminum toxicity by preventing oxidative stress in the roots of *Cassia tora* L. Plant Cell Physiol 46:1915–1923

Weber M, Trampczynska A, Clemens S (2006) Comparative transcriptome analysis of toxic metal responses in *Arabidopsis thaliana* and the Cd(2+)-hypertolerant facultative metallophyte *Arabidopsis* halleri. Plant Cell Environ 29:950–963

Wilson ID, Neill SJ, Hancock JT (2007) Nitric oxide synthesis and signaling in plants. Plant Cell Environ 31:622–631

Xiang C, Oliver DJ (1998) Glutathione metabolic genes coordinately respond to heavy metals and jasmonic acid in *Arabidopsis*. Plant Cell 10:1539–1550

Yu CC, Hung KT, Kao CH (2005) Nitricoxide reduces Cu toxicity and Cu-induced NH^{4+} accumulation in rice leaves. J Plant Physiol 162:1319–1330

Zhang H, Jiang Y, He Z, Ma M (2005) Cadmium accumulation and oxidative burst in garlic (*Allium sativum*). J Plant Physiol 162:977–984

Zhang LP, Mehta SK, Liu ZP, Yang ZM (2008) Copper-induced proline synthesis is associated with nitric oxide generation in *Chlamydomonas reinhardtii*. Plant Cell Physiol. 49:411–419

Zhao Z, Cai Y, Zhu Y, Kneer R (2005) Cadmium-induced oxidative stress and protection by l-galactono-1,4-lactone in winter wheat (*Triticum aestivum* L.). J Plant Nutr Soil Sci 168:1–5

Reactive Oxygen Species in Ozone Toxicity

Kirk Overmyer, Michael Wrzaczek, and Jaakko Kangasjärvi

Abstract The entry of ozone (O_3) to the leaf intercellular airspace is followed by its degradation to reactive oxygen species (ROS) and the induction of active ROS production by the plant itself. Using genetic and genomic tools, some of the components involved in plant O_3 responses have begun to be delineated. Mutant screens and analyses in the model plant *Arabidopsis thaliana* have revealed a picture of the O_3 response that is coming into focus in the form of recurring themes that constitute a core O_3 response, which consists of a network of ROS and hormonal interactions controlling the magnitude of O_3-induced cell death.

1 Introduction

1.1 Ozone the Pollutant

In the upper stratosphere ozone (O_3), the trimolecular allotrope of oxygen, protects life on the earth from UV irradiation. In the troposphere, O_3 is a harmful atmospheric pollutant and a potent phytotoxin. Of all the stresses discussed in this volume, O_3 is perhaps unique in that the stressor itself is a reactive oxygen species (ROS). O_3 is a potent oxidant and exerts its toxicity via the induction of oxidative stress.

1.2 O_3 Damage

There is great variation in the O_3 sensitivity of natural and cultivated plants, both within and between species. In addition to the plant sensitivity, the nature of the O_3 dose itself has implications for determining the mode of damage. Slightly elevated

K. Overmyer, M. Wrzaczek, and J. Kangasjärvi (✉)
Plant Biology, Department of Biological and Environmental Sciences,
University of Helsinki, 00014 Helsinki, Finland
e-mail: jaakko.kangasjarvi@helsinki.fi

L.A. del Río and A. Puppo (eds.), *Reactive Oxygen Species in Plant Signaling*,
Signaling and Communication in Plants, DOI 10.1007/978-3-642-00390-5_12,

O_3 during long time periods (several days to weeks), so-called chronic O_3, generally does not cause visible damage but rather metabolic changes resulting in reduced growth, altered resource allocation, and premature senescence. Acute O_3 stress is hallmarked by high O_3 concentrations lasting for short time periods (hours to a few days). Acute O_3 episodes generally result in visible damage in sensitive plants. This damage is usually in the form of lesions similar in phenotype to the hypersensitive cell death triggered in pathogen resistance responses. In this chapter, we will discuss primarily plant responses to acute ozone.

1.3 Site of Action

O_3 has little effect on the outer surfaces of plants and must traverse stomatal pores to access its major sites of action in the apolastic space. Once inside the leaf, O_3 rapidly reacts with components of the cell walls, cellular membranes and apoplastic fluids. Upon these reactions, O_3 degrades into secondary ROS (Fig. 1) such as superoxide ($O_2^{\bullet -}$), hydrogen peroxide (H_2O_2), hydroxyl radical (•OH), and singlet oxygen ($^1O_2$). These ROS then go on to further react in the apoplast and beyond. Longer-lived ROS, in essence H_2O_2, which can diffuse through the plasma membrane via certain types of aquaporins (Bienert et al. 2007), can transmit the oxidant potential

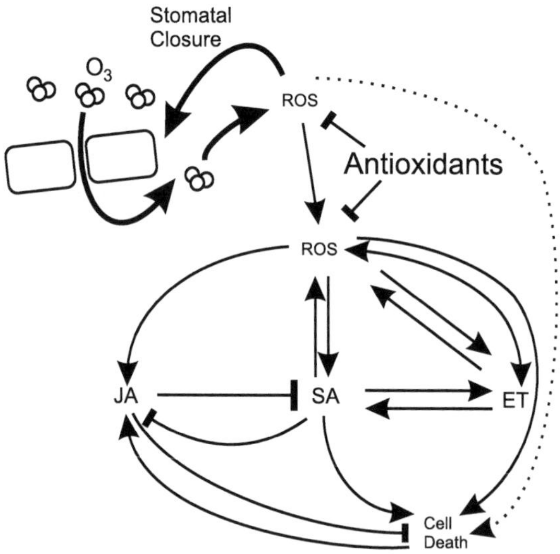

Fig. 1 Balanced signals in O_3-tolerant plants. Depicted are the ROS signaling networks active in O_3 exposed O_3-tolerant plants. Abbreviations: *SA* salicylic acid, *JA* jasmonic acid, *ET* ethylene, *ROS* reactive oxygen species. Note that ROS are split into two pools; the ROS derived from O_3 breakdown (*upper*) and the ROS actively produced by the plant (*lower*). The amplification loops between SA, ethylene, and ROS are balanced and controlled by the antagonism of JA- and ROS-scavenging by antioxidants. ROS-induced stomatal closure also limits the extent of primary ROS production from O_3 breakdown. Only a few cells undergo cell death and no visible lesions are formed

of O_3 into the cell. However, O_3 itself does not enter the cell. Thus, the apoplast is considered the primary site of O_3 action. The exact mechanism of O_3 or ROS perception in general remains unknown. However, O_3 and O_3-derived ROS have been shown to oxidize proteins, membranes, and other components in and outside of plant cells (Brosché et al. 2009).

1.4 O_3: the "Model Stress"

As discussed below, there are many similarities between O_3 responses, pathogen responses, and other stress responses; O_3 can be seen as a "signaling mimic", that co-opts signaling pathways via the production of ROS, which are perceived by the plant as signaling molecules. O_3 can be easily delivered noninvasively to a large number of plants simultaneously. Such O_3 treatment induces ROS signaling pathways in isolation to other complicating factors. For example, it is possible to induce an HR-like oxidative burst in the absence of bacteria, which manipulate defense signaling pathways via effector proteins, or to induce ABA signaling in the absence of the other physical effects associated with dehydration or osmotic shock. This makes O_3 a convenient model for the study of ROS signaling, with implications for and relevancy to a wide variety of ROS and stress signaling subspecialties.

2 O_3 Responses

2.1 The Oxidative Burst

Subsequent to the production of ROS from O_3 breakdown, these ROS further trigger the plant itself to produce an oxidative burst (Fig. 1). The O_3-induced oxidative burst is a common response that has been found in every case under study and has been demonstrated in a variety of species, including tobacco (*Nicotiana tabacum)*, *Arabidopsis thaliana*, tomato (*Lycopersicon esculentum)*, *Rumex* species and *Malva sylversteris* (Overmyer et al. 2000; Rao and Davis 1999; Schraudner et al. 1998; Wohlgemuth et al. 2002).The O_3-induced apoplastic oxidative burst is largely driven by the plasma membrane NAD(P)H oxidase and cell wall peroxidases (Overmyer et al. 2000; Pellinen et al. 1999; Rao and Davis 1999), although other ROS sources are also likely involved. ROS production in other subcellular compartments has been observed as well (Joo et al. 2004; Pellinen et al. 2002). The kinetics of the oxidative burst depends on the sensitivity of the plants. As demonstrated in tobacco and *Arabidopsis*, in sensitive plants, O_3 induces a biphasic oxidative burst that results in cell death, while in tolerant plants, a single burst of a lesser magnitude has been observed (Joo et al. 2004; Schraudner et al. 1998). Consistent with the wide variety of mechanisms determining their O_3 sensitivity or tolerance, considerable variation in O_3-induced ROS accumulation patterns, in terms of the ROS type and its spatial distribution, has been seen in naturally occurring *Arabidopsis* ecotypes

(Wohlgemuth et al. 2002) and mutants (Overmyer et al. 2000, 2008; Rao and Davis 1999). This also held true for tomato where multiple genotypes were studied side by side (Wohlgemuth et al. 2002).

2.2 Cell Death

Similar to plant–pathogen interactions, the cell death induced by O_3 can be genetically controlled and is a form of programmed cell death (PCD). In addition to the oxidative burst, there are a large number of physical, molecular, and biochemical similarities between O_3-induced cell death and the hypersensitive response (HR), which is a well-studied model of plant PCD. Beyond similarities to the HR, there are several lines of evidence of O_3-induced PCD. For example, in the *rcd1* mutant of *Arabidopsis*, O_3-induced cell death has been shown to require active metabolism and signaling. This requirement was seen as reduced or delayed cell death upon application of inhibitors of kinases, proteases, ATPases, transcription, and Ca^{+2} influx (Overmyer et al. 2005). In hybrid poplar (*Populus maximowizii xPopulus trichocarpa*), tobacco, and *Arabidopsis*, characteristic biochemical and morphological features of PCD, such as nuclear DNA fragmentation, nuclear shrinkage, chromatin condensation, membrane blebbing, and the release of cytochrome C from mitochondria, have been documented (Koch et al. 2000; Overmyer et al. 2005; Pasqualini et al. 2003). O_3 also induced protease activities in tobacco (Pasqualini et al. 2003), and caspase-like and phenylmethanesulphonyl fluoride (PMSF)-sensitive protease activities were required for O_3 induced cell death in *Arabidopsis* (Overmyer et al. 2005).

However, there are multiple forms of cell death induced by O_3. In some cases, O_3 has been shown to kill plant cells via overwhelming oxidative damage (Koch et al. 2000; Pell et al. 1997). In these cases, O_3 acted as a toxin that caused physical damage and directly killed cells. This process has been termed necrotic cell death and, in contrast to PCD, does not require active metabolism and signaling. Thus, O_3 has multiple modes of action and can act as both an oxidative toxin and a signal molecule. Interestingly, the appearance of PCD markers does not always correlate with O_3 sensitivity, as seen in hybrid poplar (Koch et al. 2000), indicating that sensitivity can be conditioned by both necrotic and programmed cell death mechanisms. It has also been suggested that more than one mode of cell death can occur even in the same plant (Overmyer et al. 2005; Pell et al. 1997).

2.3 Other Responses

O_3 induces transcriptional reprogramming of the plant cell. Much work has been devoted to the study of O_3-induced changes in gene expression both at the level of single genes and transcript profiling experiments with microarrays tracking genome level changes. This topic is beyond the scope of this chapter and has been covered elsewhere; the reader is directed to Brosché et al. (2009) and Chap. 11 of this volume.

Another of the well-studied O_3 responses is protein degradation. In potato (*Solanum tuberosum*), O_3 induced the selective degradation of certain proteins (Eckardt and Pell 1994). Additionally, the accumulation of a polyubiquitin transcript has been documented in O_3-exposed pine (*Pinus sylvestris*; Wegener et al. 1997). Together, these results suggest a role for ubiquitin-directed, proteasome-mediated protein degradation pathway in regulating O_3 responses.

Hydroxyl radicals produced from O_3-lysis can initiate lipid peroxidation. Lipid peroxidation products have indeed been observed during O_3 exposure in tobacco (Schraudner et al. 1997). Such a massive change in membrane and lipid structures is a rich source of potential signaling cues. Membrane structural changes may alter membrane protein function. Linolenic acid lipid hydroxides are intermediates on the jasmonic acid (JA) biosynthesis pathway. There is a long list of other potential signaling molecules produced, including, but not limited to, hydroxy acids, aldehydes, malondialdehyde, and ethane (reviewed in Schraudner et al. 1997).

3 O_3 Perception and Signaling

The mechanism of O_3 perception, or ROS perception in general, in the apoplast remains unknown, although some of the early events in the O_3 response have been characterized. The earliest known O_3-induced process is the activation of ion fluxes, including calcium fluxes. O_3 induced a characteristic biphasic peak in calcium (Clayton et al. 1999). Both phases of this signature were required for the induction of glutathione-S-transferase, which was used as a gene expression marker. This calcium response was very fast, occurring within seconds from the onset of exposure (Evans et al. 2005). Calcium was also required for O_3-induced cell death (Kadono et al. 2006; Overmyer et al. 2005), suggesting involvement in multiple levels of the O_3 response.

Recent reports suggest that heterotrimeric G-proteins might be critically involved in mediating the effects of O_3 in plants (Booker et al. 2004; Joo et al. 2004). One response to O_3 exposure in *Arabidopsis* is leaf curling. The heterotrimeric G-protein alpha subunit was required for this response while vegetative damage was independent of this G-protein complex (Booker et al. 2004). The alpha and beta subunits relate to more specific parts of the O_3 response (Joo et al. 2004). The Gα mutant lacked both peaks of the bi-modal oxidative burst while only the early peak was missing in the Gβ mutant. These data indicate that the late component of the oxidative burst, which is linked to tissue damage and intracellular signaling, requires only the Gα-protein. The Gα-protein mediates the activation of the membrane-bound NADPH oxidases. However, the early peak, which arises primarily from the chloroplasts, is dependent on signaling through the Gβγ-complex.

In several plant species, including tobacco, *Arabidopsis*, and hybrid poplar, exposure to O_3 has lead to a rapid activation of mitogen-activated protein kinases (MAPKs) (Ahlfors et al. 2004b; Hamel et al. 2005; Samuel et al. 2000). MAPKs are a family of protein kinases with well-defined roles in many aspects of plant life (See Chap. 2, this volume). The classical MAPK cascade is composed of a MAP kinase kinase kinase (MAP3K), a MAP kinase kinase (MAP2K) and a MAPK.

The *Arabidopsis* MAPKs AtMPK3 and AtMPK6 exhibited altered localization into the nucleus upon O_3 exposure (Ahlfors et al. 2004b). Interestingly, this O_3-dependent MAPK activation and relocalization was independent of salicylic acid (SA), JA and ethylene signaling. Activation of NtMPK4 in response to O_3 was required for JA signaling and the regulation of the stomatal aperture in tobacco (Gomi et al. 2005). Thus, MAPKs are directly involved in determining O_3 tolerance also via the regulation of the stomatal aperture.

While the activation of MAPKs in response to O_3 appears to be a common phenomenon, over-expression of the MAPK NtSIPK (*Nicotiana tabacum* salicylic acid-induced protein kinase) in tobacco lead to O_3-hypersensitivity and enhanced ethylene formation while blocking O_3-induced SA production (Samuel et al. 2005). This sensitivity could be due to prolonged SIPK activation. MAPK activation by O_3 was typically transient; a rapid activation, followed by a deactivation (Ahlfors et al. 2004b). MAPK deactivation can be achieved by dephosphorylation by so-called MAP kinase phosphatases (MKPs). A knockout of MKP2 leads to O_3-hypersensitivity (Lee and Ellis 2007). These studies demonstrate that while MAPKs are activated by O_3, their deactivation is critical to O_3 tolerance as prolonged activation has detrimental effects.

4 The Road to Ruin

4.1 Many Ways to Get There

As illustrated below, both forward genetic screens for O_3-sensitive mutants and reverse genetic analysis of mutants with knockout alleles of known genes have contributed in complementary ways. As a result of this research, a picture of the O_3 response is coming into focus in the form of recurring themes that constitute a core network of O_3 responses. In other words, as new activation points are being discovered, known signaling components are reappearing in different contexts. This presents a view of the complexity of the response and illustrates that, within one plant, multiple but converging processes determine O_3 sensitivity or tolerance. We have previously proposed a model which accounts for the role of ROS, SA, ethylene, and JA in controlling cell death (Fig. 1). This is a further development of the amplification loop between ROS, SA, and cell death, termed the oxidative cell death cycle (Overmyer et al. 2000, 2003). Figure 1 depicts the situation in tolerant plants during O_3 exposure. O_3-derived ROS accumulation is partly attenuated by antioxidants. The plant is also protected from excessive O_3 influx by ROS-induced stomatal closure. ROS induce ethylene, SA, and JA; these hormone signals interact in a balanced response resulting in a low level of cell death. This cell death involves only a few cells and is microscopic, i.e., there are no visible lesions. On one side, ethylene and SA promote ROS accumulation and cell death in a signal-amplifying feed forward loop. Counter to this, JA antagonizes SA and ethylene action keeping ROS accumulation and cell death under control. This model is based on results gained mostly with mutants of *Arabidopsis*. This work has identified several major areas that are driving current O_3 research and begins to define a core cell death pathway common to all O_3-induced cell death and some of the peripheral pathways that interact with the core pathway. We have further modified this model here to illustrate some of the mechanisms known to result in the activation of O_3-induced cell death via the oxidative cell death cycle (Fig. 2). Figure 2 depicts four different situations, each discussed in their respective sections below, where signaling imbalances have occurred, resulting in O_3 sensitivity.

4.2 Antioxidants

The first O_3-sensitive *Arabidopsis* mutant published was *soz1* (*sensitive to ozone 1*), which was subsequently renamed to *vtc1* (*vitamin C 1*) when it was discovered that the mutant was acorbate-deficient (Conklin et al. 1996). The *VTC1* gene encodes a GDP-mannose pyrophosphorylase, which is required for plant ascorbate biosynthesis (Conklin et al. 1999). This mutant underscores the importance of ascorbate and its attendant antioxidant pathway in controlling ROS levels during O_3 exposure. Reduced ascorbate levels in *vtc1* allow unregulated ROS accumulation to levels sufficient to induce cell death (Fig. 2a). There are no data available on the hormone response in *vtc1* or other antioxidant deficient plants under O_3 stress. As their roles are not known, the hormone pathways are depicted by a question mark in Fig. 2a. Ascorbate mutants have also led to the definition of ascorbate's role in pathogen stress and cell death control (Smirnoff et al. 2001). Importantly, the further application of this screen has led to the isolation of several ascorbate mutants (Conklin et al. 2000) and the definition of the ascorbate biosynthesis pathway (Smirnoff et al. 2001). Such work continues: recently, a screen for O_3 tolerance has identified a purple acid phosphatase with phytase activity which, when activated, doubles the plant ascorbate content and confers O_3 tolerance (Zhang et al. 2008). This work suggests the existence of an alternate ascorbate pathway in plants. The role of ascorbate cycling in the O_3 response was further underscored by the highly O_3 sensitive phenotype of *dhar* (*dehydroascorbate reductase*) mutant plants (Yoshida et al. 2006).

4.3 Stomata

O_3 acts in the leaf apoplast and must enter via stomata, making stomatal aperture a means of stress avoidance and the initial defense against O_3. Plants exposed to O_3 under conditions that drive stomatal opening, such as high humidity, exhibited increased O_3 damage (Kangasjärvi et al. 1994). Conversely, factors which close stomata, such as dehydration stress, exogenous ABA treatment, or increased CO_2 concentrations, protected plants from O_3 (Fiscus et al. 2005; Kangasjärvi et al. 1994).

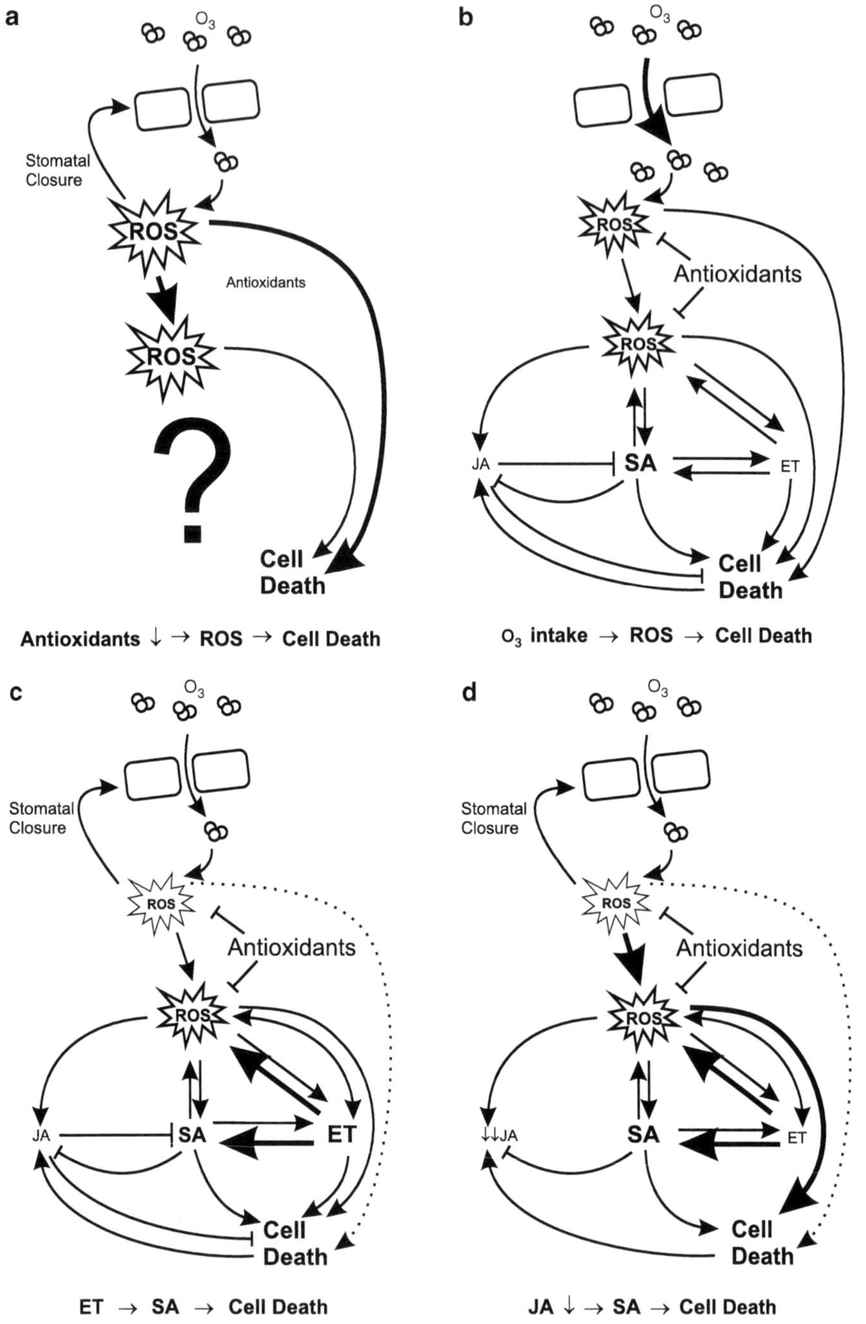

Fig. 2 Signaling imbalances leading to O_3 sensitivity. Depiction of signaling changes in some of the genotypes known to have increased O_3 damage. Abbreviations: *SA* salicylic acid, *JA* jasmonic acid, *ET* ethylene. (**a**) Loss of antioxidant capacity, as in the *vtc1* mutant, leads to increased ROS levels and cell death. (**b**) Misregulated stomata, as in the *slac1* mutant, leads to increased O_3 flux,

This has led to the concept of O_3 flux, which has now been taken into account in calculating the threshold dose for O_3 toxicity in environmental modeling (Fiscus et al. 2005).

O_3 induces stomatal closure and, while the mechanism controlling this is not known, evidence points to a role for O_3-derived ROS acting directly on stomata as signaling intermediates to promote closure. Torsethaugen et al. (1999) demonstrated that O_3 acted directly on stomata in bean (*Vicia faba)* and inhibited the guard cell K^+ channels that mediate stomatal opening. Other evidence follows the recognized roles of ROS as mediators of ABA signaling leading to stomatal closure (See Chap. 6, this volume). In *Arabidopsis*, O_3-induced stomatal closure occurred at 3 h or earlier, while O_3-induced ABA accumulation was not seen until 8 h, suggesting ABA-independent closure (Kollist et al. 2007; Overmyer et al. 2008). Kollist et al. (2007) have revealed a rapid O_3-induced stomatal response using a novel apparatus which can trace transpiration from individual untouched *Arabidopsis* rosettes. The use of this device revealed a previously unnoticed very fast closure of stomata occurring within 10 min of exposure. This response was dependent on ABI2 (ABA INSENSITIVE 2), a class II protein phosphatase, suggesting that protein phosphorylation is involved in the rapid O_3-induced stomatal closure.

O_3 sensitive mutants of *Arabidopsis* have also contributed significantly to the understanding of fundamental stomatal function. The *slac1* (*slow anion channel associated1*) mutant, which has misregulated stomata (Overmyer et al. 2008; Vahisalu et al. 2008) was originally isolated as (*radical-induced cell death 3*) *rcd3* in an O_3 sensitivity screen. Stomata in *slac1* were constitutively more open and exhibited delayed or incomplete closure in response to O_3, light, CO_2, humidity, ABA, nitric oxide, calcium, and H_2O_2 (Vahisalu et al. 2008). The *rcd3* mutant was renamed to *slac1* when it was shown that the protein encoded was required for slow (S-type) anion currents in guard cells. SLAC1 was the first component of this complex to be identified; it encodes a plasma membrane localized protein with ten predicted transmembrane domains and distant similarity to microbial dicarboxylate/malic acid transporter proteins. Taken together, this suggests that SLAC1 itself is the ion channel responsible for the S-type anion current and that SLAC1 may transport malate. However, both of these possibilities remain to be proven formally.

Work with the *slac1* mutant suggests some of the signaling processes involved in O_3 damage formation when stomata are misregulated. The inability to close stomata early in O_3 exposure (Overmyer et al. 2008; Vahisalu et al. 2008) leads to an enhanced level of ROS formation (Fig. 2b; Overmyer et al. 2008). As has been previously suggested (Kangasjärvi et al. 2005; Overmyer et al. 2008), damage caused by this

Fig. 2 (continued) which drives higher ROS accumulation. This triggers ET and SA accumulation that amplify the pro-death signal; together, these factors result increased cell death. (**c**) Ethylene overproducers, such as *eto1*, have elevated levels of SA due to the promotion of SA biosynthesis by ethylene; increased ET and SA promote ROS accumulation and cell death. (**d**) JA insensitivity, as in the *jar1* mutant, results in a loss of SA and ethylene antagonism and other potential direct anti-cell death effects of JA; increased SA levels and ethylene signaling amplify ROS and promote cell death enhanced early ROS burst likely drives the observed increased ethylene and SA accumulation, which further amplify the ROS burst and promote cell death (Fig. 2b).

4.4 Salicylic Acid

Salicylic acid (SA) plays a prominent role in O_3-induced defenses and cell death. As an outgrowth of work pioneered in the field of plant pathogen interactions (See Chap. 9, this volume), it has been established that SA and ROS each promote the other's formation, forming a feed-forward amplification loop. This process is also active during most O_3 responses (Overmyer et al. 2003). O_3 induces the synthesis of SA. The accumulation of SA became significantly increased at 5 h in *Arabidopsis* (Overmyer et al. 2008; Overmyer et al. 2005). Genetic evidence of SA signaling at earlier time points is seen in the SA dependent accumulation of superoxide and changes in gene expression as early as 1 and 0.5 h, respectively (Rao and Davis 1999). Genetic experiments involving plants deficient either in SA accumulation or SA signaling have established SA as a positive regulator of ROS accumulation, cell death, and defenses in a variety of genetic backgrounds, including both O_3-tolerant and –sensitive mutants and ecotypes (Overmyer et al. 2003). Generally, impairing SA signaling has reduced lesion formation in sensitive genotypes of tobacco and *Arabidopsis* (Örvar et al. 1997; Overmyer et al. 2005; Rao and Davis 1999). However, SA has a dual role: it is required for defense responses, and impairing SA signaling also increased the sensitivity of tolerant genotypes by promoting necrotic cell death (Rao and Davis 1999). As mentioned above, the three stress hormones SA, JA, and ethylene interact with one another at multiple levels. Generally these interactions are mutually antagonistic (Kangasjärvi et al. 2005; Overmyer et al. 2003); however, the nature of the interaction is dependent on the process and tissue under study. Importantly, SA antagonizes JA at the level of biosynthesis and signaling (Figs. 1 and 2). As is discussed below, SA has a role in the O_3 sensitivity of both mutants affected in JA signaling and ethylene signaling.

4.5 Ethylene

Similar to SA, the primary role of ethylene during O_3 exposure is in promoting damage. Ethylene is among the early responses of plants to O_3 exposure. A burst of de novo synthesized ethylene is common in the O_3 response of all plant species and genotypes. For example, the enzymatic machinery required for O_3-induced ethylene biosynthesis was rapidly induced in *Arabidopsis* and tomato (Moeder et al. 2002; Vahala et al. 1998).The magnitude of this early ethylene burst is well correlated with the level of damage that appears later in the O_3 exposure. The role of ethylene in the O_3 response was demonstrated by manipulating ethylene biosynthesis or signaling either chemically or genetically. Feeding plants 1-aminocyclopropane-l-carboxylic acid (ACC), the immediate precursor of ethylene, during O_3 or other ROS stress resulted in increased ethylene production and enhanced levels of cell death (Overmyer et al. 2000). Treatment with the ethylene signaling inhibitor norbornadiene resulted in attenuated O_3-induced ROS accumulation and cell death (Overmyer et al. 2000). Genetically these results are seen in the phenotypes of *Arabidopsis* ethylene signaling mutants. Ethylene insensitive mutants, such as *etr1* and *ein2*, are highly O_3 tolerant (Overmyer et al. 2000; Tuominen et al. 2004), while ethylene-overproducing mutants, such as the *eto1* and *eto3*, are highly O_3 sensitive (Rao et al. 2002). Also, birch trees (*Betula pendula*), transgenically modified to express the dominant mutant *Arabidopsis etr1-1* ethylene receptor, exhibited reduced ethylene sensitivity and a reduction in O_3-induced damage (Vahala et al. 2003). Significantly, there is a positive feedback between ROS accumulation and ethylene evolution (Fig. 2c); superoxide accumulation was shown to be blocked by inhibitors of ethylene signaling (Overmyer et al. 2000). The ethylene-overproducing *Arabidopsis* mutants are deficient in the control of ethylene biosynthesis and lack feedback control mechanisms. These mutants demonstrate a synergism between SA and ethylene signaling in controlling cell death during the O_3 response. These mutants had elevated SA levels, which was required for their full sensitivity to O_3 (Rao et al. 2002). This suggests that ethylene, in addition to promoting ROS accumulation, also promotes SA accumulation, which can further amplify the accumulation of ROS (Fig. 1). The signaling changes of ethylene overproducing mutants, such as *eto1*, are depicted in Fig. 2c. Uncontrolled ethylene biosynthesis results in elevated ethylene and SA levels. Together the increased level of ethylene and SA promote ROS accumulation and cell death resulting in O_3-induced cell death driven by these two interacting rampant ROS signal amplification loops.

4.6 Jasmonic Acid

In contrast to SA and ethylene, the primary role of jasmonic acid (JA) during O_3 exposure is protective. The idea for JA as a signal transducer in the O_3 response came from observations of its induction of defense genes (Kangasjärvi et al. 1994) and from the role of ROS production in wounding response, which was known to involve JA (Örvar et al. 1997). The timing of O_3-induced JA accumulation is also consistent with a protective role. JA accumulation takes place late in the exposure after damage has already occurred, in *Arabidopsis* at 5-6 h (Overmyer et al. 2005, 2008; Rao et al. 2000; Tuominen et al. 2004). The first evidence of JA protection during O_3 exposure came from JA treatment studies. Pre-treatment with JA, or the induction of JA production by wounding, reduced O_3 damage in tobacco and *Arabidopsis* (Örvar et al. 1997; Overmyer et al. 2000; Rao et al. 2000). The O_3 sensitivity of *Arabidopsis* mutants with compromised JA signaling provided genetic confirmation of JA's protective role. JA biosynthesis mutants, such as *fad3/fad7/fad8* (Rao et al. 2000), JA insensitive mutants such as *jar1* (Overmyer et al. 2000; Rao et al. 2000), as well as the mutant *rcd1* and the ecotype Cvi-0, which both exhibit some degree of JA insensitivity at the level of gene expression (Ahlfors et al. 2004a; Rao et al. 2000), were all sensitive to O_3. This was also seen in a JA-insensitive clone of hybrid poplar (Koch et al. 2000). Interestingly, not all JA insensitive *Arabidopsis* mutants are O_3 sensitive; the *jin1* mutant (Nickstadt et al. 2004), which was insensitive to JA in gene expression and in root growth assays on JA plates, was not, however, O_3 sensitive. This can be interpreted as branching in the JA pathway. The JIN1 protein encodes the MYC transcription factor AtMYC2 and together with ethylene-response-factor1 (ERF1) antagonistically regulates genes involved in pathogen defense and in the JA-mediated systemic wound response (Lorenzo et al. 2004). Thus, AtMYC2 and ERF1 might serve as an integration point in JA signaling coordinating different JA responses. JA mutants also illustrate the role of hormone interactions in the O_3 response (Fig. 2d). The loss of JA antagonism of SA and ethylene result in SA hyperaccumulation (Rao et al. 2000) and increased sensitivity to ethylene (Tuominen et al. 2004). The O_3 sensitivity of JA insensitive mutants is driven by high SA content and signaling (Fig. 2d; Rao et al. 2000). Furthermore, although ethylene does not accumulate to high levels, ethylene hypersensitivity results in high ethylene signaling flux. Similar to *eto* mutants, sensitivity is driven by uncontrolled ethylene, SA, and ROS signaling loops. Although the signaling components are the same, the mechanism driving these loops, the loss of JA antagonism, is unique.

5 The *RCD1* Gene and Family

The *radical-induced cell death 1* (*rcd1*) mutant of *Arabidopsis* was isolated in screens for plants with a HR-like lesion phenotype in response to O_3 (Overmyer et al. 2000) and paraquat tolerance (Fujibe et al. 2004). The *RCD1* gene was also isolated based on its ability to complement ROS sensitivity in the yeast strain WYT, deficient in the transcription factor YAP1(Belles-Boix et al. 2000), and in a yeast screen for interaction with SOS1, a plasma membrane Na^+/H^+ antiporter involved in the salt stress response (Katiyar-Agarwal et al. 2006). Loss of RCD1 function has pleiotrophic effects. Although *rcd1* was sensitive to O_3 and extracellular superoxide, but not to H_2O_2 or chloroplastic superoxide generated by paraquat treatment, ROS sensitivity in *rcd1* is not the result of a decrease in the expression of antioxidant genes. Additional phenotypes in *rcd1* include salt sensitivity (Katiyar-Agarwal et al. 2006), altered transcriptional response to ABA, JA, and ethylene, as well as developmental phenotypes such as altered leaf morphology and early flowering (Ahlfors et al. 2004a). The RCD1 protein is nuclear localized (Katiyar-Agarwal et al. 2006) and has been shown to interact with transcription factors (Belles-Boix et al. 2000), which suggests a mechanism for *rcd1* pleiotropism. However, under stress conditions, RCD1 is also localized in the cytosol (Katiyar-Agarwal et al. 2006). O_3 sensitivity in *rcd1* is a co-dominant trait; however, other phenotypes such as paraquat tolerance are recessive. Interestingly, over-expression of *RCD1* or *rcd1* in wild-type plants conferred a Rcd1$^-$ phenotype (Ahlfors et al. 2004a; Fujibe et al. 2006). Decreased expression of *RCD1* might also contribute the higher O_3 sensitivity in the ecotype Ws-0 (Li et al. 2006). Together, these results suggest that a balanced gene dosage is required for RCD1 function. RCD1 belongs to a family of six proteins in *Arabidopsis* and contains a WWE-domain and a putative (ADP-ribosyl)transferase domain. RCD1 alters the responses to several plant hormones, namely ethylene, ABA, and JA, and thus could serve as an integration point between hormone signaling and a coordinated O_3 response.

6 Nitric Oxide

Nitric oxide (NO) signaling in plants has recently gained much attention (Neill et al. 2008). Given the relationships between NO and ROS signaling, NO will surely play an important role in the plant O_3 response. However, results supporting this are just beginning to be published. NO production was induced quickly by O_3 in tobacco (Ederli et al. 2006), with a burst of NO appearing at 1 h, peaking by 1.5 h and returning to control levels by 2.5 h after exposure begun. In the mitochondria, O_3 exposure led to a loss of cytochrome capacity and respiration and the induction of the alternative oxidase (AOX). The induction of AOX was dependent of NO production (Ederli et al. 2006). Significantly, another study showed that over-expression, but not reduction, of AOX lead to O_3 sensitivity in tobacco (Pasqualini et al. 2007). Intriguingly, the *rcd1* mutant has elevated levels of AOX expression (Overmyer et al. 2000). The roles of AOX and NO during O_3 exposure remain unclear; however, further research in this area promises to contribute to our understanding of NO and mitchochondrial biology.

7 Chloroplast Signaling

Although the role of the chloroplast has been explored extensively for chronic O_3 response (Pell et al. 1997), this area has received less attention in the area of acute O_3 response research. However, results from *Arabidopsis* mutant studies are giving indications of an important role for chloroplasts. One such O_3-sensitive mutant was independently isolated in two different laboratories; the *soz2/lcd1* (*sensitive to ozone2/lower cell density1*) mutant (Barth and Conklin 2003) and the *rcd2* (*radical induced cell death2*) mutant (Overmyer et al. 2008). The *rcd2* mutant has been renamed to *re-8*, when it was found to be an allele of the classical mapping mutant *reticulata* (*re-1*; González-Bayón et al. 2006). This mutant has a green variegated appearance with normal-colored perivasular tissue and intervascular tissue that is pale in color due to reduced cell density (Barth and Conklin 2003; Gonzalez-Bayon et al. 2006; Overmyer et al. 2008). The ROS sensitivity and developmental phenotypes in this mutant together provide a link between ROS signaling and leaf development. The *RETICULATA* gene encodes a chloroplast-localized (Zybailov et al. 2008) protein of unknown function. Epistasis studies with *reticulata* indicate that RE functions on a genetic pathway with phophoenopyruvate/phosphate translocator1

(PPT1; González-Bayón et al. 2006; Streatfield et al. 1999). The *ppt1* mutant, formerly know as *cue1* (*chlorophyll a/b binding protein underexpressed1*), was isolated based on its altered expression of nuclear chloroplast genes and is highly sensitive to oxidative stress (Streatfield et al. 1999). This mutant was independently isolated in a NO screen and was shown to be a NO overproducer (He et al. 2004). Taken together, this suggests that RETICULATA participates in a pathway involving NO, which controls the integration of stress, environmental, and developmental signals in the chloroplast.

Further evidence for chloroplast involvement in O_3 responses comes from the pattern of ROS accumulation seen in G-protein studies. Joo et al. (2004) demonstrated that ROS accumulation during O_3 exposure began in the chloroplasts of guard cells before spreading to whole guard cells and then to surrounding tissues.

Interestingly, RCD1 may also be involved in chloroplast signaling. The *RCD1* gene was identified as a high light-response upregulated gene (Bechtold et al. 2008). Together with the paraquat tolerance of *rcd1*, this suggests possible involvement of RCD1 in signaling processes originating from the chloroplast.

8 Conclusions

O_3 research has benefited greatly from the use of the genetic model plant *Arabidopsis*. We have begun to understand how the processes regulating O_3 entry to the leaf intercellular airspace and how the subsequent signaling affects the several downstream processes, which led to the induction-adaptive processes in O_3-tolerant plants, or cell death in O_3-sensitive plants. The early apoplastic reactions followed by hormonal signaling cascades have been elucidated to the degree where the interactions between the processes can be understood and similarities with other processes have become visible. However, the perception of O_3, and ROS in general, and the role of chloroplastic processes in plant responses to O_3, are still mostly unresolved. Understanding the integration of O_3-induced responses with ROS signaling and hormonal interactions can provide powerful tools for improving plant growth, productivity, and product quality for the future plant production in changing climate. The future challenge for plant-O_3 research is to identify the components involved in O_3/ROS perception and to elucidate crosstalk pathways to understand the interaction mechanisms in the whole signal network.

Reference

Ahlfors R, Lång S, Overmyer K, Jaspers P, Brosché M, Tauriainen A, Kollist H, Tuominen H, Belles-Boix E, Piippo M, Inzé D, Palva ET, Kangasjärvi J (2004a) Arabidopsis radical-induced cell death 1 belongs to the WWE protein-protein interaction-domain protein family and modulates abscisic acid, ethylene, and methyl jasmonate responses. Plant Cell 16:1925–1937

Ahlfors R, Macioszek V, Rudd J, Brosché M, Schlichting R, Scheel D, Kangasjärvi J (2004b) Stress hormone-independent activation and nuclear translocation of mitogen-activated protein kinases (MAPKs) in *Arabidopsis thaliana* plants during ozone exposure. Plant J 40:512–522

Barth C, Conklin PL (2003) The lower cell density of leaf parenchyma in the *Arabidopsis thaliana* mutant *lcd1-1* is associated with increased sensitivity to ozone and virulent *Pseudomonas syringae*. Plant J 35:206–218

Bechtold U, Richard O, Zamboni A, Gapper C, Geisler M, Pogson B, Karpinski S, Mullineaux PM (2008) Impact of chloroplastic- and extracellular-sourced ROS on high light-responsive gene expression in *Arabidopsis*. J Exp Bot 59:121–133

Belles-Boix E, Babiychuk E, Van Montagu M, Inzé D, Kushnir S (2000) CEO1, a new protein from *Arabidopsis thaliana*, protects yeast against oxidative damage. FEBS Lett 482:19–24

Bienert GP, Moller ALB, Kristiansen KA, Schulz A, Moller IM, Schjoerring JK, Jahn TP (2007) Specific aquaporins facilitate the diffusion of hydrogen peroxide across membranes. J Biol Chem 282:1183–1192

Booker FL, Burkey KO, Overmyer K, Jones AM (2004) Differential responses of G-protein *Arabidopsis thaliana* mutants to ozone. New Phytol 162:633–641

Brosché M, Kangasjärvi S-L, Overmyer K, Wrzaczek M, Kangasjärvi J (2009) Stress signaling III: Reactive oxygen species. In: Pareek A, Sopory SK, Bohnert HJ, Govindjee (eds) Abiotic stress adaptation in plants: physiological, molecular and genomic foundation. Springer, Dordrecht, The Netherlands

Clayton H, Knight MR, Knight H, McAinsh MR, Hetherington AM (1999) Dissection of the ozone-induced calcium signature. Plant J 17: 575–579

Conklin PL, Norris SR, Wheeler GL, Williams EH, Smirnoff N, Last RL (1999) Genetic evidence for the role of GDP-mannose in plant ascorbic acid (vitamin C) synthesis. Proc Natl Acad Sci USA 96:4198–4203

Conklin PL, Saracco SA, Norris SR, Last RA (2000) Identification of ascorbate acid-deficient *Arabidopsis thaliana* mutants. Genetics 154:847–856

Conklin PL, Williams EH, Last RL (1996) Environmental stress sensitivity of an ascorbic acid-deficient *Arabidopsis* mutant. Proc Natl Acad Sci USA 93:9970–9974

Eckardt NA, Pell EJ (1994) O3-induced degradation of rubisco protein and loss of rubisco messenger-RNA in relation to leaf age in *Solanum-Tuberosum L.* New Phytol 127:741–748

Ederli L, Morettini R, Borgogni A, Wasternack C, Miersch O, Reale L, Ferranti F, Tosti N, Pasqualini S (2006) Interaction between nitric oxide and ethylene in the induction of alternative oxidase in ozone-treated tobacco plants. Plant Physiol 142:595–608

Evans NH, McAinsh MR, Hetherington AM, Knight MR (2005) ROS perception in *Arabidopsis thaliana*: the ozone-induced calcium response. Plant J 41:615–626

Fiscus EL, Booker FL, Burkey KO (2005) Crop responses to ozone: uptake, modes of action, carbon assimilation and partitioning. Plant Cell Environ 28:997–1011

Fujibe T, Saji H, Arakawa K, Yabe N, Takeuchi Y, Yamamoto KT (2004) A methyl viologen-resistant mutant of *Arabidopsis*, which is allelic to ozone-sensitive *rcd1*, is tolerant to supplemental UV-B irradiation. Plant Physiol 134:275–285

Fujibe T, Saji H, Watahiki MK, Yamamoto KT (2006) Overexpression of the *radical-induced cell death1 (RCD1)* gene of Arabidopsis causes weak rcd1 phenotype with compromised oxidative-stress responses. Biosci Biotech Biochem 70:1827–1831

Gomi K, Ogawa D, Katou S, Kamada H, Nakajima N, Saji H, Soyano T, Sasabe M, Machida Y, Mitsuhara I, Ohashi Y, Seo S (2005) A mitogen-activated protein kinase NtMPK4 activated by SIPKK is required for jasmonic acid signaling and involved in ozone tolerance via stomatal movement in tobacco. Plant Cell Physiol 46:1902–1914

González-Bayón R, Kinsman EA, Quesada V, Vera A, Robles P, Ponce MR, Pyke KA, Micol JL (2006) Mutations in the *RETICULATA* gene dramatically alter internal architecture but have little effect on overall organ shape in Arabidopsis leaves. J Exp Bot 57:3019–3031

Hamel LP, Miles GP, Samuel MA, Ellis BE, Seguin A, Beaudoin N (2005) Activation of stress-responsive mitogen-activated protein kinase pathways in hybrid poplar (*Populus trichocarpa x Populus deltoides*). Tree Physiol 25:277–288

He YK, Tang RH, Hao Y, Stevens RD, Cook CW, Am SM, Jing LF, Yang ZG, Chen LG, Guo FQ, Fiorani F, Jackson RB, Crawford NM, Pei ZM (2004) Nitric oxide represses the *Arabidopsis floral transition*. Science 305:1968–1971

Joo JH, Wang S, Chen JG, Jones AM, Fedoroff N (2004) Different signaling and cell death roles of heterotrimeric G protein α and β subunits in the Arabidopsis oxidative stress response to ozone. Plant Cell 17:957–970

Kadono T, Yamaguchi Y, Furuichi T, Hirono M, Garrec J, Kawano T (2006) Ozone-induced cell death mediated with oxidative and calcium signaling pathways in tobacco Bel W3 and Bel B cell suspension cultures. Plant Signal Behav 1:321–322

Kangasjärvi J, Jaspers P, Kollist H (2005) Signalling and cell death in ozone-exposed plants. Plant Cell Environ 28:1021–1036

Kangasjärvi J, Talvinen J, Utriainen M, Karjalainen R (1994) Plant defence systems induced by ozone. Plant Cell Environ 17:783–794

Katiyar-Agarwal S, Zhu J, Kim K, Agarwal M, Fu X, Huang A, Zhu JK (2006) The plasma membrane Na^+/H^+ antiporter SOS1 interacts with RCD1 and functions in oxidative stress tolerance in *Arabidopsis*. Proc Natl Acad Sci USA 103:18816–18821

Koch JR, Creelman RA, Eshita SM, Seskar M, Mullet JE, Davis KR (2000) Ozone sensitivity in hybrid poplar correlates with insensitivity to both salicylic acid and jasmonic acid. The role of programmed cell death in lesion formation. Plant Physiol 123:487–496

Kollist T, Moldau H, Rasulov B, Oja V, Ramma H, Huve K, Jaspers P, Kangasjärvi J, Kollist H (2007) A novel device detects a rapid ozone-induced transient stomatal closure in intact Arabidopsis and its absence in abi2 mutant. Physiol Plant 129:796–803

Lee JS, Ellis BE (2007) *Arabidopsis* MAPK phosphatase 2 (MKP2) positively regulates oxidative stress tolerance and inactivates the MPK3 and MPK6 MAPKs. J Biol Chem 282:25020–25029

Li PH, Mane SP, Sioson AA, Robinet CV, Heath LS, Bohnert HJ, Grene R (2006) Effects of chronic ozone exposure on gene expression in *Arabidopsis thaliana* ecotypes and in *Thellungiella halophila*. Plant Cell Environ 29:854–868

Lorenzo O, Chico JM, Sánchez-Serrano JJ, Solano R (2004) *Jasmonate-insensitive1* encodes a MYC transcription factor essential to discriminate between different jasmonate-regulated responses in Arabidopsis. Plant Cell 16:1938–1950

Moeder W, Barry CS, Tauriainen AA, Betz C, Tuomainen J, Utriainen M, Grierson D, Sandermann H, Langebartles C, Kangasjärvi J (2002) Ethylene synthesis regulated by bi-phasic induction of ACC synthase and ACC oxidase genes is required for H_2O_2 accumulation and cell death in ozone-exposed tomato. Plant Physiol 130:1918–1926

Neill S, Barros R, Bright J, Desikan R, Hancock J, Harrison J, Morris P, Ribeiro D, Wilson I (2008) Nitric oxide, stomatal closure, and abiotic stress. J Exp Bot 59:165–176

Nickstadt A, Thomma BPHJ, Feussner I, Kangasjärvi J, Zeier J, Loeffler C, Scheel D, Berger S (2004) The jasmonate-insensitive mutant *jin1* shows increased resistance to biotrophic as well as necrotrophic pathogens. Mol Plant Pathol 5:425–434

Örvar BL, McPherson J, Ellis BE (1997) Pre-activating wounding response in tobacco prior to high-level ozone exposure prevents necrotic injury. Plant J 11:203–212

Overmyer K, Brosché M, Kangasjärvi J (2003) Reactive oxygen species and hormonal control of cell death. Trends Plant Sci 8:335–342

Overmyer K, Brosché M, Pellinen R, Kuittinen T, Tuominen H, Ahlfors R, Keinanen M, Saarma M, Scheel D, Kangasjärvi J (2005) Ozone-induced programmed cell death in the Arabidopsis *radical-induced cell death1* mutant. Plant Physiol 137:1092–1104

Overmyer K, Kollist H, Tuominen H, Betz C, Langebartels C, Wingsle G, Kangasjärvi S-L, Brader G, Mullineaux P, Kangasjärvi J (2008) Complex phenotypic profiles leading to ozone sensitivity in *Arabidopsis thaliana* mutants. Plant Cell Environ 31:1237–1249

Overmyer K, Tuominen H, Kettunen R, Betz C, Langebartels C, Sandermann H, Jr, Kangasjärvi J (2000) The ozone-sensitive *Arabidopsis rcd1* mutant reveals opposite roles for ethylene and jasmonate signaling pathways in regulating superoxide-dependent cell death. Plant Cell 12:1849–1862

Pasqualini S, Paolocci F, Borgogni A, Morettini R, Ederli L (2007) The overexpression of an alternative oxidase gene triggers ozone sensitivity in tobacco plants. Plant Cell Environ 30:1545–1556

Pasqualini S, Piccioni C, Reale L, Ederli L, la Torre G, Ferranti F (2003) Ozone-induced cell death in tobacco cultivar Bel W3 plants. The role of programmed cell death in lesion formation. Plant Physiol 133:1122–1134

Pell EJ, Schlagnhaufer CD, Arteca RN (1997) Ozone-induced oxidative stress: mechanisms of action and reaction. Physiol Plant 100:264–273

Pellinen R, Palva T, Kangasjärvi J (1999) Subcellular localization of ozone-induced hydrogen peroxide production in birch (*Betula pendula*) leaf cells. Plant J 20:349–356

Pellinen RI, Korhonen MS, Tauriainen AA, Palva ET, Kangasjärvi J (2002) Hydrogen peroxide activates cell death and defense gene expression in birch. Plant Physiol 130:549–560

Rao MV, Davis KR (1999) Ozone-induced cell death occurs via two distinct mechanisms in *Arabidopsis*: the role of salicylic acid. Plant J 17:603–614

Rao MV, Lee H, Davis KR (2002) Ozone-induced ethylene production is dependent on salicylic acid, and both salicylic acid and ethylene act in concert to regulate ozone-induced cell death. Plant J 32:447–456

Rao MV, Lee HI, Creelman RA, Mullet JA, Davis KR (2000) Jasmonic acid signaling modulates ozone-induced hypersensitive cell death. Plant Cell 12:1633–1646

Samuel MA, Miles GP, Ellis BE (2000) Ozone treatment rapidly activates MAP kinase signalling in plants. Plant J 22:367–376

Samuel MA, Walia A, Mansfield SD, Ellis BE (2005) Overexpression of SIPK in tobacco enhances ozone-induced ethylene formation and blocks ozone-induced SA accumulation. J Exp Bot 56:2195–2201

Schraudner M, Langebartels C, Sandermann H, Jr (1997) Changes in the biochemical status of plants cells induced by the environmental pollutant ozone. Physiol Plant 100:274–280

Schraudner M, Moeder W, Wiese C, Van Camp W, Inzé D, Langebartels C, Sandermann H, Jr (1998) Ozone-induced oxidative burst in the ozone biomonitor plant, tobacco Bel W3. Plant J 16:235–245

Smirnoff N, Conklin PL, Loewus FA (2001) Biosynthesis of ascorbic acid in plants: a renaissance. Annu Rev Plant Physiol Plant Mol Biol 52:437–467

Streatfield SJ, Weber A, Kinsman EA, Häusler RE, Li J, Post-Beittenmiller D, Kaiser WM, Pyke KA, Fluegge UI, Chory J (1999) The phosphoenolpyruvate/phosphate translocator is required for phenolic metabolism, palisade cell development, and plastid-dependent nuclear gene expression. Plant Cell 11:1609–1621

Torsethaugen G, Pell EJ, Assmann SM (1999) Ozone inhibits guard cell K^+ channels implicated in stomatal opening. Proc Natl Acad Sci USA 96:13577–13582

Tuominen H, Overmyer K, Keinänen M, Kollist H, Kangasjärvi J (2004) Mutual antagonism of ethylene and jasmonic acid regulates ozone-induced spreading cell death in *Arabidopsis*. Plant J 39:59–69

Vahala J, Ruonala R, Keinänen M, Tuominen H, Kangasjärvi J (2003) Ethylene insensitivity modulates ozone-induced cell death in birch. Plant Physiol 132:185–195

Vahala J, Schlagnhaufer CD, Pell EJ (1998) Induction of an ACC synthase cDNA by ozone in light-grown *Arabidopsis thaliana* leaves. Physiol Plant 103:45–50

Vahisalu T, Kollist H, Wang Y-F, Nishimura N, Chan W-Y, Valerio G, Lamminmäki A, Brosché M, Moldau H, Desikan R, Schroeder JI, Kangasjärvi J (2008) SLAC1 is required for plant guard cell S-type anion channel function in stomatal signalling. Nature 452:487–493

Wegener A, Gimbel W, Werner T, Hani J, Ernst D, Sandermann H (1997) Sequence analysis and ozone-induced accumulation of polyubiquitin mRNA in *Pinus sylvestris*. Can J For Res 27:945–948

Wohlgemuth H, Mittelstrass K, Kschieschan S, Bender J, Weigel HJ, Overmyer K, Kangäsjarvi J, Sandermann H, Langebartels C (2002) Activation of an oxidative burst is a general feature of sensitive plants exposed to the air pollutant ozone. Plant Cell Environ 25:717–726

Yoshida S, Tamaoki M, Shikano T, Nakajima N, Ogawa D, Ioki M, Aono M, Kubo A, Kamada H, Inoue Y, Saji H (2006) Cytosolic dehydroascorbate reductase is important for ozone tolerance in *Arabidopsis thaliana*. Plant Cell Physiol 47:304–308

Zhang W, Gruszewski HA, Chevone BI, Nessler CL (2008) An Arabidopsis purple acid phosphatase with phytase activity increases foliar ascorbate. Plant Physiol 146:431–440

Zybailov B, Rutschow H, Friso G, Rudella A, Emanuelsson O, Sun Q, van Wijk KJ (2008) Sorting signals, N-terminal modifications and abundance of the chloroplast proteome. PLoS ONE 3:e1994; doi:10.1371/journal.pone.0001994

ROS and Plant Membrane Rafts

Françoise Simon-Plas and Sébastien Mongrand

Abstract Although excess reactive oxygen species (ROS) are toxic, physiological concentrations of ROS may function as signaling molecules to mediate various responses. However, given that ROS are diffusible and short-lived, localizing the ROS signal at a precise subcellular location is essential for stimulation of specific redox signaling. In animals, recent studies have indicated lipid microdomain platforms or lipid rafts may be importantly implicated in redox signaling of a variety of cells in response to agonists or stimuli (for a review, see Li and Gulbins 2007). The plant plasma membrane (PM) is in charge of sensing the various environmental modifications faced by the plant cell and triggering the appropriate physiological responses. It thus exemplifies this requirement for an extremely fine-tuning of ROS production in plants, which has been evidenced as a mediator in many different biotic or abiotic stresses leading to significantly different responses. The spatial compartmentalization of ROS-producing enzymes in specialized domains of the plant PM could be one key element of such a regulation.

1 Evidence for Lateral Segregation of Plasma Membrane Components in Cells, the Case of Membrane Rafts

1.1 Definition for Lipid Rafts and Estimation of their Size

None would doubt that membranes of living cells are laterally subcompartmented and that this heterogeneity play major a role in cell biology. Nevertheless, characterizing this heterogeneity has proven to be difficult. In plant cells, experimental evidence provides clues on the existence of membrane domains of various size from the large domain (ca. 10-20 μm) of auxin transporter proteins (Galweiler et al. 1998;

F. Simon-Plas and S. Mongrand (✉)
Laboratoire de Biogenèse Membranaire, UMR 5200 CNRS-Université de Bordeaux, 146 rue Léo Saignat, 33076, Bordeaux, France
e-mail: sebastien.mongrand@biomemb.u-bordeaux2.fr

L.A. del Río and A. Puppo (eds.), *Reactive Oxygen Species in Plant Signaling*, Signaling and Communication in Plants, DOI 10.1007/978-3-642-00390-5_13,

Swarup et al. 2001) to rather small domains of less than 1 μm evidenced in plasma membrane (PM) domains of guard cell protoplasts, thought to act as perception site for the phytohormone abscisic acid (ABA) (Yamazaki et al. 2003). The tips of pollen tubes (Kost et al. 1999) and root hairs (Takeda et al. 2008) are also good examples of domains segregating lipids and proteins essential for the cell polarized growth. Another example of microdomains is given by the in vivo tagging strategy of the K^+ channel KAT1 targeted to the PM within positionally stable microdomains of ca. 0.5 μm in diameter (Sutter et al. 2006).

As stated before, the biochemical or biophysical approaches to study these membrane domains turned out to be tedious. Almost 20 years ago, the concept of "lipid rafts" arose from various studies going from virus biology, sphingolipid sorting, and membrane protein solubility to detergent and model membranes (see Powell 2006, for a historical perspective). The Keystone Symposium held in 2006 helped to clarify current thinking in the raft field by providing a definition for lipid rafts. First, although lipid-lipid interactions provide a driving force for the formation of membrane domains, undoubtedly proteins also play a major role in contributing to the genesis of these membrane microdomains. Hence, the term "lipid raft" was abandoned in favor of "membrane rafts". Second, membrane rafts are defined as "small (10–200 nm), heterogeneous, highly dynamic, sterol-, and sphingolipid-enriched domains that compartmentalize cellular processes. Small rafts can sometimes be stabilized to form larger platforms through protein–protein and protein–lipid interactions" (Pike 2006). We must emphasize that the size of rafts remains a subject of intense debates (Jacobson et al. 2007).

In mammalian cells, the sizes of rafts have been estimated, by immunogold labeling-electronic microscopy strategy, to be 25–70 nm in diameter (Prior et al. 2003). The size of raft thus cannot be resolved in living cells by conventional light microscopy (Glebov and Nichols 2004; Simons and Toomre 2000). Nevertheless, a punctuated distribution of raft markers in mammalian cells is light-microscopically visible after stimulation in various cell types (Gekara and Weiss 2004; Tanimura et al. 2003; Triantafilou et al. 2004), suggesting a stimulus-dependent aggregation of rafts.

In yeast, large raft-based membrane domains, more likely raft clusters, can be visualized in living cells (Malinska et al. 2003, 2004). In plant cells, up to now, only indirect evidence was provided. For example, the hexose-proton symporter HUP1 of the green alga *Chlorella kessleri* shows a punctuated localization in the plane of the PM. When expressed in yeast mutants lacking ergosterol and sphingolipids, confocal images showed a homogenous distribution of HUP1-GFP within the PM (Grossmann et al. 2006), a first experimental evidence of a lipid-dependant segregation of a PM protein.

This could be connected with the fact that biological membranes are composed of many types of lipids, including phospholipids, sphingolipids, and sterols. Numerous studies performed with artificial membranes with a composition similar to PM supported the formation of domains enriched in sterols and sphingolipids, largely excluding phospholipids (Niemela et al. 2007). The main forces enabling the formation of such domains would result from the ability of sphingolipids to associate with each other. Cholesterol molecules tightly interact with sphingolipids by hydrogen bonding and serve as spacers to fill voids between sphingolipids (Simons and Ikonen 1997). This would lead to the formation of tightly packed subdomains, corresponding to a "liquid ordered" phase, surrounded by the fluid phase of the membrane, enriched in fatty acids with unsaturated acyl chains, and qualified as "liquid disordered", e.g., for plant lipids (Beck et al. 2007; Dufourc 2008).

1.2 Biochemical Characterization: Lipidomic and Proteomic Approaches on Detergent Insoluble Membranes

A widely used feature of lipid rafts is their resistance to solubilization at low temperature in mild nonionic detergents such as Triton X-100 (Brown and Rose 1992; Rietveld and Simons 1998). It can be explained by the high degree of acyl-chain ordering, which prevents the diffusion of detergent molecules inside these domains. Detergent-insoluble membranes (DIMs) are further purified by density gradient centrifugation; raft-associated proteins are found floating in the low-density fraction (top of the gradient) due to the lipids remaining attached to the proteins, while the solubilized fluid phase is recovered in high-density fractions (see Fig. 1a). The insoluble fractions are also called detergent-resistant membranes (DRM), detergent-insoluble glycolipid-enriched microdomains (DIG), or glycolipid-enriched membranes. For clarity, in this chapter, we choose the acronym DIM.

Do DIMs obtained by this procedure selectively recruit specific membrane proteins, whereas others are excluded? It has to be noticed that, even widely used, this procedure of DIM isolation raises a profound question: to what extent do these DIMs correspond to microdomains present in vivo on the biological membrane? There is no definite answer to that question. However, two types of arguments can be put forward to justify the use of DIMs: (1) the lipid composition of DRMs is similar to the one of domains formed spontaneously, in complex lipid mixtures; and (2) in many studies, the association of proteins with DIMs has been correlated, using imaging techniques, with their clustering on the PM of living cells (Jacobson et al. 2007). It is important to note that compositionally distinct lipid microdomains may coexist on the membrane. They can differ both in their resistance towards different detergents and in the types of associated proteins (Roper et al. 2000).

In term of lipids, cholesterol, sphingomyelin, and glycosphingolipids are the major lipids of DIM in mammalian cells (Rietveld and Simons 1998). In yeast, these are ergosterol, inositolphosphoceramide, and its mannosylated derivatives (Bagnat et al. 2000; Kubler et al. 1996). In plants cells, free phytosterols, sterol conjugates and sphingolipids are found to be enriched in DIMs (Borner et al. 2005; Lefebvre et al. 2007; Mongrand et al. 2004). Sphingolipids were mainly indirectly measured by quantitation of LCB by HPLC (Borner et al. 2005; Lefebvre et al. 2007). This measurement suggests that shingolipids in DIMs are mainly glucosylceramides and glycosyl inositol phosphoceramide. Free sterols, but not steryl-conjugates, can be depleted from plant PM vesicles by the cyclic oligosaccharide methyl-beta-cyclodextrin, commonly used in animal cells to decrease cholesterol levels. After such treatment,

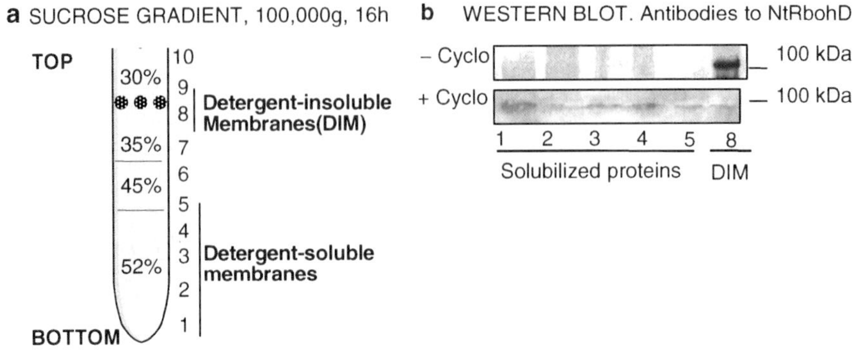

Fig. 1 Biochemical purification of detergent-insoluble membranes (DIMs) by flotation on discontinous sucrose gradient. Western-blotting indicating that NtrbohD is mainly associated with DIMs, in a sterol dependant manner. (**a**) Schematic representation of the centrifuge tube used for DIM isolation. (**b**) PMs extracted from BY-2 cells, treated or not with methyl-β cyclodextrin 20 mM (–/+ cyclo) were solubilized with Triton X-100 at low temperature and centrifuged on a sucrose gradient, as described in (**a**). An equal amount of proteins present in the fractions collected were submitted to SDS-PAGE and blotted with an anti-NtrbohD antibody. Adapted with permission from Roche et al. (2008)

leading to a reduction of 50% in the PM total free sterol, DIMs were no longer extracted from PM, showing that phytosterols are key compounds for the formation of plant PM microdomains (Roche et al. 2008).

In term of proteins, studies using mass spectrometry gave an extensive description of *Arabidopsis* (Shahollari et al. 2004; Borner et al. 2005) and tobacco BY-2 cell PM DIM proteome (Morel et al. 2006). From a functional point of view, this analysis indicated that if a primary function of the PM, such as transport, seems under-represented in the raft fraction, others undergo a significant increase of their relative importance, such as those involved in signaling and response to biotic and abiotic stress, cellular trafficking, and cell wall metabolism. Recently, a similar approach was performed on *Medicago truncatula* root PM rafts, in which a large number of proteins (i.e., 270) have been identified. Interestingly, evidence was given for the presence of a complete PM redox system in *Medicago* DIMs (see below). All these studies suggest that raft domains are likely to constitute signaling platforms involved in various physiological functions.

1.3 Role(s) for Membrane Rafts in Plant Cell

Several roles were attributed to membrane rafts in animal and yeast biology. Rafts form platforms for lipid (Simons and Ikonen 1997) and protein sorting and trafficking (Simons and van Meer 1988; Galbiati et al. 2001; Ikonen 2001) and cell signaling (Dykstra et al. 2001; Field et al. 1997; Simons and Toomre 2000; Stauffer and

Meyer, 1997). Membrane rafts also provide the cells with a mechanism for functional and spatial control of exocytosis (Chamberlain et al. 2001) and are involved in immune cell activation (Dykstra et al. 2001; Galbiati et al. 2001; Katagiri et al. 2001). Different models proposed that the small size of rafts in resting cells is important to keep raft-associated proteins in an inactive state. Upon stimulation, rafts may cluster to form a larger platform where functionally-related proteins can interact (Simons and Toomre 2000).

In plants, rather little evidence suggests the role of rafts in cell biology processes, some of them coming from plant–pathogen interaction. Recently, two studies reported the pathogen–triggered focal accumulation of components of plant defence pathway in the PM, a process reminiscent of lipid rafts (Assaad et al. 2004; Bhat et al. 2005). It concerns the *Arabidopsis* syntaxin AtPEN1, originally identified in a genetic screen for "non-host" resistance mutants (Collins et al. 2003), and the barley homolog, HvROR2, the heptahelical HvMLO; putative positive and negative regulators of a basal defence pathway against powdery mildew attack and Cytb561-containing protein, respectively (Bhat and Panstruga 2005). These authors showed that fluorescently-tagged versions of these proteins were evenly distributed in the PM in healthy leaf epidermal cells. When challenged with fungal pathogens, the fluorescence accumulates at fungal pathogen entry sites. It defines a stable and circular PM microdomain of approximately 3–10 μm diameter, labeled by filipin, a sterol dye, a result which strengthens the link between these proteins and raft domain (Bhat and Panstruga 2005).

In the next section, we will discuss a hypothesis, already obtained through experimental evidence, for the role of membrane rafts in regulating reactive oxygen species (ROS) production in plant cells.

2 Membrane Localization of ROS-Producing Enzymes and Compartmentalization of ROS Production

2.1 ROS-Producing Enzymes are Associated with DIMs

2.1.1 NADPH Oxidase (NtrbohD) in Tobacco Cells

Studies performed on different plant species showed that enzymes (named Rboh for respiratory burst oxidase homologue), similar to the respiratory burst oxidase of mammalian neutrophil cells, were an essential ROS-producing system during plant–microorganism interaction (for a review see Chap. 1 of this volume). Cell fractionation and membrane purification of plant tissue indicates that Rboh proteins are localized in PM (Sagi and Fluhr 2001; Simon-Plas et al. 2002). In particular, NtrbohD (for *Nicotiana tabacum* respiratory burst oxidase homolog, isoform D), located on the PM, was shown to be responsible for the oxidative burst triggered by the fungal elicitor cryptogein, since its downregulation by an antisense strategy on

Table 1 Enzymes and regulators found in plant raft plasma membrane, putatively involved in ROS formation

Protein name	Plant specie/accession number	Reference
NADPH oxidase NtrbohD	Nt, Q8RVJ9	Mongrand et al. (2004); Morel et al. (2005)
RAC5	Nt024142	Mongrand et al. (2004); Morel et al. (2005)
NADH-ubiquinone oxidoreductase	MtC00504_1	Lefebvre et al. (2006)
NADH-ubiquinone oxidoreductase	MtC00519.2_1	Lefebvre et al. (2006)
NADH-ubiquinone oxidoreductase	MtC00544_1	Lefebvre et al. (2006)
NADH-ubiquinone oxidoreductase	MtC20003_1	Lefebvre et al. (2006)
L-Acorbate Oxidase	MtC20240_1	Lefebvre et al. (2006)
Peroxidase precursor (haem peroxidase)	AC123575_10.1	Lefebvre et al. (2006)
Cytochrome b561, domon domain	AC147406_8.1	Lefebvre et al. (2006)
Domon domain	AC126793_22.1	Lefebvre et al. (2006)
Germin; cupin region	AC149038_25.1	Lefebvre et al. (2006)
Plastocyanin like	MtD04199_1	Lefebvre et al. (2006)
Ubiquinol-cytochrome c reductase	MtC50199_1	Lefebvre et al. (2006)
Cytochrome bd ubiquinol oxidase (14 kDa-subunit)	MtD22966_1	Lefebvre et al. (2006)
1,4-Benzoquinone reductase	MtC00067_1	Lefebvre et al. (2006)
Cytochrome p450, trans-cinnamate 4-monooxygenase	S36878	Lefebvre et al. (2006)
Blue copper-binding protein, plastocyanin like	MtC30578_1	Lefebvre et al. (2006)
Cytochrome c1	AC138464_11.1	Lefebvre et al. (2006)

tobacco bright-yellow 2 (BY-2) cells completely abolished ROS production upon challenging with this elicitor (Simon-Plas et al. 2002).

Studies performed on this model, using either immunological or mass spectrometry experiments (see Table 1), revealed the presence of NtrbohD in DIMs of tobacco BY-2 cells (Mongrand et al. 2004; Morel et al. 2006). Furthermore, it has been confirmed that all the protein was associated with the DIM fraction, and that this segregation was abolished by methyl-β-cyclodextrin, which indicates its sterol-dependency (Fig. 1b).

2.1.2 Redox System Around a Cytochrome b_{561} in *Medicago* Root?

In 2001, the ascorbate-reducible cytochrome b561 was found to be differentially solubilized by Triton X-100 in different plant PMs (Berczi et al. 2001). The authors speculated that the differential Triton X-100 solubility could be the consequence of the formation of sterol-containing rafts in the *Arabidopsis* membrane (Berczi and Horvath 2003). A few years later, biochemical characterization and extensive proteomic analysis of DIMs from *Medicago truncatula* root PM not only allowed the identification of a specific set of proteins common to other lipid rafts, but also the presence of a putative redox system around a cytochrome b561 not previously identified in lipid rafts of either plants or animals (Lefebvre et al. 2007; for review, see Furt et al. 2007) (see Table 1).

Cytb561 proteins belong to a family of proteins located in both PM and endomembranes; one subclass of this family containing an extracellular domain which is also found in dopamine β-monooxygenase N-terminal domain (DOMON). Proteins found in *Medicago* DIMs contain both DOMON and Cytb561 domains. In animals, Cytb561 proteins are able to transfer electrons across membranes from an ascorbate to a dehydro-ascorbate to allow the biosynthesis of noradrenaline from dopamine by dopamine β-monooxygenase. In addition, there is increasing evidence that Cytb561 proteins can act as ferric reductases, by reducing ferric ions (Ponting 2001; Berczi et al. 2007). Thus these proteins in *M. truncatula* are likely to be involved in redox control by shuttling electrons across the PM. It must be noted that its e^- acceptor is unknown.

On the apoplastic side, both ascorbate oxidase and class III heme peroxydase, both identified in DIMs, could also be involved in the redox system. In good agreement, strong illumination of watermelon to produce an excess of electron leads to a coregulation of both ascorbate oxidase and Cytb561 proteins (Nanasato et al. 2005). Class III heme peroxidase (PODs) could reduce hydrogen peroxide (involved in ROS degradation). Alternatively, depending on the apoplastic redox conditions, PODs could produce hydrogen peroxide (Mika et al. 2004) participating in ROS generation. Such ROS accumulation could have several functions, and ROS production may be a mean of controlling the potential symbiont (see Chap. 10 in this volume). Indeed, *Rhizobium* mutants with impaired ROS detoxification machinery present abnormal nodulation phenotypes (Pauly et al. 2006). On the plant side, it is involved in signaling, as accumulation of ROS is dependent on Nod factors, signal molecules secreted by *Rhizobium*.

The functional analysis of the PM cytochrome b_{561} redox system found in *M. truncatula* root DIM need to be substantiated for evaluating its role in the redox balance, and also determining whether it is involved in the establishment of the legume–*Rhizobium* symbiosis. Interestingly, the NADPH oxidase found in tobacco BY2 cell PM DIM has not been found in *M. truncatula* roots. Therefore, PODs in DIMs might be considered as other candidates, in addition to NADPH oxidase, to participate in ROS production. This model presented in Furt et al. 2007, albeit speculative, provides a good hypothesis for further studies.

2.2 *ROS Production is Observed in Distinct Areas of Plant PM*

To determine both the dynamics and subcellular location of ROS accumulation in cryptogein-elicited tobacco suspension cells and plants, cerium chloride, a capture agent of H_2O_2 widely employed in TEM studies in animal (Briggs et al. 1975; Warren et al. 1989; Soares et al. 1994; Miura et al. 2003) and plant cells and tissues (Bestwick et al. 1997; Pellinen et al. 1999), was used. As soon as 5 min after elicitation,

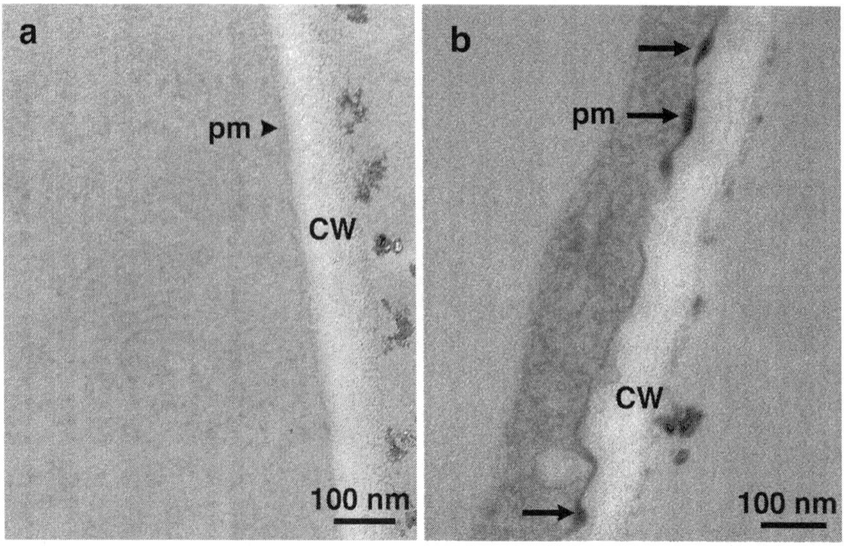

Fig. 2 Electron microscope localization of H_2O_2 accumulation triggered by cryptogein in BY-2 cells after staining by $CeCl_3$. $CeCl_3$ precipitates with H_2O_2 at reaction sites, forming electron-dense cerium perhydroxides that are visualized as black spots. (**a**) Experiment on nonelicited BY-2 cells: PM is free from cerium precipitates (*arrowhead*). (**b**) Aliquots of BY-2 cells sampled 5 min after the addition of 50 nM cryptogein. Patches (80 nm) of cerium perhydroxides are located along the PM (*arrows*). *CW* Cell wall, *pm* plasma membrane

H_2O_2 is clearly shown on the PM of cryptogein-elicited BY-2 cells with a maximum of precipitates visualized on PM of cells sampled 15 min after elicitation (Fig. 2). All along the time-course of the experiment (30 min), H_2O_2 was only visualized on the cell PM as distinct and local patches (Lherminier et al. 2009). This constitutes one of the very first correlations between the association of a protein with particular domains of specific lipidic composition, and the detection of its biological activity on distinct areas of the PM. It also strengthens the validity of using DIMs as biochemical counterparts of specialized domains present on the PM of living cells. Furthermore, the size of these precipitates was around 80 nm, which is within the size range (10–200 nm) generally admitted for lipid rafts.

Chemiluminescence experiments have failed to detect ROS accumulation in BY-2 cells transformed with antisense constructs of the PM NtrbohD (Simon-Plas et al. 2002). Consistently, these cells do not exhibit any precipitates of cerium on the PM, which leads to the assumption that the patches along the PM represent the accumulation of H_2O_2 generated by the NADPH oxidase. The same technique allowed the detection of hydrogen peroxide accumulation in cultured cells of *Rubia tinctorum* challenged with a fungal elicitor, and revealed that H_2O_2 generation occurred locally along the PM (Boka et al. 2007). In the context of cell growth and development, (Foreman et al 2003) reported that localized ROS production by NADPH oxidase in the growing tip of cells is required for polarized hair growth. Deregulation of such a process leading to a broader distribution of ROS led to mislocalization of root hair cells and ectopic hair formation sites (Carol et al. 2005).

These observations corroborate experimental evidence obtained in animal cells. The use of scanning near-field optical microscopy, combined with the immunodetection of PM NADPH oxidase, produced high resolution imaging indicating the occurrence of a clustered localization of these proteins on the PM of haematopoietic stem cells (Frassanito et al. 2008). Similarly, in vascular smooth muscle cells, Nox1 colocalized with caveolin in punctuate patches on the PM, whereas Nox4 was observed in particular intracellular sites (Hilenski et al. 2004). This differential localization in separate signaling domains may support opposing functions and differential regulation by growth factors and hormones. Conversely, an interesting study demonstrated that ROS could promote raft formation in T lymphocytes (Lu and Cederbaum 2007).

Another interesting feature is the demonstration that the functionality of animal NADPH oxidase depends on the presence of cholesterol, and that this effect is explained by the recruitment of the different subunits of this oxidase in cholesterol-rich microdomains to assemble the active enzymatic complex (Vilhardt and van Deurs 2004). This observation of microdomains acting as platforms gathering different proteins involved in a redox signaling pathway fully supports the hypothesis developed above concerning the "redox system" identified in *Medicago truncatula* roots.

Finally, both from experiments performed on animal and plant cells, NADPH oxidases appear to be activated within discrete membrane compartments, thereby facilitating localized ROS production. These findings provide insights into our understanding of the temporal and spatial organization of agonist stimulated ROS production and activation of specific redox signaling pathways. It remains to be determined how the subcellular localization and targeting of NADPH oxidases are regulated. Future experiments should reveal the functional importance of localized ROS production, and how its activity is controlled to produce ROS in a regulated manner.

References

Assaad FF, Qiu JL, Youngs H, Ehrhardt D, Zimmerli L, Kalde M, Wanner G, Peck SC, Edwards H, Ramonell K (2004) The PEN1 syntaxin defines a novel cellular compartment upon fungal attack and is required for the timely assembly of papillae. Mol Biol Cell 15:5118–5129

Bagnat M, Keranen S, Shevchenko A, Simons, K (2000) Lipid rafts function in biosynthetic delivery of proteins to the cell surface in yeast. Proc Natl Acad Sci USA 97:3254–3259

Beck J, Mathieu D, Loudet C, Buchoux S, Dufourc E (2007) Plant sterols in "rafts": a better way to regulate membrane thermal shocks, FASEB J 21:1714–1723

Berczi A, Horvath G (2003) Lipid rafts in the plant plasma membrane? Acta Biol Szeged 47:7–10

Berczi A, Luthje S, Asard H (2001) b-Type cytochromes in plasma membranes of Phaseolus vulgaris hypocotyls, *Arabidopsis thaliana* leaves, and Zea mays roots. Protoplasma 217:50–55

Berczi A, Su D, Asard H (2007) An *Arabidopsis* cytochrome b561 with trans-membrane ferrireductase capability. FEBS Lett 581:1505–8

Bestwick CS, Brown IR, Bennett MH, Mansfield JW (1997) Localization of hydrogen peroxide accumulation during the hypersensitive reaction of lettuce cells to *Pseudomonas syringae* pv phaseolicola. Plant Cell 9:209–221

Bhat RA, Miklis M, Schmelzer E, Schulze-Lefert P, Panstruga R (2005) Recruitment and interaction dynamics of plant penetration resistance components in a plasma membrane microdomain. Proc Natl Acad Sci USA 102:3135–3140

Bhat RA, Panstruga R (2005) Lipid rafts in plants. Planta 223:5–19

Boka K, Orban N, Kristof Z (2007) Dynamics and localization of H2O2 production in elicited plant cells. Protoplasma 230:89–97

Borner GH, Sherrier DJ, Weimar T, Michaelson LV, Hawkins ND, Macaskill A, Napier JA, Beale MH, Lilley KS, Dupree P (2005) Analysis of detergent-resistant membranes in Arabidopsis. Evidence for plasma membrane lipid rafts. Plant Physiol 137:104–116

Briggs RT, Dratch DB, Karnovsky ML, Karnovsky MJ (1975) Localization of NADH oxidase on the surface of human polymorphonuclear leukocytes by a new cytochemical method. J Cell Biol 67:566–596

Brown DA, Rose JK (1992) Sorting of GPI-anchored proteins to glycolipid-enriched membrane subdomains during transport to the apical cell surface. Cell 68:533–544

Carol RJ, Takeda S, Linstead P, Durrant MC, Kakesova H, Derbyshire P, Drea S, Zarsky V, Dolan L (2005) A RhoGDP dissociation inhibitor spatially regulates growth in root hair cells. Nature 438:1013–1016

Chamberlain LH, Burgoyne RD, Gould GW (2001) SNARE proteins are highly enriched in lipid rafts in PC12 cells: implications for the spatial control of exocytosis. Proc Natl Acad Sci USA 98:5619–5624

Collins NC, Thordal-Christensen H, Lipka V, Bau S, Kombrink E, Qiu JL, Huckelhoven R, Stein M, Freialdenhoven A, Somerville SC, Schulze-Lefert P (2003) SNARE-protein-mediated disease resistance at the plant cell wall. Nature 425:973–977

Dykstra ML, Cherukuri A, Pierce SK (2001) Floating the raft hypothesis for immune receptors: access to rafts controls receptor signaling and trafficking. Traffic 2:160–166

Dufourc E (2008) The role of phytosterols in plant adaptation to temperature, Plant Signal Behav 3:133–134

Field KA, Holowka D, Baird B (1997) Compartmentalized activation of the high affinity immunoglobulin E receptor within membrane domains. J Biol Chem 272:4276–4280

Foreman J, Demidchik V, Bothwell JH, Mylona P, Miedema H, Torres MA, Linstead P, Costa S, Brownlee C, Jones JD, Davies JM, Dolan L (2003) Reactive oxygen species produced by NADPH oxidase regulate plant cell growth. Nature 422:442–446

Frassanito MC, Piccoli C, Capozzi V, Boffoli D, Tabilio A, Capitanio N (2008) Topological organization of NADPH-oxidase in haematopoietic stem cell membrane: preliminary study by fluorescence near-field optical microscopy. J Microsc 229:517–524

Furt F, Lefebvre B, Cullimore J, Bessoule JJ, Mongrand S (2007) Plant lipid rafts fluctuat nec mergitur. Plant Signal Behav 2:1011–1013

Galbiati F, Razani B, Lisanti MP (2001) Emerging themes in lipid rafts and caveolae. Cell 106:403–411

Galweiler L, Guan C, Muller A, Wisman E, Mendgen K, Yephremov A, Palme K (1998) Regulation of polar auxin transport by AtPIN1 in Arabidopsis vascular tissue. Science 282:2226–2230

Gekara NO, Weiss S (2004) Lipid rafts clustering and signalling by listeriolysin O. Biochem Soc Trans 32:712–714

Glebov OO, Nichols BJ (2004) Distribution of lipid raft markers in live cells. Biochem Soc Trans 32:673–675

Grossmann G, Opekarova M, Novakova L, Stolz J, Tanner, W (2006) Lipid raft-based membrane compartmentation of a plant transport protein expressed in Saccharomyces cerevisiae. Eukaryot Cell 5:945–953

Hilenski LL, Clempus RE, Quinn MT, Lambeth JD, Griendling KK (2004) Distinct subcellular localizations of Nox1 and Nox4 in vascular smooth muscle cells. Arterioscler Thromb Vasc Biol 24:677–683

Ikonen E (2001) Roles of lipid rafts in membrane transport. Curr Opin Cell Biol 13(4):470–477

Jacobson K, Mouritsen OG, Anderson RG (2007) Lipid rafts: at a crossroad between cell biology and physics. Nat Cell Biol 9:7–14
Katagiri YU, Kiyokawa N, Fujimoto J (2001) A role for lipid rafts in immune cell signaling. Microbiol Immunol 45:1–8
Kost B, Lemichez E, Spielhofer P, Hong Y, Tolias K, Carpenter C, Chua, NH (1999) Rac homologues and compartmentalized phosphatidylinositol 4,5-bisphosphate act in a common pathway to regulate polar pollen tube growth. J Cell Biol 145:317–330
Kubler E, Dohlman HG, Lisanti MP (1996) Identification of Triton X-100 insoluble membrane domains in the yeast *Saccharomyces cerevisiae*. Lipid requirements for targeting of heterotrimeric G-protein subunits. J Biol Chem 271:32975–32980
Lefebvre B, Furt F, Hartmann MA, Michaelson LV, Carde JP, Sargueil-Boiron F, Rossignol M, Napier JA, Cullimore J, Bessoule JJ, Mongrand S (2007) Characterization of lipid rafts from Medicago truncatula root plasma membranes: a proteomic study reveals the presence of a raft-associated redox system. Plant Physiol 144:402–418
Lherminier J, Elmayan T, Fromentin J, Tantaoui Elaraquia K, Vesa S, Morel J, Verrier JL, Cailleteau B, Blein JP, Simon-Plas F (2009) NADPH oxidase mediated ROS production: subcellular localization and reassesment of its role in plant defense. MPMI in press
Li PL, Gulbins E (2007) Lipid rafts and redox signaling. Antioxid. Redox Signal 9:1411–1415
Lu Y, Cederbaum A (2007) The mode of cisplatin-induced cell death in CYP2E1-overexpressing HepG2 cells: modulation by ERK, ROS, glutathione, and thioredoxin. Free Radic Biol Med 43:1061–1075
Malinska K, Malinsky J, Opekarova M, Tanner W (2003) Visualization of protein compartmentation within the plasma membrane of living yeast cells. Mol Biol Cell 14:4427–4436
Malinska K, Malinsky J, Opekarova M, Tanner W (2004) Distribution of Can1p into stable domains reflects lateral protein segregation within the plasma membrane of living *S. cerevisiae* cells. J Cell Sci 117:6031–6041
Mika A, Minibayeva F, Beckett R, Lüthje S (2004) Possible functions of extracellular peroxidases in stress-induced generation and detoxification of active oxygen species. Phytochem Rev 3:173–193
Miura H, Bosnjak JJ, Ning G, Saito T, Miura M, Gutterman DD (2003) Role for hydrogen peroxide in flow-induced dilation of human coronary arterioles. Circ Res 92:e31–40
Mongrand S, Morel J, Laroche J, Claverol S, Carde JP, Hartmann MA, Bonneu M, Simon-Plas F, Lessire R, Bessoule JJ (2004) Lipid rafts in higher plant cells: purification characterization of Triton X-100-insoluble microdomains from tobacco plasma membrane. J Biol Chem 279:36277–36286
Morel J, Claverol S, Mongrand S, Furt F, Fromentin J, Bessoule JJ, Blein JP, Simon-Plas F (2006) Proteomics of plant detergent resistant membranes. Mol Cell Proteomics 5:1396–1411
Niemela PS, Ollila S, Hyvonen MT, Karttunen M, Vattulainen I (2007) Assessing the nature of lipid raft membranes. PLoS Comput Biol 3(2):e34
Nanasato Y, Akashi K, Yokota A (2005) Co-expression of cytochrome b561 and ascorbate oxidase in leaves of wild watermelon under drought and high light conditions. Plant Cell Physiol 46:1515–1524
Pauly N, Pucciariello C, Mandon K, Innocenti G, Jamet A, Baudouin E, Hérouart D, Frendo P, Puppo A (2006) Reactive oxygen and nitrogen species and glutathione: key players in the legume-Rhizobium symbiosis. J Exp Bot 57:1769–1776
Pellinen R, Palva T, Kangasjarvi J (1999) Short communication: subcellular localization of ozone-induced hydrogen peroxide production in birch (*Betula pendula*) leaf cells. Plant J 20:349–356
Pike LJ (2006) Rafts defined: a report on the Keystone Symposium on Lipid Rafts and Cell Function. J Lipid Res 47:1597–1598
Powel L (2006) From the archive; Lipid raft idea is floated. J Cell Biol 172:166–167
Ponting CP (2001) Domain homologues of dopamine beta-hydroxylase and ferric reductase: roles for iron metabolism Hum Mol Genet 10:1853–1858

Prior IA, Muncke C, Parton RG, Hancock JF (2003) Direct visualization of Ras proteins in spatially distinct cell surface microdomains. J Cell Biol 160:165–170
Rietveld A, Simons K (1998) The differential miscibility of lipids as the basis for the formation of functional membrane rafts. Biochim Biophys Acta 1376:467–479
Roper K Corbeil D, Huttner WB (2000) Retention of prominin in microvilli reveals distinct cholesterol-based lipid micro-domains in the apical plasma membrane. Nat Cell Biol 2:582–592
Roche Y, Gerbeau-Pissot P, Buhot B, Thomas D, Bonneau L, Gresti J, Mongrand S, Perrier-Cornet JM, Simon-Plas F (2008) Role of phytosterols in the structuration of plant plasma membrane. FASEB J 22:3980–3991
Sagi M, Fluhr R (2001) Superoxide production by plant homologues of the gp91(phox) NADPH oxidase. Modulation of activity by calcium and by tobacco mosaic virus infection. Plant Physiol 126:1281–1290
Shahollari B, Peskan-Bergho T, Oelmuller R (2004) Receptor kinases with leucine-rich repeats insoluble plasma membrane microdomains Physiol. Plantarum 122:397–403
Simon-Plas F, Elmayan T, Blein JP (2002) The plasma membrane oxidase NtrbohD is responsible for AOS production in elicited tobacco cells. Plant J 31:137–147
Simons K, Ikonen E (1997) Functional rafts in cell membranes. Nature 387:569–572
Simons K, van Meer G (1988) Lipid sorting in epithelial cells. Biochemistry 27(17): 6197–6202
Simons K, Toomre D (2000) Lipid rafts and signal transduction. Nat Rev Mol Cell Biol 1:31–39
Soares FA, Shaughnessy SG, MaClarkey WR, Orr FW (1994) Quantification and morphologic demonstration of reactive oxygen species produced by Walker-256 Tumor-Cells In-Vitro and during metastasis in-vivo. Lab Invest 71:480–489
Stauffer TP, Meyer T (1997) Compartmentalized IgE receptor-mediated signal transduction in living cells. J Cell Biol 139:1447–1454
Sutter JU, Campanoni P, Tyrrell M, Blatt MR (2006) Selective mobility and sensitivity to SNAREs is exhibited by the Arabidopsis KAT1 K+ channel at the plasma membrane. Plant Cell 18:935–954
Swarup R, Friml J, Marchant A, Ljung K, Sandberg G, Palme K, Bennett M (2001) Localization of the auxin permease AUX1 suggests two functionally distinct hormone transport pathways operate in the Arabidopsis root apex. Genes Dev 15:2648–2653
Takeda S, Gapper C, Kaya H, Bell E, Kuchitsu K, Dolan L (2008) Local positive feedback regulation determines cell shape in root hair cells. Science 319:1241–1244
Tanimura N, Nagafuku M, Minaki Y, Umeda Y, Hayashi F, Sakakura J, Kato A, Liddicoat DR, Ogata M, Hamaoka T, Kosugi A (2003) Dynamic changes in the mobility of LAT in aggregated lipid rafts upon T cell activation. J Cell Biol 160:125–135
Triantafilou M, Manukyan M, Mackie A, Morath S, Hartung T, Heine H, Triantafilou K (2004) Lipoteichoic acid and toll-like receptor 2 internalization and targeting to the Golgi are lipid raft-dependent. J Biol Chem 279:40882–40889
Vilhardt F, van Deurs B (2004) The phagocyte NADPH oxidase depends on cholesterol-enriched membrane microdomains for assembly. EMBO J 23:739–748
Warren JS, Kunkel RG, Simon RH, Johnson KJ, Ward PA (1989) Ultrastructural cytochemical analysis of oxygen radical-mediated immunoglobulin A immume complex induced lung injury in the rat. Lab Invest 60:651–658
Yamazaki D, Yoshida S, Asami T, Kuchitsu K (2003) Visualization of abscisic acid-perception sites on the plasma membrane of stomatal guard cells. Plant J 35:129–139

ROS in Retrograde Signalling from the Chloroplast to the Nucleus

Philip M. Mullineaux

Abstract This chapter reviews the roles of reactive oxygen species (ROS), such as singlet oxygen and hydrogen peroxide, in initiating and transducing retrograde signals from the chloroplast to the nucleus in response to environmental cues. Recent research is extensively cited that show ROS-directed retrograde signalling triggering both programmed cell death and acclimation to changed conditions. The implications for signal transduction are discussed in which ROS, by virtue of their reactivity or due to cellular antioxidant systems, are contained within chloroplasts, or under more extreme conditions, diffuse out to promote oxidative damage in other subcellular compartments. Finally, a discussion on the imperatives that will drive future ROS-related research in plants is provided that sets out why, and perhaps how, such research must be translated into crop improvement strategies for the benefit of global agriculture.

1 Introduction

This is one of many reviews on the subject of chloroplast-to-nucleus retrograde signalling (hereafter referred to simply as retrograde signalling) in plants and the author is acutely aware of the following sentiment: -"Given the number of recent reviews written in this area of plastid-to-nucleus signalling, there is considerable interest in these signalling pathways, yet very little is actually known" (Nott et al. 2006). Since these words were published, we are a little further forward, mainly due to a major contribution from the authors of this statement (Koussevitzky et al. 2007), but not in all areas. The preponderance of reviews reflects the importance of

P.M. Mullineaux
Department of Biological Sciences, University of Essex,
Colchester, CO4 3SQ, United Kingdom
e-mail: mullin@essex.ac.uk

L.A. del Río and A. Puppo (eds.), *Reactive Oxygen Species in Plant Signaling*, Signaling and Communication in Plants, DOI 10.1007/978-3-642-00390-5_14,

this subject, since the communication between the chloroplast and the nucleus is clearly essential for plant cell function and a detailed justification for reviewing this subject need not be expounded here. For different perspectives on this subject, the reader is referred to recent excellent reviews covering this subject (Nott et al. 2006; Pesaresi et al. 2007; Pfannschmidt et al. 2009).

In an attempt to be different and perhaps to be more speculative, this chapter will focus on one specific aspect of retrograde signalling, viz., how the signal exits the chloroplast and how reactive oxygen species (ROS) can play a role in this process. In this context, two mutually exclusive roles of ROS can be perceived: either the initiation of a retrograde signalling pathway by ROS leading to the exit of a non-ROS signal from the chloroplast (Mullineaux et al. 2006), or the exit of ROS from the chloroplast to propagate the signal in the cytosol and beyond (Mullineaux and Karpinski 2002; Baier and Dietz 2005; Pesaresi et al. 2007). These two roles will hopefully become apparent to the reader when examples of actual or potential retrograde signalling pathways are considered. Integrated into these mutually exclusive signalling roles is the notion of containment (or not) of ROS at their site of production as the means of distinguishing between two different outcomes – either the death of the cell or acclimation leading to its survival. This may allow a means of classifying the roles of ROS at two levels. First, the distinction can be made between promotion by ROS of uncontrolled oxidative stress and when the same species act as signalling molecules (Mullineaux et al. 2006). The outcome of ROS-directed signalling provides a second level of distinction, in which ROS signalling triggers either programmed cell death or acclimation of the cell to adverse conditions. It should be emphasised that, at the level of the organism, the programmed death of specific cells or tissues involving ROS is a normal part of plant development or in mounting resistance to environmental challenges (Pennell and Lamb 1997; Grant and Loake 2000; Bechtold et al. 2005). Only in extremis, where cell death is triggered over much large areas of the plant, does this become a problem, as in exposure to acute levels of ozone which inappropriately triggers an extensive hypersensitive response (HR; Langebartels et al. 2002, Joo et al 2005) or over-production of the ROS singlet oxygen ($^1O_2$) in the *flu1* mutant (Op den Camp et al. 2003; see Sect. 14.2.4.1).

Retrograde signalling in plants has been studied most extensively from damaged chloroplasts and is beginning to be studied in situations where this mode of signalling occurs as part of a plant's normal response to a physiologically relevant environmental cue. There are also examples that appear to fall between these neat categories, mainly because there is little information on the physiological state of the chloroplast. Thus, experimental systems range from studying retrograde signalling in bleaching, and therefore dying, whole seedlings, through to cells which undergo cell death as part of a protective mechanism for the whole organism. At the other end of this spectrum of responses, retrograde signalling drives cellular adjustment to the new conditions and consequent acclimation of the plant. Considering the aims of this chapter, it is hoped that the role of ROS in each situation will be apparent or can be pointed to as a potential involvement.

2 Mechanisms of Retrograde Signalling from Damaged Chloroplasts and the Involvement of ROS

In situations where chloroplasts malfunction, it is clearly important that this situation be communicated to the rest of the cell. Signal transduction from damaged chloroplasts is the best defined retrograde signalling pathway in plants. This experimental system is predicated on the observation that malfunctioning chloroplasts in seedlings generally leads to downregulation in the expression of many genes coding for components of the photosynthetic apparatus. Recent research has highlighted three distinct pathways distinguished by whether these signalling pathways are activated by disruptions to porphyrin biosynthesis, chloroplast gene expression, or redox signals originating from photosynthetic linear electron flux (LEF). As will be described briefly below, these three signalling pathways may be coordinately regulated by two genes, *GUN1* and *ABI4* (Koussevitzky et al. 2007). In addition, studies on *Arabidopsis* mutants that accumulate $^1O_2$ in chloroplasts and promote oxidative stress can be argued as signalling from compromised chloroplasts that elicit a number of signalling pathways that lead to cell death (see Sect. 14.2.4).

2.1 The Mg-ProtoIX Signalling Pathway: a Brief Description

When compounds or mutants are used that promote inhibition of porphyrin biosynthesis and accumulation of Mg-Protoporphyrin IX (Mg-ProtoIX) and its methyester derivatives, a distinct retrograde signalling pathway is activated (Kittsteiner et al 1991; Susek et al 1993; Oster et al 1996; Strand et al 2003; Mochizuki et al 2001; Nott et al; 2006; Koussevitzky et al. 2007). The cumulative evidence presented in these papers strongly suggests that Mg-ProtoIX accumulation specifically triggers this signalling pathway. Genetic dissection of this signalling pathway has been greatly facilitated by the isolation of so-called *genomes uncoupled* (*gun*) mutants of *Arabidopsis thaliana*. These mutants were identified from an elegant screen of norflurazon-treated photo-bleached seedlings in which there was a search for higher levels of LIGHT-HARVESTING CHLOROPHYLL *a/b*-BINDING PROTEIN gene (*Lhc*) expression than would occur in wild-type seedlings under these conditions (Susek et al. 1993; Nott et al. 2006; Koussevitzky et al. 2007). Most of these mutant loci (*gun2* to *gun5*) encode plastid-localised enzyme subunits that are part of the porphyrin biosynthetic pathway leading to chlorophyll or phytochromobilin (the chromophore for phytochrome; Nott et al. 2006). All these mutants fail to accumulate Mg-ProtoIX and therefore fail to repress the expression of nuclear photosynthetic genes (Nott et al. 2006). Arguments have been made that Mg-ProtoIX could exit the chloroplast and be the transducer of the signal (Strand et al. 2003; Nott et al. 2006). This may well be the case, reflecting a better established extra-plastidial role for Mg-ProtoIX in *Chlamydomonas reinhardtii* in which the induction of nuclear photosynthetic genes and stress-protective genes occurs in cultures that undergo a dark-to-light transition (Kropat et al. 1997, 2000).

The mutant *gun1* was characterised more recently (Koussevitzky et al. 2007), and the *GUN1* gene was found to encode a plastid-localised pentatricopeptide repeat (PPR)-containing protein. This family of proteins is involved in the maturation, translation and turnover of messenger RNAs. However, GUN1 contains a second domain called a *mutS*-related (SMR) domain. This highly conserved domain across the kingdoms may reflect an ability of GUN1 to bind DNA and thus GUN1 may be distinct from other PPR-proteins (Koussevitzky et al. 2007). Microarray analysis comparing *gun* mutant and wild-type seedlings in the presence of norflurazon revealed an enrichment of differentially expressed genes whose promoters contained a combined light and abscisic acid (ABA) regulated *cis* element. However, testing of a range of ABA-deficient and -insensitive mutants for *gun* mutant phenotypes was negative apart from mutants in *ABI4*, indicating that ABA signalling was not involved in this retrograde signalling pathway (Koussevitzky et al. 2007). This gene encodes an Apetala-2 (AP-2) class transcription factor which had previously been implicated in sugar signalling (Rolland et al. 2006), which also promotes downregulation of the expression of nuclear genes encoding components of photosynthesis (Oswald et al. 2001). Genetic analysis showed that GUN1 acts above ABI4 in the same signalling pathway, and that this GUN1/ABI4 combination co-ordinately regulates the three signalling pathways identified to date. ABI4 was shown to bind to the *Lhc* promoter and this transcription factor is known to act as a repressor in ABA-, ethylene- and jasmonic acid (JA)-mediated signalling and is therefore consistent with a role in the downregulation of nuclear photosynthetic genes described here.

2.1.1 A Role for ROS in the Mg-ProtoIX Signalling Pathway?

Might there be a role for ROS, especially $^1O_2$, in the Mg-ProtoIX signalling pathway in norflurazon-treated photobleaching seedlings? Many porphyrin- and haem-containing compounds are photodynamic compounds that generate $^1O_2$ in the presence of light (Duke et al. 1991; Jimenez-Banzo et al. 2008). However, there is specificity of Mg-ProtoIX in repressing light-induced nuclear gene expression in both *Chlamydomonas* (Kropat et al. 1997) and *Arabidopsis* because closely related compounds, such as protoporphyrin IX or protochlorophyllide, do not work in this system (Kropat et al. 1997; Strand et al. 2003). Light induction of *HSP70* in *Chlamydomonas* requires Mg-ProtoIX, but addition of this compound to dark-grown cultures achieves the same effect (Kropat et al. 1997). This rules out the possibility of light promoting $^1O_2$ production as part of the signalling process, and instead the role of light has been suggested to promote the export of Mg-ProtoIX to the cytosol from the chloroplast (Kropat et al. 1997). Therefore, it is unlikely that the required accumulation of Mg-ProtoIX is linked to the production of $^1O_2$. However, in the *Arabidopsis* experimental system, the presence of light plus norflurazon will generate $^1O_2$

(Thomas et al. 1998; Jung et al. 2000) and is an additional variable factor not emphasised in these studies. Given the proposed role of $^1O_2$ in signalling (see Sect. 14.2.4), there would seem to be potential to explore this question further, especially when one considers that *GUN1/ABI4* may also regulate the expression of genes that respond to photobleaching induced by exposure to intense light (see Sect. 14.2.3).

2.2 A Plastid Gene Expression-Dependent Signalling Pathway

The same *GUN1/ABI4* genes also mediate the downregulation of nuclear photosynthetic gene expression when seedlings are treated with inhibitors of chloroplast translation such as lincomycin and chloramphenicol (Nott et al. 2006; Koussevitzky et al. 2007). This plastid gene expression (PGE)-dependent pathway is light independent, which was concluded from the observation that constitutively photomorphogenic mutants of pea and *Arabidopsis* grown in the dark accumulate *Lhc* mRNA, which could be inhibited by treatment with lincomycin (Sullivan and Gray 1999). Unlike most *gun* mutants, seedlings harbouring mutant alleles of *GUN1* show increased levels of *Lhc* mRNA when treated with lincomycin, arguing that *GUN1* exerts control over this PGE-dependent pathway as well as the Mg-ProtoIX pathway (Koussevitzky et al. 2007). The nature of the signal from chloroplasts that reports disruption of translation remains elusive, but the data provided seem to rule out a role for ROS. However, lincomycin treatment of leaves in the light can promote $^1O_2$ production in the chloroplast that causes oxidative damage to photosystem II (PSII) reaction centres (Hideg et al. 2007). This observation might suggest that there is some connectivity provided by $^1O_2$ in initiating these GUN1/ABI4-regulated pathways in light-exposed plants.

2.3 A Photo-Oxidative Stress-Dependent Pathway: ROS and Redox Signalling

The third pathway involving *GUN1/ABI4* may well involve ROS at a number of levels. In the studies reporting the convergent regulation of retrograde signalling by *GUN1/ABI4* (Koussevitzky et al. 2007), low light-grown seedlings were exposed to >40-fold intense light over their growth light intensity. The consequent photobleaching of the seedlings was associated with a downregulation of *Lhc* expression as well as the induction of the photo-oxidative stress inducible genes, *Zat10* and *Zat12* (Koussevitzky et al. 2007; Rossel et al. 2007; Davletova et al. 2005). In contrast, in both *gun1* and *abi4* mutants, these effects were at least partially reversed i.e. *Lhc* expression was maintained for longer and the *Zat* genes were slower to be induced in the mutants. Under these intense light conditions, chloroplasts will produce high levels of ROS including $^1O_2$, superoxide anion radical ($O_2{\bullet}^-$), hydrogen peroxide (H_2O_2) and lipid (hydro)peroxides (Sofo et al. 2004; Flors et al. 2006; Karpinski et al. 1997; Havaux et al. 2006; Scarpecci et al. 2008) which could initiate signalling (Levine et al. 1994; Karpinski et al. 1997, 1999; Jabs et al. 1996; Op den Camp et al. 2003; Rusterucci et al. 2001; Fryer et al. 2003; Apel and Hirt 2004; Mateo et al. 2004; Joo et al. 2005; Gadjev et al. 2006; Chang et al. 2004; Sattler et al. 2006). Under these extreme conditions, seedlings bleach and are unlikely to recover. From damaged photobleaching chloroplasts, the secretion of stable ROS such as H_2O_2 and lipid peroxidation products could initiate a cell death programme. In this respect, while ROS diffusing from chloroplasts could be regarded as a retrograde signal (Mullineaux and Karpinski 2002; Baier and Dietz 2005; Nott et al. 2006; Pfannschmidt et al. 2009), specificity is not an issue since this degree of oxidative stress could be argued to elicit the same type of programmed cell death irrespective of the source of ROS. Identical outcomes are observed when, for example, peroxisomal catalase-deficient plants are exposed to moderately high light (Willekens et al. 1997; Vanderauwera et al. 2005) or when pathogen elicitors induce cell death (Rusterérucci et al. 1999). However, while numerous ROS must be present in these seedlings exposed to intense light, the role of the GUN1/ABI4 pathway in regulating a ROS-associated signalling pathway is difficult to discern from the data currently available (Koussevitzky et al. 2007), but it is hard to imagine that ROS are not involved.

2.4 *Multiple Retrograde Signalling Initiated by* $^1O_2$

The production of $^1O_2$ in PSII reaction centres occurs under a very wide range of light intensities, and its rapid accumulation upon sudden exposure of leaves to high light or treatment with herbicide inhibitors of photosynthetic electron transport, such as DCMU (3-(3,4-dichlorophenyl)-1,1-dimethylurea), promotes photoinhibition (Fischer et al. 2002; Hideg et al. 2007; Fufezan et al. 2002; Flors et al. 2006). Protection against oxidative damage to thylakoid membranes by $^1O_2$ is afforded by both carotenoids and α-tocopherol (Havaux et al. 2005; Sattler et al. 2006). From a signalling perspective, it is generally accepted that the reactivity and consequent in vivo half-life of $^1O_2$ is so short (Halliwell and Gutteridge 1985) that some secondary ROS would have to transduce the signal as a secondary messenger. In particular, the (per)oxidation products of polyunsaturated fatty acids that would arise from the reaction of $^1O_2$ with predominant fatty acids in the chloroplast such as α-linolenic acid would be the likely candidates (Apel and Hirt 2004; Montillet et al. 2004). In turn, further free radical catalysed reactions of α-linolenic acid generate cyclic oxylipins (isoprostanes) which are potent inducers of nuclear gene expression (Thoma et al. 2003). Recently, it has been shown that isoprostane-dependent signalling requires TGA class B-Zip transcription factors (Mueller et al. 2008). Therefore, from general $^1O_2$-initiated lipid peroxidation, a number of potential retrograde signalling pathways could be activated in which lipid peroxides or their derivatives could act as the retrograde signal from chloroplasts, although no specific molecule or process has been identified that can achieve this.

Recently, however, a contrary view has been offered, suggesting that $^1O_2$ may be more stable in vivo than thought (Fischer et al 2007). In the cytosol as well as the chloroplast of high light-exposed *Chlamydomonas*, $^1O_2$ was detected and has been suggested to signal the induction of the antioxidant gene *GPXH*. This observation needs to be verified, but it is known that fatty acid peroxide radicals can precipitate autocatalytic cascades of further peroxidation that can generate more $^1O_2$ (Halliwell and Gutteridge 1985; Flors et al. 2006). In other words, it may be that cytosol-located $^1O_2$ did not originate directly from chloroplasts but was formed in the cytosol from fatty acid peroxides that had been formed in thylakoid membranes and then diffused from chloroplasts. This is potentially a retrograde signalling route, since this process of $^1O_2$ to lipid peroxide back to $^1O_2$ in the chloroplast and then the cytosol does trigger nuclear gene expression and warrants further investigation.

$^1O_2$ can also be generated in the chloroplast by diphenyl ether herbicides that inhibit protoporphyrinogen IX oxidase (Duke et al. 1991), the last common step for chlorophyll and haem biosynthesis (Nott et al. 2006). The accumulation of this potent photodynamic compound promotes $^1O_2$ accumulation, oxidative stress and photobleaching. Again, the treatment with levels of herbicide which promotes non-enzymatic lipid peroxidation can lead to changes in nuclear gene expression (Jung and Black 2005).

2.4.1 $^1O_2$ and the Overlap of Cell Death and Disease Resistance Pathways

That chloroplast-generated $^1O_2$ may initiate the activation of cell death-disease resistance pathways has been reinforced comprehensively with work on the *flu1* mutant of *Arabidopsis* (Meskauskiene et al. 2001; Op den Camp et al. 2003; Apel and Hirt 2004; Wagner et al. 2004; Gadjev et al. 2006; Przybyla et al. 2008). This mutant is defective in the regulation of enzymes of the chlorophyll biosynthetic pathway and consequently accumulates protochlorophyllide in the dark up to ten times greater than wild type plants (Przybyla et al 2008). The attraction of using *flu1* is that the timing of accumulation and the amount of $^1O_2$ can be controlled by growing plants as seedlings or to maturity under a continuous 24 h photoperiod to prevent accumulation of protochlorophyllide. Shifting such plants to one cycle of darkness brings about accumulation of protochlorophyllide which upon re-introduction of plants back to the light leads to production of $^1O_2$ (Meskauskiene et al. 2001). Mutant *flu1* plants or seedlings treated in this way undergo bleaching which is the end result of the $^1O_2$-mediated activation of a genetically programmed cell death response (Op den Camp et al. 2003; Wagner et al. 2004; Przybyla et al. 2008). In contrast, *flu1* etiolated seedlings grown heterotrophically produce four times more protochlorophyllide than the same seedlings grown in 24 h light and then transferred to the dark (Przybyla et al. 2008). In this case, the photobleaching of the *flu1* seedlings is due to massive oxidative stress caused by auto-catalysed lipid peroxidation (Przybyla et al 2008).

In the genetically controlled $^1O_2$ initiated cell death response, enzyme-catalysed lipid peroxidation via the activation of plastidial lipoxygenase is important (Op den Camp et al. 2003). One major product of enzyme-catalysed lipid peroxidation is 13-hydroperoxy octadecatrienoic acid, a biosynthetic precursor for jasmonic acid (JA; Przybyla et al. 2008). JA is a well-studied stress responsive signalling molecule (Devoto and Turner 2005; Balbi and Devoto 2008), and the stimulation of its biosynthesis and signalling by $^1O_2$ means that this is one route in which this ROS can stimulate a retrograde signalling pathway from the chloroplast to the nucleus. However, despite clear evidence of interaction between JA- and $^1O_2$-mediated signalling, the $^1O_2$-induced cell death programme *per se* does not depend on JA signalling (Przybyla et al. 2008). A similar situation exists for $^1O_2$ activation of salicylic acid (SA)-mediated signalling; i.e., there is a clear connection but the SA pathway is not required for $^1O_2$-mediated cell death (Ochsenbein et al. 2006). The novel nature of this signalling pathway is underscored by the identification of two genes, *EXECUTOR1* (*EXE1*) and *EXE2*, whose chloroplast-localised products are required for the $^1O_2$-induced cell death (Wagner et al. 2004; Lee et al. 2007). Precisely how EXE1 and EXE2 carry out their role in this signalling pathway is not clear and no retrograde signal specific for this cell death pathway has yet been identified (Pfannschmidt et al. 2009). However, it is clear that in *flu1* at least, $^1O_2$ production elicits a whole network of signalling pathways.

The elegant exploitation of *flu1* has clearly demonstrated that $^1O_2$-directed retrograde signalling pathways exist, but the question arises of how and under what circumstance do these pathways operate in wild-type plants? Treatment of plants with high light and/or DCMU evokes $^1O_2$ production (Fufezan et al. 2002; Flors et al. 2006), but genes whose expression was induced in *flu1* producing $^1O_2$ were not upregulated in DCMU-treated plants (Bechtold et al. 2008). This impression is reinforced by an extensive meta-analysis of data from plants producing different ROS, in which *flu1* showed alterations in expression of a highly specific group of genes (Gadjev et al. 2006). This has been argued to be because the production of $^1O_2$ in DCMU/high light-treated plants is in PSII reaction centres, whereas in *flu1* this will not be the site of $^1O_2$ production (Bechtold et al. 2008). However, this potentially important difference between wild-type and *flu1* may not be due to the precise origin of $^1O_2$, since accumulation of protochlorophyllide in *flu1* has recently been shown to be associated with thylakoid membranes (Przybyla et al. 2008) and the photobleaching effects of low concentrations of DCMU are ameliorated in *exe1* null mutants (Wagner et al. 2004).

The observation that sub-lethal doses of diphenyl ether herbicides that generate $^1O_2$ can invoke SA-mediated pathogen resistance (Molina et al. 1999), mimicking the situation in *flu1* (Ochsenbein et al. 2006), might offer a clue to resolving this conundrum. High light-exposed "sun" leaves have been suggested to depend more on resistance signalling pathways mediated by the chloroplast than those in shade (Bechtold et al. 2005). Might it be that $^1O_2$-mediated retrograde signalling has an important role in sun leaves when challenged with a pathogen and can use increased $^1O_2$ production to accelerate the induction of resistance? The role of *ENHANCED DISEASE SUSCEPTIBILITY1* (*EDS1*), a gene that integrates ROS into SA-mediated disease resistance pathways (Rusterérucci et al. 2001), in $^1O_2$-mediated signalling (Ochsenbein et al. 2006) does argue for the involvement of $^1O_2$ in disease resistance in wild-type plants in their natural environment.

2.5 *Multiple Pathways for H_2O_2-Directed Signalling from Chloroplasts*

A number of scenarios can be described in which H_2O_2, generated from enhanced photoreduction of O_2 at photosystem I (PSI; Asada 1999) plays a central role in chloroplast-to-nucleus signalling that activates responses ranging from acclimation to cell death, and in which this stable ROS (Halliwell and Gutteridge 1985) can be argued to be contained within the chloroplast and exert its signalling function there, or to have been secreted from it and elicit signalling in the cytosol or beyond (Mullineaux et al. 2006).

2.5.1 High Light-Induced Retrograde Signalling from Chloroplasts: H_2O_2 or Plastiquinone Redox State?

Prior to photobleaching, the exposure of seedlings to intense light as described in Sect. 14.2.3 would, in addition to bringing about increased ROS production, induce substantial over-reduction of LEF chain components as the amount of light energy absorbed would rapidly exceed the capacity of photosynthetic metabolism to use it (Karpinski et al. 1997, 1999; Ort and Baker 2002; Nott et al. 2006; Pfannschmidt et al. 2009). Therefore, redox changes around LEF may initiate signalling as well as, or in place of, ROS. Discriminating between these two possibilities is important.

The redox state of plastoquinone (PQ) and other LEF chain components can also be manipulated experimentally by using the LEF inhibitors DCMU and DBMIB (2,5-dibromo-6-isopropyl-3-methyl-1,4-benzoquinone), which, among their other effects, promote oxidation and reduction of the PQ pool, respectively (summarised in Pfannschmidt et al. 2009). A similar effect can be achieved by exposing plants to wavelengths of light which are preferentially absorbed by PSII or PSI and would promote reduction and oxidation of the PQ pool, respectively (summarised in Pfannschmidt et al. 2009). In these somewhat artificial experimental arrangements, it has been argued that changes in the redox state of PQ initiates signalling leading to changes in both plastid and nuclear gene expression (Escoubas et al. 1995; Karpinski et al. 1997, 1999; Pfannshmidt et al. 1999; Oswald et al. 2001; Fey et al. 2005). These PQ-driven signalling pathways do not appear to involve ROS, suggesting that PQ redox changes *per se* are indeed important in retrograde signalling (Fey et al. 2005).

The role of PQ as an initiator of retrograde signalling was also initially supported by studies on the high light-mediated induction of *ASCORBATE PEROXIDASE2* (*APX2*). Exposure of *Arabidopsis* leaves to ca. tenfold high light increases the reduction state of Quinone A (Q_A) consistent with a concomitant increased redox state of PQ (Russell et al. 1995; Fryer et al. 2003). The same degree of increased light intensity rapidly induces *APX2* expression in the bundle sheath cells of *Arabidopsis* leaves (Karpinski et al. 1997, 1999; Ball et al. 2004; Chang et al. 2004; Nishizawa et al. 2006). By using DCMU, it has been established that there is a requirement for LEF for *APX2* induction under high light conditions (Karpinski et al. 1997, 1999; Rossel et al. 2002; Chang et al. 2004; Bechtold et al. 2008). Combined with observations that under growth light conditions DBMIB-treated leaves show increased *APX2* expression, this has been taken as evidence for the importance of a PQ-mediated redox signal in regulating the expression of this gene (Karpinski et al. 1997, 1999; Yabuta et al. 2004). However, there is a problem when induction of *APX2* expression in a range of high light intensities is measured (Fryer et al. 2003). Both the speed and degree of *APX2* induction are essentially the same from 2.5- to 10-fold increased light intensities above growth PPFD. For example, at a 2.5-fold increased light intensity, the redox state of the PQ pool in bundle sheath cells was calculated to have changed by 4% compared with the growth light conditions and yet the degree and speed of *APX2* induction were hardly affected (Fryer et al. 2003). This is because, under moderate increases in light intensity, increased photochemical and non-photochemical quenching act to maintain a constant rate of photosynthetic electron transport and consequently an unchanged redox state of LEF components. This situation argues against PQ redox state being a determinant of *APX2* expression under all light regimes, especially since exposure of leaves to light intensities that subsequently lead to photobleaching (and would be associated with a strong increase in PQ redox state) actually blocks induction of *APX2* expression in damaged sectors (Karpinski et al. 1999). Thus, under circumstances where both *APX2* expression and the redox state of the PQ pool change, the two processes may be unconnected. The alternative explanation for the requirement for LEF for *APX2* induction is that it reflects the role of LEF in the photoreduction of O_2 at PSI (Asada 1999; Ort and Baker 2002), thus generating H_2O_2 for signalling (Fryer et al. 2003).

Notwithstanding the above conclusions, there is some evidence that LEF does play an additional role in *APX2* regulation which does not directly involve H_2O_2. Infiltration of this ROS into leaves under ambient light conditions can partly induce *APX2* expression that can still be inhibited by DCMU (Karpinski et al. 1999), demonstrating a requirement for active photosynthesis in a situation where H_2O_2 has been supplied exogenously. These data could indicate that LEF is the source of some other non-ROS signal, but this does not necessarily mean that PQ redox state is the alternative explanation; other chloroplast localised functions could be important (see Sects. 14.2.5.3 and 14.2.5.4). In summary, PQ redox state may not be an important determinant of *APX2* expression, although this does not exclude it being a regulator of the expression of other genes, and there is some evidence that *APX2* is not typical of high light-induced genes in terms of its effectors (Bechtold et al. 2008).

2.5.2 The Role of H_2O_2 in the Induction of Systemic Acclimation to High Light

When leaves on one side of an *Arabidopsis* rosette, covering up to 30% of its area, are exposed to a >tenfold increase over growth light, it can be shown that unexposed leaves become more tolerant of a subsequent exposure to high light, implying the establishment of systemic photo-acclimation, termed systemic acquired acclimation (SAA; Karpinski et al. 1999; Rossel et al. 2007). It is not the intention to go into details of the SAA phenomenon here; the reader is referred to the two previously cited papers and Ball et al. (2004). However, the accumulation of ROS in chloroplasts may initiate this process and, with the levels of increased light used, the photo-oxidative stress that would be elicited could mean that the leaching of this ROS from the chloroplasts goes as far as exiting the cell and initiating long-range signalling (Mullineaux and Karpinski 2002). The infiltration of bovine catalase into intracellular spaces of detached leaves was used to inhibit high light induction of *APX2* expression and was taken as evidence that H_2O_2 could be part of the systemic signal (Karpinski et al. 1999). Since then, however, extracellular/plasma membrane-sourced H_2O_2 has been shown to be important for *APX2* expression (Bechtold et al. 2008), and the H_2O_2 dismutated by catalase (Karpinski et al. 1999) almost certainly did not originate from chloroplasts.

The ≥tenfold increased light intensities used to elicit SAA (Karpinski et al. 1999; Ball et al. 2004; Rossel et al. 2007) would promote a degree of irreversible photo-inhibition (Russell et al. 1995) associated with increased production of $^1O_2$ in PSII reaction centres (Flors et al. 2006). The possible retrograde signalling role of $^1O_2$ is discussed above (see Sect. 14.2.4), but it is noticeable that, as with $^1O_2$-overproducing plants, both exposed and distal leaves in SAA-elicited plants show a preponderance of genes associated with responses to pathogens that are associated with cell death (Rossel et al. 2007). This implies that establishment of SAA in partial exposure to excess light is akin to the induction of systemic acquired resistance (SAR) to biotrophic pathogens (Bechtold et al. 2005). SAR and SAA share many features in common, including the appearance of micro-HR-like lesions (Zaier et al. 2004; Bechtold et al. 2005). Interestingly, *Arabidopsis* exposed to conditions that promote photo-oxidative stress can show some degree of enhanced resistance to a bacterial pathogen (Bechtold et al. 2005). This view is further supported by observations that, when HR is induced (which leads to SAR), there is a strong inhibition of photosynthetic metabolism that brings about increased H_2O_2 production in chloroplasts that in turn activates a mitogen-activated protein (MAP) kinase cascade associated with activation of a programmed cell death pathway responsible for HR-associated lesions (Liu et al. 2007).

In summary, H_2O_2 is implicated in initiating signalling under excess light conditions that promote photo-oxidative stress, and in the establishment of HR. However, it is not clear whether in these conditions the H_2O_2 is secreted from chloroplasts to propagate a signal or is contained at its site of production. In leaves exposed to lower increases in light intensity, the situation is clearer and described in the next section.

2.5.3 H_2O_2 and ABA-Mediated Retrograde Signalling

When *Arabidopsis* leaves are exposed to a three- to fivefold increase in light intensity, bundle sheath cell chloroplasts accumulate H_2O_2, but these cells do not suffer oxidative stress and consequent cell death. Rather, they induce antioxidant defence gene expression, which is associated with increased photochemical and non-photochemical quenching capacity in these cells (Fryer et al. 2003). Induction of antioxidant defences may be associated with increased acclimation to the high light conditions (Karpinski et al. 1999). Given the lack of evidence of cellular damage, it has been proposed that, in such circumstances, H_2O_2 is contained within bundle sheath cell chloroplasts by an extensive and subsequently enhanced cytosolic antioxidant network (Mullineaux et al. 2006). Thus, in these circumstances, since any H_2O_2 that does exit the chloroplast is effectively scavenged, the only way in which H_2O_2 could engage in signalling would be to be transduced to a non-ROS signal in the chloroplast (Mullineaux et al. 2006). How this could be achieved is unknown, but in this and the following section (Sect. 14.2.5.4), two potential signalling events are described.

APX2 and many other high light responsive genes can be induced by treatment of leaves with ABA (Fryer et al. 2003; Rossel et al. 2006; Bechtold et al. 2008). The ABA signalling mutants, *abi1-1* and *abi2-1* (*abi* stands for *ABA insensitive*) show hampered induction of *APX2* expression under high light conditions (Fryer et al. 2003), and the selection of a mutant, *alx8-1*, for constitutive expression of *APX2* was revealed to have higher foliar levels of ABA (Rossel et al. 2006). This dependency of high light responsive genes upon ABA suggest a model for a retrograde signalling route, since the ABA biosynthetic pathway is split such that the first steps from the precursor xanthophyll carotenoid to the oxidative cleavage step to produce xanthoxin occurs in the plastid, the xanthoxin is exported to the cytosol where the remaining steps to ABA are completed (Nambara and Marion-Poll 2005; North et al. 2007). The possibility therefore exists that increased oxidising conditions in the chloroplast stroma, brought about by increased H_2O_2, could stimulate ABA biosynthesis in a manner similar to that for glutathione biosynthesis (see Sect. 14.2.5.4) by targeting a key enzyme of the biosynthetic pathway. This is a compelling hypothesis and some support is provided by the recent observation that exposure of *Arabidopsis* to high light does bring about an increase in foliar ABA content (Rossel et al. 2006), but the means whereby this increase occurs has not been described.

2.5.4 Glutathione-Mediated Retrograde Signalling

In mutants with lowered levels of the thiol antioxidant reduced glutathione (GSH), altered steady state levels of some stress-responsive genes have been observed either prior to or during exposure to abiotic or biotic stress (Ball et al. 2004; Parisy et al. 2007). These mutants harbour defective alleles of the gene *GSH1*, which encodes the enzyme γ-glutamylcysteine synthetase (γ-ECS) that catalyses the first of two reactions leading to the biosynthesis of GSH from its constituent amino acids cysteine, glutamate and glycine (Mullineaux and Rausch 2005; Meyer and Hell 2005). These mutant alleles are *cadmium deficient2-1* (*cad2-1*; Cobbett et al. 1998), *root meristemless1-1* (*rml1-1*; Vernoux et al. 2000), *regulator of APX2 1-1* (*rax1-1*; Ball et al. 2004) and *phytoalexin deficient2-1* (*pad2-1*; Parisy et al. 2007). It is not the intention to go into the full details of the phenotypes, molecular characterisation or all of the roles that GSH can play in cells, these have recently been extensively reviewed (Mullineaux and Rausch 2005; Meyer and Hell 2005). In the context of retrograde signalling, one key feature of glutathione biosynthesis is that it has been recently established in *Arabidopsis* that the two-step pathway is split between the plastid and the cytosol such that γ-ECS is in the plastid and the enzyme catalysing the second step of GSH biosynthesis, glutathione synthetase (GS), is primarily located in the cytosol (Wachter et al. 2005; Pasternak et al. 2008). Since GSH levels are responsive to environmental challenges such as exposure to high light (Muller-Moulé et al. 2003; Kopriva and Rennenberg 2004) and infection with pathogens (Mou et al. 2003; Parisy et al. 2007), it follows that the changes in GSH levels can carry a signal out of the chloroplast to induce a redox change in the cytosol, to which regulatory proteins such as NPR1 could be responsive (Mou et al. 2003; Mullineaux and Rausch 2005). This scheme does not immediately invoke ROS, especially since, in the one case where it has been examined, exposure of *rax1-1* to high light led to no greater photo-inhibition or accumulation of H_2O_2 in veinal tissue than occurred in wild-type plants and yet *APX2* expression was substantially higher (Ball et al. 2004). This was taken as evidence of a redox- as opposed to a ROS-mediated signal regulating *APX2* expression (Ball et al. 2004). However, it is possible that the rise in H_2O_2 levels in bundle sheath chloroplasts create the conditions that stimulate γ-ECS activity, since, in vitro, the activity of the recombinant enzyme is stimulated by oxidising conditions, most likely by formation of an intra-molecular disulphide bond (Jez et al. 2004). Therefore, in the context of this article, H_2O_2 by promoting oxidising conditions within the chloroplast could play a role in strengthening or amplifying a retrograde redox signal.

3 The Future: Major Opportunities and Challenges

In late spring and early summer, the time of the year when this chapter was being written, the English suburban garden this author often gazed on is a riot of fresh greenery reflecting the accelerated growth rates of over-wintered plants as environmental conditions become more benign. This vision of calm and beauty belies the continual balancing act that comes with carrying out oxygenic photosynthesis in a highly fluctuating environment. Even in a protected environment such as a garden, the divide between growth and successful reproduction on the one hand, and accelerated senescence and premature death on the other hand, is a fine one. The processes that govern this highly dynamic situation revolve around the multiple roles of ROS in every aspect of plants' lives and are reflected in every chapter of this book. However, the understanding of these processes is not some esoteric academic exercise but is becoming increasingly urgent as part of the world-wide research effort that is needed to increase global crop productivity in the face of a decline in available arable land and associated environmental services, increased population size, urbanisation and the depredations of climate change (Morison et al. 2008). I am not going to attempt to pick out specific technologies and predict these will have importance, I will certainly be wrong. However, it is pertinent to ask what are the major challenges and opportunities which will allow a community of researchers working on the role of ROS in plants to move the subject on and begin to generate crop germplasm of use to world agriculture.

The chapters described in this book cover the roles of ROS in initiating or perpetuating signalling that elicit changes in growth and metabolism of plants in response to changes in their abiotic environment, attack by pathogens and the establishment of symbiotic associations, all of which involve wholesale changes in metabolism governing every aspects of a plant's life from photosynthesis to respiration and from the turnover and synthesis of cell walls to hormones, to name but a few. For example, in response to drought, because of their involvement in many signalling processes, ROS are implicated in whole organism responses ranging from controlling the changes in the turgor of guard cells to the stimulation of root growth.

The sheer range of cellular processes in which ROS are now implicated are reflected in the great progress that has been made in understanding the underlying molecular mechanisms that allow ROS to function in so many ways. Recent endeavours in our field have revolved around describing, under many situations, the many hundreds of genes that appear to be regulated by, or responsive to, ROS. This ever-increasing volume of data generated using high throughput technologies such as microarrays, and now massively parallel sequencing technology, has created a set of new problems. How do we make sense of all these data? Notwithstanding the valiant efforts using meta-analysis approaches (Gadjev et al. 2006), at the time of writing we do not have any gene network models that are robust enough to withstand experimental analysis in several laboratories. There are glimmers of progress though; common features are beginning to emerge such as the role of heat shock transcription factors in high light responses (Davletova et al. 2005; Nishizawa et al. 2006). This suggests that we are not far off unified models appearing that describe so-called local networks of genes, although how constellations of such networks cooperate to produce a physiological response at the level of an organ, such as the leaf, is some way off. To achieve these ends, the field is going to have to take a so-called Systems Biology approach and go onto a more structured footing, which allows the construction of in silico gene network models that guide re-iterative targeted experimentation. To achieve this paradigm shift in research requires a re-think of classical approaches, abandonment of cherished beliefs, a strong commitment to multi-disciplinary interactions and access to expanded resources, such that one laboratory can conduct such projects under one roof, or develop long-term multi-centre collaborations. In a UK context, where this author resides, the norm in plant science research will be multi-institutional collaborations, which will demand new practices from participating laboratories. For example, such projects will have to devise and manage the coordinated growing of plants under near-identical conditions, agree and execute defined *physiologically relevant* changes in environmental conditions (e.g. just what is "high light" defined as?) and carry out a common set of physiological analyses. This is only the beginning. In addition, high throughput platforms to analyse the transcriptome, proteome and metabolome need to be maintained and managed efficiently, and data analysis and bioinformatics pipelines created that allow efficient production of testable network models. This unprecedented level of organisation, especially for academic laboratories, must occur against a backdrop where the different disciplines – physiologists, molecular biologists, bio-informaticians and mathematicians – need to learn to communicate and understand each others' limitations and strengths. Therefore, the challenges are daunting to produce functioning ROS-responsive network models that can be generated and productively interrogated for an organ as complex as a leaf. Such models will ultimately have to contain both temporal and spatial (tissue-specific) components and form the basis for the development of the "virtual" plant, a project which has begun in the USA.

Given the progress that has been made in even the last 10 years in this specific research area, the future suggests that one or more of us (or more likely our successors) will succeed in overcoming these challenges, but despite the great achievements these would be, the imperatives do not stop there. These activities with model plants systems will have to be coupled, as they develop, to translational research that can quickly take advantage of any gene or genes that might promote improved crop performance. The development of Systems Biology-based model plant research and exploitation of its outputs for crop improvement will have to occur in parallel and involve dialogue with crop physiologists and agronomists (Morison et al. 2008). This is happening in one or two institutions, but is not universal. The times-cales are likely to be too long for crop research to be attendant on the maturation of Systems Biology in model plant research and, as has often formerly hap-pened, be considered as an afterthought after academic success has been achieved. Thus, the current situation must compel us to bring crop research into our ROS research programmes even if they are not fully developed and are sure of immediate delivery, because time is not on our side. Therefore, I see this as a future development that is necessary if we are to continue to ask for publicly funded resources to continue our research.

Finally, for all of these positive thoughts and the hope that our research field engenders, this author cannot help wondering and worrying with the question – how do we know we are focussing in the laboratory on the right genes and processes which will be relevant to plants in their cropping or natural environments? The answer is we do not know and we could be in for some (nasty?) surprises should we decide to analyse the function of apparently key genes in even model species grown outside the laboratory. What if our inducible genes in the laboratory are in effect constitutively expressed in the natural environment? What if high levels of ROS are always present? If we have focussed on the wrong genes, in a world increasingly desperate for improvements in crop performance, will we be forgiven?

References

Apel K, Hirt H (2004) Reactive oxygen species: Metabolism, oxidative stress and signal transduction. Annu Rev Plant Biol 55:373–399.

Asada K (1999). The water–water cycle in chloroplasts: Scavenging of active oxygens and dissipation of excess photons. Annu Rev Plant Physiol Plant Mol Biol 50:601–639.

Baier M, Dietz K-J (2005) Chloroplasts as source and target of cellular redox regulation: A discussion on chloroplast redox signals in the context of plant physiology. J Exp Bot 56:1449–1462.

Balbi V, Devoto (2008) Jasmonate signalling network in *Arabidopsis thaliana*: crucial regulatory nodes and new physiological scenarios. New Phytol 177:301–318

Ball L, Accotto G-P, Bechtold U, Creissen G, Funck D, Jimenez A, Kular B, Leyland N, Mejia-Carranza J, Reynolds H, Karpinski S, Mullineaux P (2004) Evidence for a direct link between glutathione biosynthesis and stress defence gene expression in Arabidopsis. Plant Cell 16:2448–2462.

Bechtold U, Karpinski S, Mullineaux PM (2005) The influence of the light environment and photosynthesis oxidative signalling responses in plant-biotrophic pathogen interactions. Plant Cell Environ 28:1046–1055.

Bechtold U, Richard O, Zamboni A, Gapper C, Geisler M, Pogson B, Karpinski S, Mullineaux PM (2008) Impact of chloroplastic- and extracellular-sourced ROS on high light-responsive gene expression in Arabidopsis. J Exp Bot 59:121–133.

Chang CC-C, Ball L, Fryer MJ, Baker NR, Karpinski S, Mullineaux PM (2004) Induction of ASCORBATE PEROXIDASE 2 expression in wounded Arabidopsis leaves does not involve known wound-signaling pathways but is associated with changes in photosynthesis. Plant J 38:499–511.

Cobbett CS, May MJ, Howden R, Rolls B (1998) The glutathione deficient, cadmium-sensitive mutant, *cad2-1* of *Arabidopsis thaliana* is deficient in γ-glutamylcysteine synthetase. Plant J 16:73–78.

Davletova S, Rizhsky L, Liang H, Shengqiang Z, Oliver DJ, Coutu J, Shulaev V, Schlauch K, Mittler R (2005) Cytosolic Ascorbate peroxidase 1 is a central component of the reactive oxygen gene network in Arabidopsis. Plant Cell 17:268–281.

Devoto A, Turner JG (2005) Jasmonate-regulated Arabidopsis stress signalling network. Physiol Plant 123:161–172.

Duke SO, Becerril JM, Sherman TD, Lehnen LP, Matsumoto H (1991) Photosensitizing porphyrins as herbicides. ACS Symp Ser 449:371–386.

Escoubas JM, Lomas M, LaRoche J, Falkowski PG (1995) Light intensity regulation of cab gene transcription is signaled by the redox state of the plastoquinone pool. Proc Nat Acad Sci USA 92:10237–10241.

Fey V, Wagner R, Bräutigam K, Wirtz M, Hell R, Dietzmann A, Leister D, Oelmüller R, Pfannschmidt T (2005) Retrograde plastid redox signals in the expression of nuclear genes for chloroplast proteins of *Arabidopsis thaliana*. J Biol Chem 280:5318–5328.

Fischer BB, Krieger-Liszkay A, Hideg E, Snyrychova I, Weisendanger M, Eggen RIL (2007) Role of singlet oxygen in chloroplast to nucleus retrograde signalling in *Chlamydomonas reinhardtii*. FEBS Lett 581:5555–5560.

Flors C, Fryer MJ, Waring J, Reeder B, Bechtold U, Mullineaux PM, Nonell S, Wilson MT, Baker NR (2006) Imaging the production of singlet oxygen in vivo using a new fluorescent sensor, Singlet Oxygen Sensor Green. J Exp Bot 57:1725–1734.

Fryer MJ, Ball L, Oxborough K, Karpinski S, Mullineaux PM, Baker NR (2003) Control of Ascorbate Peroxidase 2 expression by hydrogen peroxide and leaf water status during excess light stress reveals a functional organisation of Arabidopsis leaves. Plant J 33:691–705.

Fufezan C, Rutherford AW, Krieger-Liszkay A (2002) Singlet oxygen production in herbicide-treated photosystem II. FEBS Lett 532:407–410.

Gadjev I, Vanderauwera S, Gechev TS, Laloi C, Minkov IN, Shulaev V, Apel K, Inzé D, Mittler R, Van Breusegem F (2006) Transcriptomic footprints disclose specificity of reactive oxygen species signaling in Arabidopsis. Plant Physiol 141:436–445.

Grant JJ, Loake GJ (2000) Role od reactive oxygen intermediates and cognate redox signaling in disease resistance. Plant Physiol 124:21–29.

Halliwell B, Gutteridge JMC (1985) Free Radicals in Biology and Medecine (3rd edition). Oxford University Press, Oxford, UK

Havaux M, Eymery F, Porfirova S, Rey P, Dormann P (2005) Vitamin E protects against photoinhibition and photooxidative stress in *Arabidopsis thaliana*. Plant Cell 17:3451–3469.

Havaux M, Triantophyllidès C, Genty B (2006) Autoluminescence imaging: a non-invasive tool for mapping oxidative stress. Trends Plant Sci 11:1360–1385.

Hideg E, Kos PB, Vass I (2007) Photosystem II damage induced by chemically generated singlet oxygen in tobacco leaves. Physiol Plant 131:33–40.

Jabs T, Dietrich RA, Dangl JL (1996) Initiation of runaway cell death in an *Arabidopsis* mutant by extracellular superoxide. Science 273:1853–1856.

Jez JM, Cahoon RE, Chen S (2004) *Arabidopsis thaliana* glutamate-cysteine ligase. J Biol Chem 279:33463–33470.

Jimenez-Banzo A, Sagrista ML, Mora M, Nonell S (2008) Kinetics of singlet oxygen photosensitization in human skin fibroblasts. Free Rad Biol Med 44:1926–1934.

Joo JH, Wang S, Chen JG, Jones AM, Federoff NV (2005) Different signaling and cell death roles of heterotrimeric G protein α and β subunits in the Arabidopsis oxidative stress response to ozone. Plant Cell 17:957–970.

Jung SY, Black KW (2005) Herbicidal and antioxidant responses of transgenic rice overexpressing *Myxococcus xanthus* protoporphyrinogen oxidase. Plant Physiol Biochem 43:423–430.

Jung SY, Kim JS, Cho KY, Tae GS, Kang BG (2000) Antioxidant responses of cucumber (*Cucumis sativus*) to photoinhibition and oxidative stress induced by norflurazon under high and low PPFDs. Plant Sci 153:145–154.

Karpinski S, Escobar C, Karpinska B, Creissen G, Mullineaux PM (1997) Photosynthetic electron transport regulates the expression of cytosolic ascorbate peroxidase gene in Arabidopsis during excess light stress. Plant Cell 9:627–640.

Karpinski S, Reynolds H, Karpinska B, Wingsle G, Creissen G, Mullineaux PM (1999) Systemic signalling and acclimation in response to excess excitation energy in *Arabidopsis*. Science 284:654–657.

Kittsteiner U, Brunner H, Rüdiger W (1991) The greening process in cress seedlings. II. Complexing agents and 4-aminolevulinate inhibit accumulation of cab messenger RNA coding for the light-harvesting chlorophyll a/b protein. Physiol Plant 81:190–196.

Kopriva S, Rennenberg H (2004) Control of sulphate assimilation and glutathione synthesis: interaction with N and C metabolism. J Exp Bot 55:1831–1842.

Koussevitzky S, Nott A, Mockler TC, Hong F, Sachetto-Martins G, Surpin M, Lim J, Mittler R, Chory J (2007) Multiple signals from damaged chloroplasts converge on a common pathway to regulate nuclear gene expression. Science 316:715–719.

Kropat J, Oster U, Rüdiger W, Beck CF (1997) Chlorophyll precursors are signals of chloroplast origin involved in light induction of nuclear heat shock genes. Proc Natl Acad Sci USA 94:14168–14172.

Kropat J, Oster U, Rüdiger W, Beck CF (2000) Chloroplast signalling in the light induction of nuclear *HSP70* genes requires the accumulation of chlorophyll precursors and their accessibility to cytoplasm/nucleus. Plant J 24:523–531.

Langebartels C, Wohlgemuth H, Kschiechan S, Grün S, Sandermann H (2002) Oxidative burst and cell death in ozone-exposed plants. Plant Physiol Biochem 40:567–575.

Lee KP, Kim C, Landgraf F, Apel K (2007) EXECUTER1- and EXECUTER2-dependent transfer of stress-related signals from the plastid to the nucleus of *Arabidopsis thaliana*. Proc Natl Acad Sci USA 104:10270–10275.

Levine A, Tenhaken R, Dixon R, Lamb C (1994) H_2O_2 from the oxidative burst orchestrates the plant hypersensitive disease resistance response. Cell 79:583–593.

Liu Y, Ren D, Pike S, Pallardy S, Gassmann W, Zhang S (2007) Chloroplast-generated reactive oxygen species are involved in hypersensitive response-like cell death mediated by a mitogen-activated protein kinase cascade. Plant J 51:941–954.

Mateo A, Mühlenbock P, Rustérucci C, Chang CC-C, Miszalski Z, Karpinska B, Parker JE, Mullineaux PM, Karpinski S (2004) *LESION SIMULATING DISEASE 1* is required for acclimation to conditions that promote excess excitation energy. Plant Physiol 135:2818–2830

Meskauskiene R, Nater D, Goslings D, Kessler F, op den Camp R, Apel K (2001) FLU: A negative regulator of chlorophyll biosynthesis in *Arabidopsis thaliana*. Proc Natl Acad Sci USA 98:12826–12831.

Meyer AJ, Hell R (2005) Glutathione homeostasis and redox-regulation by sulfyhdryl groups. Photosynth Res 86:435–457.

Mochizuki N, Brusslan JA, Larkin R, Nagatani A, Chory J (2001) *Arabidopsis* genomes uncoupled 5 (GUN5) mutant reveals the involvement of Mg-chelatase H subunit in plastid-to-nucleus signal transduction. Proc Natl Acad Sci USA 98:2053–2058.

Molina A, Volrath S, Guyer D, Maleck K, Ryals J, Ward E (1999) Inhibition of protoporphyrinogen oxidase expression in *Arabidopsis* causes a lesion-mimic phenotype that induces systemic acquired resistance. Plant J 17:667–678.

Montillet J-L, Cacas J-L, Garnier L, Montané M-H, Douki T, Bessoule J-J, Polkowska-Kowalczyk L, Maciejewska U, Agnel J-P, Vial A, Triantophyllidès C (2004) The upstream oxylipin profile of *Arabidopsis thaliana*: a tool to scan oxidative stresses. Plant J 40:439–451.

Morison JIL, Baker NR, Mullineaux PM, Davies WJ (2008) Improving water use in crop production. Phil Trans Royal Soc B 363:639–658.

Mou Z, Fan W, Dong X (2003) Inducers of plant systemic acquired resistance regulate NPR1 function through redox changes. Cell 113:935–944.

Mueller S, Hilbert B, Dueckershoff K, Roitsch T, Krischke M, Mueller MJ, Berger S (2008) General detoxification and stress responses are mediated by oxidized lipids through TGA transcription factors in *Arabidopsis*. Plant Cell 20:768–785.

Mullineaux P, Karpinski S (2002) Signal transduction in response to excess light: Getting out of the chloroplast. Curr Opin Plant Biol 5:43–48.

Mullineaux PM, Rausch T (2005) Glutathione, photosynthesis and the redox regulation of stress-responsive gene expression. Photosynth Res 86:459–474.

Mullineaux PM, Karpinski S, Baker NR (2006) Spatial dependence for hydrogen-peroxide-directed signaling in light-stressed plants. Plant Physiol 141:346–350.

Müller-Moulé P, Havaux M, Niyogi KK (2003) Zeaxanthin deficiency enhances the high light sensitivity of an ascorbate-deficient mutant of Arabidopsis. Plant Physiol 133:748–760.

Nambara E, Marion-Poll A (2005) Abscisic acid biosynthesis and catabolism. Annu Rev Plant Biol 56:165–185.

Nishizawa A, Yabuta Y, Yoshida E, Maruta T, Yoshimori K, Shigeoka S (2006) Arabidopsis heat shock transcription factor A2 as a key regulator in response to several types of environmental stress. Plant J 48:535–547.

North HM, De Almeida A, Boutin JP, Frey A, To A, Botran L, Sotta B, Marion-Poll A (2007). The Arabidopsis ABA-deficient mutant *aba4* demonstrates that the major route for stress-induced ABA accumulation is via neoxanthin isomers. Plant J 50:810–824.

Nott A, Jung H-S, Koussevitzky S, Chory J (2006) Plastid-to-nucleus retrograde signaling. Annu Rev Plant Biol 57:739–759.

Ochsenbein C, Przybyla D, Danon A, Landgraf F, Göbel C, Imboden A, Feussner I, Apel K (2006) The role of *EDS1* (enhanced disease susceptibility) during singlet oxygen-mediated stress responses of Arabidopsis. Plant J 47:445–456.

Op den Camp RGL, Przybyla D, Ochsenbein C, Laloi C, Kim C, Danon A, Wagner D, Hideg E, Göbel C, Feussner I, Nater M, Apel K (2003) Rapid induction of distinct stress responses after the release of singlet oxygen in Arabidopsis. Plant Cell 15:2320–2332.

Ort DR, Baker NR (2002) A photo-protective role for O_2 as an alternative electron sink in photosynthesis. Curr Opin Plant Biol 5:193–198.

Oster U, Brunner H, Rüdiger W (1996) The greening process in cress seedlings. V. Possible interference of chlorophyll precursors, accumulated after thujaplicin treatment, with light regulated expression of *Lhc* genes. J Photochem Photobiol 36:255–261.

Oswald O, Martin T, Dominy PJ, Graham IA (2001) Plastid redox state and sugars: interactive regulators of nuclear-encoded photosynthetic gene expression. Proc Natl Acad Sci USA 98:2047–2052.

Parisy V, Poinssot B, Owsianowski L, Buchala A, Glazebrook J, Mauch F (2007) Identification of PAD2 as a γ-glutamylcysteine synthetase highlights the importance of glutathione in disease resistance of Arabidopsis. Plant J 49:159–172.

Pasternak M, Lim B, Wirtz M, Hell R, Cobbett CS, Meyer AJ (2008) Restricting glutathione biosynthesis to the cytosol is sufficient for normal plant development. Plant J 53:999–1012.

Pennell RI, Lamb C (1997) Programmed cell death in plants. Plant Cell 9:1157–1168.

Pesaresi P, Schneider A, Kleine T, Leister D (2007) Interorganellar communication. Curr Opin Plant Biol 10:600–606.

Pfannschmidt T, Nilsson A, Allen JF (1999) Photosynthetic control of gene expression. Nature 397:625–628.

Pfannschmidt T, Bräutigam K, Wagner R, Dietzel L, Schröter Y, Steiner S, Nykytenko A (2009) Potential regulation of gene expression in photosynthetic cells by redox and energy state: approaches towards a better understanding. Ann Bot 103:599–607.

Przybyla D, Göbel C, Imboden A, Hamberg M, Feussner I, Apel K (2008) Enzymatic, but not non-enzymatic, $^1O_2$-mediated peroxidation of polyunsaturated faty acids forms part of the EXECUTER1-dependent stress response program in the *flu* mutant of *Arabidopsis thaliana*. Plant J 54:236–248.

Rolland F, Baena-Gonzalez E, Sheen J (2006) Sugar sensing and signaling in plants: Conserved and novel mechanisms. Annu Rev Plant Biol 57:675–709.

Rossel JB, Wilson IW, Pogson BJ (2002) Global changes in gene expression in response to high light in Arabidopsis. Plant Physiol 130:1109–1120.

Rossel JB, Walter PB, Hendrickson L, Chow WS, Poole A, Mullineaux PM, Pogson BJ (2006) A mutation affecting *ASCORBATE PEROXIDASE2* gene expression reveals a link between responses to high light and drought tolerance. Plant Cell Environ 29:269–281.

Rossel JB, Wilson PB, Hussain D, Woo NS, Gordon MJ, Mewett OP, Howell KA, Whelan J, Kazan K, Pogson BJ (2007) Systemic and intracellular responses o photooxidative stress in *Arabidopsis*. Plant Cell 19:4091–4110.

Russell AW, Critchley C, Robinson SA, Franklin LA, Seaton GGR, Chow WS, Anderson JM, Osmond CB (1995) Photosystem II regulation and dynamics of the chloroplast D1 protein in *Arabidopsis* leaves during photosynthesis and photoinhibition. Plant Physiol 107: 943–952.

Rusterérucci C, Montillet J-L, Agnel J-P, Battesti C, Alonso B, Knoll A, Bessoule J-J, Etienne P, Suty L, Blein J-P, Triantophylidès C (1999) Involvement of lipoxygenase-dependent production of fatty acid hydroperoxides in the development of the hypersensitive cell death induced by cryptogein on tobacco leaves. J Biol Chem 274:36446–36455.

Rusterérucci C, Aviv DH, Holt BF, Dangl JL, Parker JE (2001) The disease resistance signaling components *EDS1* and *PAD4* are essential regulators of the cell death pathway controlled by LSD1 in Arabidopsis. Plant Cell 13:2211–2224.

Sattler SE, Mène-Saffrané, Farmer EE, Krischke M, Mueller MJ, DellaPenna D (2006) Nonezymatic lipid peroxidation reprograms gene expression and activates defense markers in *Arabidopsis* tocopherol-deficient mutants. Plant Cell 18:3706–3720.

Scarpecci TE, Zanor MI, Carrillo N, Mueller-Roeber B, Valle EM (2008) Generation of superoxide anion in chloroplasts of *Arabidopsis thaliana* during active photosynthesis: a focus on rapidly induced genes. Plant Mol Biol 66:361–378.

Sofo A, Dichio B, Xiloyannis C, Masia A (2004) Effects of different irradiances on some antioxidant enzymes and on malondialdehyde content during rewatering in olive trees. Plant Sci 166:293–302.

Sullivan JA, Gray JC (1999) Plastid translation is required for the expression of nuclear photosynthesis genes in the dark and in roots of the pea *lip1* mutant. Plant Cell 11:901–910.

Strand A, Asami T, Alonso J, Ecker JR, Chory J (2003) Chloroplast to nucleus communication triggered by accumulation of Mg-protoporphyrinIX. Nature 421:79–83.

Susek RE, Ausubel FM, Chory J (1993) Signal transduction mutants of *Arabidopsis* uncouple nuclear *CAB* and *RBCS* gene expression from chloroplast development. Cell 74:787–799.

Thoma, I, Loeffler C, Sinha AK, Gupta M, Krischke M, Steffan B, Roitsch T, Mueller MJ (2003) Cyclopentanone isoprostanes induced by reactive oxygen species trigger defense gene activation and phytoalexin accumulation in plants. Plant J 34:363–375.

Thomas DJ, Avenson TJ, Thomas JB, Hebert SK (1998) A cyanobacterium lacking iron superoxide dismutase is sensitized to oxidative stress induced with methyl viologen but is not sensitized to oxidative stress induced with norflurazon. Plant Physiol 116:1593–1602.

Vanderauwera S, Zimmermann P, Rombauts S, Vandenabeele S, Langebartels C, Gruissem W, Inzé D, Van Breusegem F (2005) Genome-wide analysis of hydrogen peroxide-regulated gene expression in Arabidopsis reveals a high light-induced transcriptional cluster involved in anthocyanin biosynthesis. Plant Physiol 139:806–821.

Vernoux T, Wilson RC, Seeley KA, Reicheld JP, Muroy S, Brown S, Maughan SC, Cobbett CS, Van Montagu M, Inzé D, May MJ, Sung ZR (2000) The *ROOT MERISTEMLESS 1/CADMIUM SENSITIVE 2* gene defines a glutathione dependent pathway involved in initiation and maintenance of cell division during postembryonic root development. Plant Cell 12:97–110.

Wachter A, Wolf S, Steininger H, Bogs J, Rausch T (2005) Differential targeting of GSH1 and GSH2 is achieved by multiple transcription initiation: implications for the compartmentation of glutathione biosynthesis in the *Brassicaceae*. Plant J 41:15–30.

Wagner D, Przybyla D, op den Camp R, Kim C, Landgraf F, Lee KP, Würsch M, Laloi C, nater M, Hideg E, Apel K (2004). The genetic basis of singlet oxygen-induced stress responses of *Arabidopsis thaliana*. Science 306:1183–1185.

Willekens H, Chamnongpol S, Davey M, Schraudner M, Langebartels C, Van Montagu M, Inze D, Van Camp W (1997) Catalase is a sink for H_2O_2 and is indispensable for stress defence in C3 plants. EMBO J 16:4806–4816.

Yabuta Y, Maruta T, Yoshimura K, Ishikawa T, Shigeoka S (2004) Two distinct redox signaling pathways for cytosolic APX induction under photooxidative stress. Plant Cell Physiol 45:1586–1594.

Zeier J, Pink B, Mueller MJ, Berger S (2004) Light conditions influence specific defence responses in incompatible plant-pathogen interactions: uncoupling systemic resistance from salicylic acid and PR-1 accumulation. Planta 219:673–683.

Index